Technological Change and the United States Navy, 1865–1945

JOHNS HOPKINS STUDIES IN THE HISTORY OF TECHNOLOGY

Merritt Roe Smith, *Series Editor*

Technological Change
and the
United States Navy,
1865–1945

William M. McBride

The Johns Hopkins University Press
Baltimore & London

© 2000 The Johns Hopkins University Press
All rights reserved. Published 2000
Printed in the United States of America on acid-free paper
9 8 7 6 5 4 3 2 1

The Johns Hopkins University Press
2715 North Charles Street
Baltimore, Maryland 21218-4363
www.press.jhu.edu

A catalog record for this book is available from the British Library.

Library of Congress Cataloging-in-Publication Data

McBride, William M.
Technological change and the United States Navy, 1865–1945 /
William M. McBride.
    p. cm. — (Johns Hopkins studies in the history of technology)
    Includes bibliographical references and index.
    ISBN 0-8018-6486-0 (hardcover : acid-free paper)
    1. United States. Navy — Officers — Attitudes.   2. United States. Navy —
Civilian employees — Attitudes.   3. Naval art and science — Technological
innovations — United States.   I. Title.   II. Series.
VA55 .M33 20000
359'.00973'09034 — dc21   00-009313

For Sally, Billy, Emily, and Kylie,
and in memory of Robert Glenn Thrasher (1952–89),
a courageous shipmate who maintained his dignity
and dry sense of humor during our plebeian year,
and also for Guy Henry Brown (1950–77), who was lost at sea

We may now picture this great Fleet, with its flotillas and cruisers, steaming slowly out of Portland Harbour, squadron by squadron, scores of gigantic castles of steel wending their way across a misty, shiny sea, like giants bowed in anxious thoughts.

—Winston S. Churchill
on the sortie of the fleet, 29 July 1914, *The Modern Crisis*

Contents

     Task Forces and "Three-Plane" Warfare   182

9   Castles of Steel: *Technological Change and*
     the Modern Navy   211

     Notes   243
     Note on Sources   319
     Index   325

Acknowledgments

This study would not have been possible without the strong support, vision, and continuing encouragement of Stuart W. Leslie, who introduced me to the history of technology at the Johns Hopkins University well over a decade ago. I am indebted to Robert H. Kargon, Willis K. Shepherd Professor of the History of Science, for shedding light on the institutional development of science and technology in the United States and for introducing me to the work of the philosopher Ludwik Fleck. I would also like to thank Alex Roland, of Duke University, for his continued endorsement of my investigation into the history of naval technology. I am also grateful to Louis Galambos, editor of the Dwight D. Eisenhower Papers, whose seminars in American political and business history at Johns Hopkins allowed me to place American naval technology within an appropriate political context. I am in debt to Paul M. Kennedy, Dilworth Professor of History at Yale University, for providing the opportunity to extend my research on American naval policy into the post-1945 period as a John M. Olin Fellow in Military and Strategic History in the International Security Program at Yale. I also appreciate the friendship of a fine colleague, James Sadkovich, who "internationalized" my perceptions of naval power and strategy during our year in New Haven. I also owe debts to Robert O'Connell and Jon Sumida for the insights their works have offered regarding military technology and the battleship in particular.

I am grateful to Rameswar Bhattacharyya, Roger C. Compton, and Paul Van Mater for their interest and pedagogical efforts during my undergraduate education in naval architecture at the U.S. Naval Academy three decades ago. I also benefited a great deal from the two years I spent as a senior naval architect at John J. McMullen and Associates in northern Virginia. I especially value my interactions with Dr. Volf Asinovsky, whose years of design experience in Leningrad provided me with a unique window into transcultural technological and engineering practices.

This study has only been possible due to the financial support I received from various institutions. I would like to thank the Naval Historical Center of Washington, D.C., specifically Dean Allard and Ronald Spector, now of George Washington University, for selecting me as the inaugural Rear Admiral John D. Hayes Fellow in 1987. My research was also aided by the Hoover Presidential Library Association, which designated me a 1988 Hoover Presidential Scholar and the 1988–89 U.S. Senator (Ret.) and Mrs. Roman L. Hruska Fellow. I am indebted to the Franklin and Eleanor Roosevelt Institute of Hyde Park, New York, for a grant to conduct research at the Franklin D. Roosevelt Library, and to the Center for the History of Business, Technology, and Society of the Hagley Museum and Library of Wilmington, Delaware, for an additional grant to explore the Elmer A. Sperry Papers.

Some later portions of this study were undertaken at James Madison University, where I was blessed with a pleasant group of colleagues on the history faculty. I am grateful to the head of the history department, Michael Galgano, for granting me release from some of my teaching responsibilities over several semesters to devote time to this book. I also appreciate the external financial support for my research that resulted from my selection as inaugural Edna T. Shaeffer Distinguished Humanist at James Madison in 1993.

I would also like to thank my colleagues in the Department of History at the United States Naval Academy for their support. I have benefited a great deal from conversations on naval history with Fred Harrod, Bob Love, Bill Roberts, and Craig Symonds. I am especially indebted to Eric Reed for numerous thought-provoking theoretical discussions, grounded firmly in naval realities. Many of my students over the years have been exposed to my ideas, and I appreciate their responses. I would like to thank Bill Roberts for his patient advice and Bob Love for the insight he has provided into the personalities who populated the twentieth-century navy. I also owe thanks to the military members of the history faculty whose questions as new instructors have forced me to clarify and refine many of my conceptualizations of American naval history. I also appreciate the readings of various sections of this study by my colleagues, David Appleby, Mary DeCredico, Allison Fuss, and Brian Van De Mark.

Any historical study dealing with the federal government requires the sifting of extensive archival records. If not for the assistance I received from the staff members at various archives, I would still be lost amid miles of manuscripts and records. I would like to thank Dwight Miller, senior archivist at the Hoover Presidential Library, West Branch, Iowa, for his generous assistance during my review of the Hoover Papers pertaining to the U.S. Navy and disarmament questions. I would also like to thank the very professional and helpful staff at the Manuscript Division of the Library of Congress, who cheerfully dragged out innumerable dusty file boxes from the Naval Manuscript Collection for my review. The staff of the Franklin D. Roosevelt Library at Hyde Park, New York, was also most cooperative in finding material about the navy during the New Deal. Dr. Michael Nash, of the Hagley Museum and Library, kindly provided me the pertinent papers of Elmer A. Sperry, and of the Sperry Gyroscope Company, for my investigation of one facet of U.S. naval-industrial relations. Dr. Evelyn Cherpak, director of the Naval Historical Collection, U.S. Naval War College, was most helpful in directing me to pertinent archival holdings. I would be remiss if I did not also thank the staff of the Nimitz Library at the U.S. Naval Academy, including the staff of the Special Collections Division, for their assistance in locating the appropriate Naval Academy records in their vast holdings. The staff of the Seeley G. Mudd Manuscript Library at Princeton University was most helpful in guiding me through material related to my study. Finally, I acknowledge the assistance from the staff of the Military Reference Branch of the National Archives in Washington.

Preparing books for publication is a lengthy process and I appreciate the efforts of the staff at the Johns Hopkins University Press, especially those of my editor, Robert J. Brugger, and my copy editor, Maria E. denBoer.

This project has spawned several articles which have received peer recognition, encouraging me when my energy was ebbing. I am grateful for the Society for the History of Technology's I.E.E.E. Life Members' Prize, the Society of Historians of the Gilded Age and Progressive Era's Inaugural Biennial Prize, the Society for Military History's Moncado Prize for Excellence, and the Naval Historical Center and Naval Historical Foundation's joint Ernest M. Eller Prize, Honorable Mention.

Most important, I treasure the selflessness of my wife, who eschewed a post-doctoral fellowship at Yale Medical School for a five-year sojourn in a section of Virginia where Bobby Lee still lives and Stonewall Jackson look-a-likes lurk behind almost every tree. I also appreciate her tolerance of my periodic discourses on technology and history.

Many individuals and organizations contributed to this study. The interpretations and conclusions herein, and any shortcomings that may exist, are mine alone.

Technological Change and the United States Navy, 1865–1945

Introduction

A little over a quarter-century ago I was a naval officer newly assigned to a Pacific Fleet destroyer. During my warfare qualification I was introduced to a piece of electronic gear designated ULQ-6. One of its features was the ability to increase the radar reflection of our ship. In this "blip-enhance mode" the ULQ-6 theoretically would trick radar-guided missiles into mistaking our tiny destroyer for a huge aircraft carrier. When I expressed surprise at such a suicidal device, I was told not to worry because the blip-enhance mode was "always broken." The notion of hierarchical sacrifice the ULQ-6 entailed was not new to me. The Naval Academy's "Sea Power" course had acquainted me with the sacrifice of destroyers and destroyer escorts, and the deaths of 526 of their crews, off Samar in 1944 to protect aircraft carriers.[1] What bothered me was *my* expendable status thirty years later in what many nonaviators considered a futile sacrifice to protect a vulnerable capital ship.

What intrigued me as I later took up the study of the history of technology was how the naval profession developed the intellectual and institutional framework necessary for the conceptualization, design, testing, and deployment of such a device. In retrospect, I am amazed at the ULQ-6 designers' optimism that sailors would use it and its apparent easy circumvention by personnel unwilling to play capital ship during an attack. I never found out if the account of constant disablement of the ULQ-6 was apocryphal. I spent the next thirty-three months far below the air-conditioned domain of the ULQ-6, in a world dominated by boilers and steam turbines and populated by colorful characters drawn from Georgius Agricola's 1556 treatise, *De Re Metallica*.

My introduction to the ULQ-6 occurred within the context of the "hollow force" navy of the late Vietnam era. Aviators dominated the service, and the navy contained large numbers of aging World War II ships such as mine. In general, the navy measured its health in the numbers of attack carriers able to project power ashore and, theoretically, able to control the seas in the tradition of Alfred Thayer Mahan. While my shipmates and I were busy serving as lesser cogs within the technological framework that supported the aircraft carrier, I perceived some weakness in the claims of its dominance of war at sea.

Nuclear-powered submarines posed a serious challenge. During periodic exercises, submarines routinely and successfully attacked us and the aircraft carriers we escorted. Most times they did so undetected, even by newer antisubmarine ships. Since the Soviet navy had over three hundred submarines, I wondered about the aircraft carrier's ability to survive even a limited conventional conflict. I was not alone and recall one illustration in the U.S. Naval Institute *Proceedings* that featured an aviator admiral scanning the heavens while centered in the crosshairs of a submarine periscope.

Two years after my introduction to the ULQ-6, another occurrence caused me to question the navy's apparent singular focus on air warfare and became the nascent catalyst for this study. My old World War II–era destroyer sank the nuclear-powered supercarrier *Enterprise* in two of three phases of a fleet exercise. Despite local umpire agreement with the success of our attacks, we were informed that the final exercise report did not mention them. I assume the admiral who approved the report had a broader view of the exercise and had valid reasons to discount our success. Or perhaps he was just partial to aircraft carriers and *knew* that a nuclear-powered supercarrier like *Enterprise* was only vulnerable to an airborne assailant. The incident paralleled a successful British submarine attack on a battleship during a pre-1914 exercise. The battleship admiral's signal to the submarine captain was dismissive: "You be damned!"[2] The whole *Enterprise* issue seemed to repeat the stereotyped myopia of pre–Pearl Harbor battleship admirals.

I can think of no other modern weapon that, in the popular view, so embodies obsolescence as the battleship. The 1965 Random House *American College Dictionary* on my bookshelf even used the battleship as an example in its definition of "obsolete."[3] This view originated with postwar perceptions that the battleship had failed to fulfill its pre-1914 billing as a decisive weapon. Images of the 1921 battleship bombing trials lingered in the public and political consciousness and contributed to questions about the battleship's relevance. During the 1920s, air power advocates such as Billy Mitchell portrayed the battleship as passé and aviation as a technology and weapon of the future. Popular literature, including science fiction, reinforced this and was part of what H. Bruce Franklin characterized as the American fascination with "wonder weapons."[4] The images of burning and sinking battleships at Pearl Harbor went the rest of the way to discredit the battleship and its champions. "Battleship admiral" became a pejorative term, connoting a reactionary stubbornly clinging to tradition like the generals who blundered through World War I.

In lumping carrier admirals and battleship admirals together after the *Enterprise* affair, I naively had accepted an aviation-biased history that forward-thinking aviators had been oppressed by reactionary battleship admirals before Pearl Harbor. This facile, but common, view has obscured the realities of the battleship era. Intrinsic to this view was the idea that the prewar conservatism of the naval profession was so pervasive that only Japanese use of aircraft carriers at Pearl Harbor brought about a new, vigorous U.S. Navy run by aviator visionaries. Quite simply, the aircraft carrier, used by an enemy not constrained by barnacle-encrusted conservatism, provided the catalyst for a change in the navy's technological, social, and cultural hierarchies as well as a fundamental change in the way the U.S. Navy would organize itself and fight.

What bothered me as I delved into military history, and naval history specifically, was the pervasive, simplistic technological determinism that new technologies were "better" and the cause of new ways of war. The relationship between military technology and strategic doctrine was portrayed as cause and effect. Military historians, like their colleagues in business history, had succumbed too frequently to technological determinism, making technology an exogenous factor—a "black box"—that has guided the evolution of the military arts. A good example is Martin van Creveld's *Technology and War* (1989), which reads like a biblical account in which the sword begat the musket, the musket the cannon, and so on, down to nuclear weapons.[5] Such interpretations ignore any cultural or social framework for technological innovation, development, cultivation, or rejection.

Over a decade ago, Merritt Roe Smith observed that military historians have ignored technological enterprise in their attempts to "weave" military history "more tightly into the fabric of American history."[6] Not a great deal has changed. In addition to more emphasis on technology, military history would also benefit from more sophisticated interpretations that blend social constructivism with more refined interpretations of technological determinism.

Navies have always been technical. The warship *Sovereign of the Seas*, built by Charles I, was arguably the most complex, large artifact in England in 1637. The same can be said of the superdreadnought battleships in service on the eve of World War I and the modern supercarrier. All these ships required officers who could master the technologies involved. Naval officers of the seventeenth and eighteenth centuries — while not natural philosophers, engineers, or mechanics — were members of a technically conversant society. By the late nineteenth century, technology, science, and society had changed to the point that naval officers had become members of a technology-based profession.

In such a professional society, bound by a strong, traditional warrior ethos, new technologies could be destabilizing and were not glibly accepted as improvements. A few historians — John Ellis, Robert O'Connell, Alex Roland, and Tim Travers — have described how traditional warrior ethos have been powerful filters against adoption of new military technologies.

Military hierarchies seek stability, and when a new technology challenges that stability, the reaction can be sharp and hostile. For example, the American naval profession generally accepted steam engines before 1865. After the war, there was a significant backlash against steam engines and the engineers who ran them. This technology, useful during the Civil War, challenged the idealized postwar cultural, social, and technical self-images of the naval profession.

The antisteam reaction occurred amid the rise of science-based engineering during the 1870s. This engineering, not artisanal practice, was responsible for the new technological measure of first-rate naval power. By century's end, warships were complex systems that bore little resemblance to those of fifty years earlier. Seagoing line officers were leery of the importance of engineers and their power over the technological basis of the profession. As a result, line officers spent almost four decades trying to dominate engineers and the technology they controlled. In the end, all naval officers became hybrid warrior-engineers capable of operating *and* fighting the ship, just as in the age of sail.

The purpose of this study is to understand the dynamics through which social groups, in this case the American naval profession, have addressed technological change. In addition to fostering a greater understanding of the specifics of military technological change, this analysis also informs the current debate within the history of technology between those favoring social constructivism and those partial to technological determinism. I have found that neither of these analytical frameworks solely explains the technological dynamics within the American navy since the Civil War.

Questions relevant to technological change within the U.S. Navy include: How did changes in the navy's technological trajectory occur? Can change only result from war and an exogenous demonstration of a new way to fight? Did battleship admirals comprise an ancien régime that wrongfully suppressed naval aviation and the submarine? Is the navy's post-1945 focus on naval aviation waning?

In searching for answers, I have framed this study around the concepts of technological and strategic paradigms. A technological paradigm, according to Edward Constant, involves an exemplary artifact and a cultural framework devoted to sustaining that artifact.[7] For approximately five decades before 1945, the U.S. Navy's technological paradigm was based on the battleship. This battleship technological paradigm had its own technological "momentum" and technological "trajectory."[8]

The naval technological paradigm is distinct from Constant's technological paradigm and its dependence on cultural or market factors. The naval technological paradigm, also influenced by culture, interacts with an intellectual paradigm — the strategic philosophy of the naval profession. This strategic philosophy is comparable to scientific theory in Thomas Kuhn's traditional definition of a scientific paradigm.[9]

The U.S. naval officer corps did not invent the strategy of guerre d'escadre (fleet versus fleet warfare). Nor did it create the battleship technological paradigm. Both were copied from the British Royal Navy in an attempt to emulate it. In doing so in 1890, the United States turned its back on a more individualistic strategy of guerre de course (commerce raiding) and what Kenneth Hagan has termed a continentalist naval tradition of "great virtue."[10] There were many reasons for rejecting the old strategic paradigm, but perhaps the simplest was that it was not appropriate. It was too "weak" for a first-rate naval power, exactly what many influential people were intent on the United States becoming as the twentieth century approached.

I also prefer to view the naval profession in the terms developed by the Viennese philosopher, Ludwig Fleck. Fleck's concepts are especially valuable in describing the profession's intellectual and social dynamics, including the pre–World War I conflict between naval engineers and line officers and also the technologically defined subgroups within the postwar naval profession: aviators, submariners, and battleship/surface ship sailors. These could be characterized as lobbies or factions, but Fleck's definitions of a *thought collective* as "a community of persons mutually exchanging ideas or maintaining intellectual interaction" which provides the "special 'carrier' " for the *thought style*, the "given stock of knowledge and level of culture," are more precise and applicable.[11]

It is almost axiomatic that, for a navy, technological and strategic paradigms are inextricably linked. A profound shift in the technological paradigm would affect the naval profession significantly. Nevertheless, technological paradigm shifts typically occur within the framework of the existing strategic paradigm.[12] In his history of the turbojet, Constant postulated two types of technological paradigm change. The first was functional failure (in terms of this study, this would parallel significant military setbacks or defeat for a navy) and the hunt for a solution. The second involved technological co-evolution and the interaction of competing technologies (in this study, battleships, airplanes, and submarines) within the larger "macro-system."[13]

I find merit in, and make use of, Constant's significant conceptualization of a "presumptive anomaly" as a useful bridge between scientific and technological paradigmatic analysis. The naval profession would evaluate any alternative technology that was presumed to be superior to the existing technological paradigm (a presumptive anomaly), for example, a navy organized around aircraft carriers instead of battleships. However, as Constant cogently observed, a "normal technology, even if plagued with problems, is not easily abandoned."[14] Adoption of a new technological paradigm is dependent on "perceived costs, efficiency, and risks . . . the ease with which the new system can be explained . . . and the extent to which it can be easily tried."[15] In addition to these requirements, a new naval technological paradigm must also serve the prevailing strategic paradigm and be consonant with the profession's thought style and warrior ethos.

Paradigms, presumptive anomalies, thought collectives, and thought styles all usefully describe an American naval profession that became more complex and fragmented as its technology became more complicated and diverse, its strategic philosophy shifted, and the nation it served took on a global role between 1865 and 1945. During this period, the battleship technological paradigm was dominant, and the battleship retained strategic importance even after the attack on Pearl Harbor in December 1941.

In 1945 Fleet Admiral Ernest J. King attributed the U.S. Navy's victory over Japan to the "flexibility and balanced character of our naval forces."[16] Surface forces, which before the war were defined myopically in terms of the battle fleet arranged around the battleship, were characterized in 1945 as "fast task forces comprised of aircraft carriers, fast battleships, cruisers, and destroyers."[17] Almost thirty years later, there seemed to me to be little "balanced character" in a navy built around an aviation technological paradigm and its supercarrier and typified by the ULQ-6.

In his introduction to the 1986 edition of *Makers of Modern Strategy*, Peter Paret observed that "The past — even if we could be confident of interpreting it with high accuracy — rarely offers direct lessons."[18] On the whole, Paret is correct, yet the dynamics of technological change within the battleship navy offer direct insight into the post-1945 aviation navy as well as into the post–cold war navy as it faces the uncertainties of a new century. On a broader scale, this study also offers a historical lesson on the dynamics of change pertinent to nonmilitary societies as well.

CHAPTER 1

The Postbellum Naval Profession
From Discord to Amalgamation

During the late nineteenth century, rapidly evolving, science-based technology posed a challenge to the established values of the American naval profession. The sail-powered ship of the line, whose basic attributes had changed little in two hundred years, defined first-rate naval power in 1800. By mid-century, technological change resulted in a shift from a purely quantitative measure of perceived naval greatness — the number of ships of the line a country possessed — to a nascent qualitative measure of naval power based on significant technological differences. These included steam propulsion, ironcladding of wooden hulls, iron hulls, the use of propellers instead of paddlewheels, and advances in naval ordnance such as rifled barrels and exploding projectiles.

Naval officers, then and now, draw their professional identity from the technology they operate, and new technologies have the potential to destabilize the existing "sociotechnical" framework. As a result, naval officers historically have shared technological skepticism with other warrior societies.[1]

One option for the hierarchies of such societies is simply to ignore technological change. If warfare remains within a common framework, one in which all the combatants base their war fighting on the same technical foundation, military strength can be defined quantitatively. However, technical divergence, whether evolutionary or discontinuous, can result in an effective "counterweapon" — to use Robert O'Connell's term. Because of a counterweapon's destabilizing nature, established warrior sociotechnical systems work against its adoption and use. Defenders of the status quo re-

Admiral David Farragut watches from the rigging of his steam-powered flagship Hartford *as his crew pounds the Confederate ironclad* Tennessee *in Mobile Bay in 1864. This idealized view reflected line officers' extension of the traditional naval warrior ethos into the steam age.* (William Heysham Overrend, *An August Morning with Farragut: The Battle of Mobile Bay, August 5, 1864.* Reproduced by permission of the Wadsworth Atheneum, Hartford. Gift of Citizens of Hartford by Subscription)

ject radical technological innovation in favor of nonthreatening, symmetrical technological responses (tank versus tank, battleship versus battleship).[2] Intercultural and technologically discontinuous warfare, however, can shatter an existing sociotechnical framework. The victories by Portuguese "Atlantic"-style ships over Moslem galleys and smaller sailing ships off India in 1509 is a good example.[3]

A second option for hierarchies chary of new technology is to try to control the course and rate of technological change. This is what some American naval officers sought to do, citing valid geostrategic restraints, after the Civil War. This served their attempts to restore the social and cultural homogeneity of the profession by suppressing the engineer officer pathogens which had infected it during the Civil War. To control technological change successfully, seagoing line officers had to master the officers of the

Engineering Corps and their politically well-connected and powerful technical bureaus within the Navy Department.[4] Unfortunately for the line, naval innovation abroad demanded a technological response that would empower naval engineers.

THE PRE-1865 STEAM NAVY

The development of explosive-shell guns, such as that of Henri Paixhans, led to ironcladding of wooden warships — the French *La Gloire* of 1857 being the first.[5] The combination of shell guns and ironcladding was a destabilizing technical amalgamation. All but the most dull-witted naval officers at mid-century could see how their profession had been changed irretrievably by applications in metallurgy and chemistry.

Steam propulsion, which augmented wind power, was more subtle than ironcladding and weapon developments in its disruption. To some, early steam engines presented an emerging presumptive anomaly to the technological paradigm based on the sail-powered warship.[6] Robert Fulton's 1815 "steam frigate" *Demologos* (twenty guns) was the world's first steam-propelled warship. She was designed as a floating steam battery for harbor defense, had a centerline paddlewheel, and steamed at 5 miles per hour. Her $320,000 cost rivaled that of the heavy frigate *Constitution*, built for $302,719 in 1797.

Demologos also had the distinction of being the first in a long line of steamships modified to conform better to the prevailing technological paradigm. Captain David Porter, given command of *Demologos* after losing the frigate *Essex* to the British, balked at *Demologos*'s mode of propulsion. He forced the addition of two masts and two bowsprits to carry sails. As a result, *Demologos*'s completion was delayed further to build bulwarks to protect the additional crew handling the sailing rig. The war ended before these modifications were complete. Nevertheless, the pertinent subcommittee of the Coast and Harbor Defense Association recommended her continued operation to provide a trained cadre familiar with a steam vessel. Despite this recommendation, *Demologos* — renamed *Fulton* in honor of her designer — only served as a receiving ship (a floating barracks) at Brooklyn until destroyed by a gunpowder explosion in 1829.[7]

Naval officers had no consensus that this early foray into steam propulsion was in any way successful. *Demologos* was a "brown-water" defensive platform and not a "blue-water" ship that could serve the navy's primary mission of commerce protection in far oceans. Edward Constant observed correctly that "few practitioners will abandon a highly successful normal technology in the absence of a convincing alternative."[8] It was almost a quarter-century before the navy built another steam-propelled ship. Meanwhile, the shallow-draft galliot *Sea Gull* was purchased in 1822 for use in David Porter's Caribbean efforts to suppress piracy and illegal privateers during the wars for Latin American independence. The navy laid *Sea Gull* up in 1825 and sold her in 1840. In contrast to the navy's cautious forays into steam propulsion, almost seven hundred commercial steam vessels were in use in American waters by the 1830s while the navy had none.

The 1816 Act for the Gradual Increase of the Navy budgeted $1 million per year for eight years to build nine ships of the line, twelve heavy frigates, and three "steam batteries" similar to *Demologos*. This grandiose plan was hurt by the 1819 economic panic and no steamships were built until Secretary of the Navy Mahlon Dickerson directed the three captains on the Board of Navy Commissioners to begin construction of a steam vessel in 1835. This request caught the Board off-balance and they were forced to admit, writing in the third person, that "they are incompetent themselves, and have no person under their direction who could furnish them with the necessary information to form a contract for steam engines that may secure the United States from imposition, disappointment, and loss." In February 1836 Charles Haswell was hired to assist the Board and was appointed chief engineer of the steam warship *Fulton (the Second)* then under construction.[9]

The sail-steam hybrid *Fulton* was designed as a harbor defense vessel and entered service in 1837. Matthew C. Perry used her effectively to suppress pirates in the Caribbean even though the ship was not an open-ocean steamer. However, the general inefficiency of marine steam engines offered no advantages to a navy operating commerce-protecting squadrons as far away as the East Indies.[10]

Several histories describe a nineteenth-century naval profession populated with antisteam reactionaries.[11] Some naval officers objected to steam propulsion on grounds ranging from aesthetic to military. Steam engines were large and noisy and their lubricating oil and coal fuel were intrusions into the clean and orderly world of a sailing man-of-war characterized so well by Herman Melville in *White-Jacket* in 1844.[12] Side-mounted paddle-wheels had significant military disadvantages: they were vulnerable to gunfire and displaced many broadside guns.

More surprising than opposition to technological change were the nonengineer naval officers of the line, such as Matthew C. Perry, who championed steam propulsion. By 1842, in spite of the aesthetically motivated disdain of Secretary of the Navy James Paulding, the navy had two paddlewheel frigates, *Missouri* and *Mississippi*.[13] The attraction of steam propulsion for naval officers — independence from natural forces — was similar to that felt by early users of waterwheels and windmills. The weaknesses of the engines — the inefficient transmission of engine power and the engines' voracious appetites for coal — limited the potential of these early steam propulsion systems to providing added speed during pursuit, evasion, or battle.

For naval officers, the ultimate validation of technology was its performance at sea and in war. While useful in combat, inefficient, early steam propulsion offered few advantages to a far-flung navy patrolling a growing American "empire of commerce."[14] Perry again made good use of steam propulsion when he commanded U.S. naval forces in the Gulf of Mexico during the Mexican War (1846–47), just as the British had done a few years earlier during the First Opium War with China.[15] However, keeping ships coaled in Mexico, more than 800 miles from the nearest U.S. naval base, was a significant logistical problem. In 1851 the 2,450-ton side-wheel steamer *Susquehanna* departed Norfolk on her maiden voyage for duty as flagship of the East India Squadron. The 8-month trip covered 18,500 miles and the ship consumed its weight in coal and 1,100 sticks of wood. Twenty-five percent of the voyage was spent coaling the ship.[16]

By the mid-1850s paddlewheel steamers had given way to new classes of propeller-driven ships with improved steaming and sailing qualities, but the problem of coal was far from resolved. The demand for it translated into a need for overseas possessions that would later feed into a circular argument, akin to William McNeill's "self-perpetuating feedback loop" in favor of imperialism to provide naval coaling bases to protect U.S. trade.[17]

ENGINE-DRIVERS AS WELL AS SAILORS

The rapid expansion of the U.S. Navy during the Civil War and its extensive use of steam-propelled vessels changed the professional landscape of the navy. Before the 1861 rebellion, the cultural purity of the naval profession was threatened little by the few naval engineers required by the fledgling steam navy — a navy with approximately the same number of ships (primarily sail-powered) America possessed at the beginning of the century.[18] However, engineering grew increasingly important in the maintenance and employment of naval power to achieve victory over the Confederacy.

Prewar engineers were typically drawn from the "higher" end of the antebellum mechanical culture and more easily acculturated into their appropriate nook within the naval profession. This was not the case during the war.[19] The dearth of qualified marine engineers led to reduced naval engineering entrance qualifications and the navy accepted many individuals without mechanical experience as engineering officers. The case of former Confederate lieutenant colonel Don Carlos Hasseltino is illustrative. Hasseltino abandoned the rebellion and entered the federal navy as an acting first assistant engineer after just a few weeks of book study. He was typical of many wartime naval engineers who "did not know a steam engine from a horse power."[20] As a gentleman, Hasseltino at least was socially acceptable.

Line officers resented the intrusion of nongentlemanly mechanics into their wardrooms and the resulting contamination of their "aristocratic" officer corps.[21] The Civil War navy differed from the sailing navy in which seamen officers of the line controlled the weapons *and* means of propulsion. Reliance on these professionally and socially inferior mechanics to produce the correct technological voodoo to fight the ship must have galled many line officers.

Divisiveness between engineers and line officers posed serious problems in a navy building and fighting a large number of steam-propelled and mechanically complex warships, typified by John Ericsson's *Monitor*. Naval engineers worked below decks, out of sight, and, most often, out of mind of the line officers who commanded, maneuvered, and fought these ships. Commander Alfred Thayer Mahan's pejorative dismissal of "those who snored away below while line officers fought the ship" may have been a representative view.[22] Many ships' captains commended their engineers for bravery. However, the charges pressed by Rear Admiral Samuel F. Du Pont against Chief Engineer Alban C. Stimers, in the wake of the failed attack on Charleston, reflected the "internal strife in the service" attending the introduction of "mastless war vessels" noted by Frank Bennett in his 1896 history of the steam navy.[23] Such divisiveness was dangerous since fighting efficiency depended upon the technical understanding of the captain and his relationship with his engineers. It would probably not be a long step between most Civil War command-engineering relationships and Captain Kirk's *Star Trek* dealings with Chief Engineer Scott, in which the captain demanded more power, speed, or shield strength with no interest in how Scott's engineers provided it.

Engineers suffered reduced status when they were left out of the reorganization of the navy's rank structure in 1862. The navy created the line officer ranks of rear admiral, commodore, lieutenant commander, and ensign. No corresponding engineer ranks meant that most engineers, tied to the older rank structure specified by Secretary of the Navy Isaac Toucey in 1859, lost ground compared with their line colleagues. Naval engineers complained to Congress and to Secretary of the Navy Gideon Welles. The importance of engineers to the naval war effort led Welles to redress this disparity in ranks in an executive order in 1863.[24]

Secretary Welles and Assistant Secretary of the Navy Gustavus Fox desired to restructure the officer corps and to formalize engineer officer education and this worried officers of the line. In his 1863 annual report, Welles threatened a radical recasting of the navy with line officers subordinated to engineers. The report laid the basis of "our true position as a naval power" on the line officers and engineering officers who "conduct the varied operations of the service." Naval engineers were portrayed as the "men combining science with mechanical skill and genius" on whom naval operations depend. Because of this, "Steam engineering should indeed be one of the important studies of all naval officers. . . . It is a question, indeed, as *sails are subordinate to steam* [emphasis added], whether every officer of the line ought not to be educated to and capable of performing the duties that devolve upon engineers."[25]

In his 1864 annual report, Secretary Welles argued that all Naval Academy students should study steam engineering "in view of the radical changes which have been wrought by steam as a motive power for naval vessels." Welles's comments on the future of the naval officer corps boded ill for officers of the old line, who like Rear Admiral Du Pont, remained in the "wooden age":

in our future navy every line officer will be a steam engineer, and qualified to have complete command and direction of his vessel. Hereafter every vessel of war must be a steam vessel. Those designed for ocean service will be furnished with sails to economize fuel. . . . The officers to sail and navigate a ship and the officers to run the steam-engine are about equal in number. . . . But half the officers of a steamship cannot keep watch, cannot navigate her, cannot exercise the great guns or small arms, nor, except as volunteers under a line officer, take part in any expedition against an enemy. On the other hand, the other half of the officers are incapable of managing the steam motive power, or of taking charge of the engine-room in an emergency, nor can the commander of a vessel . . . understand, of his own knowledge, whether the engineers and firemen are competent or not.

The remedy for this very simple, [we should make] our officers engine-drivers as well as sailors. It would not be expedient to interfere with the present status of line officer or engineers, — the change would be too radical; but we should begin by teaching each midshipman to be able to discharge the duties of line officers and engineers, to combine the two into one profession, so that officers so educated can take their watch alternately in the engine-room and on deck.[26]

A first step toward a new hybrid engineer-line officer was the assignment of several assistant engineers to the Naval Academy — relocated to Newport, Rhode Island, during the war — as "acting assistant professors of natural and experimental philosophy." Legislation enacted on 4 July 1864 authorized the secretary of the navy to appoint up to fifty cadet engineers to the Academy. When the Academy returned to Annapolis in 1865, the Department of Steam-Enginery (later renamed Steam Engineering) went to work acquainting midshipmen with the basics of steam engines. Congress appropriated $20,000 for a department building that housed a complete marine steam engine, including boilers, engine, propeller shaft, and propeller.[27] The secretary of the navy's annual report for 1866 lauded the new facility at Annapolis and warned that a line officer untrained in steam engineering would be "taking a secondary position" within the profession.[28]

While the secretary of the navy annual reports painted a steamy future, only four individuals had applied to be cadet engineers through 1866. The failure to attract young cadet engineers (who had to be 18 and have two years' mechanical experience before appointment) forced another tack: recruitment from the ranks of engineering colleges. Fifty young men, most graduate engineers, were tested at Annapolis in 1866. The top sixteen were appointed acting third assistant engineers and began a two-year course separate from the midshipmen. This program was never repeated and after the graduation of two of the four cadet engineers in 1868, the cadet-engineering curriculum was dropped.[29]

POST-1865 ROCKS AND SHOALS

At the conclusion of hostilities in 1865, the U.S. Navy returned to its antebellum mission of commerce protection and a strategy of guerre de course. These far-flung missions were not ones in which "sail was subordinate to steam." The Engineer Corps entered a difficult time since legislation of 4 July 1864 fixed the number of chief engineers at "one for each first and second rate steam vessel of war." The number of assistant engineers was set vaguely to the "actual needs of the service," a phrase that allowed line officer reductions of engineers in the postwar years. In January 1865 the navy list had 474 regular and 1,803 volunteer engineer officers. All these volunteer engineers were discharged over the next several years, with the bulk separated soon after the war with three months' severance pay and an "ornamental discharge paper."[30]

The termination of the Naval Academy cadet engineer program in 1868 paralleled the navy's postwar downsizing from a wartime high of 671 ships. The naval force bought, borrowed, and built to suppress the rebellion was ill-suited for a blue-water force returning to a mission of worldwide commerce protection. Mahanian navalists later decried this reduction, but most of the ships decommissioned were tied to calmer inland and coastal waters by marginal stability, limited steaming radius, and common sense.[31]

Vice Admiral David D. Porter was master of the postbellum navy, serving under the figurehead secretary of the navy, Alexander Borie.[32] Porter, the victor of the steam-powered western river campaign with Ulysses S. Grant, seemed to turn against steam propulsion. Elting Morison maintained that nostalgia for the sail-powered ships of the prewar era drove Porter and his senior line colleagues to minimize steam propulsion and to marginalize engineers after 1865.[33] Given naval officers' traditional affinity for their ships and Porter's lengthy service in steam-propelled ironclads, a knee-jerk rejection of a technology that brought him fame and high rank seems unlikely. Lance Buhl has argued that Porter's antisteam crusade merely reflected the strategic and tactical difficulties attending inefficient, steam-propelled ships serving worldwide.[34] Robert Albion maintained Porter resented the "growing importance of the engineer corps" along with "the machinery they operated," and with Porter running the navy, the engineers "caught the full brunt of line displeasure," much of it directed against the lightning-rod chief of the Bureau of Steam Engineering, Benjamin Isherwood.[35]

Porter's policies were more antiengineer than antitechnology. Porter and his line colleagues feared the social changes steam engineering and engineers would bring to the naval profession with its clear warrior ethos and aristocracy. The Porterist notion of the naval profession was Admiral David Farragut clinging to the rigging of his flagship, damning the torpedoes as he *steamed* into Mobile Bay under fire. This was a contemporary representation of the American naval warrior ideal that stretched back to John Paul Jones's exploits during the Revolutionary War.

This ideal contrasted markedly with engineers' general perception of a warship as a collection of machines — a weapons system in today's jargon. This subservience to the machine and the dangerous reorganizational imperative that the machine had on the navy was anathema to the traditional naval thought style. In their distrust of the engineers' version of the future, Porterists were no different from those who decried the coming of mechanical time during the Middle Ages or workers who resented the role of the machine in eclipsing the craft tradition at Harpers Ferry Armory before the Civil War.[36]

Line officers, such as Porter, resented their inability to control technology, and the hostility of the postwar line officer boards, convened under Porter, was the prime manifestation. The line also supported the 1867 legislation that divided the naval officer corps into line and staff officers. The latter group included naval engineers, naval constructors, paymasters, and medical officers.

In 1869 Porter and his senior line officer colleagues rejected USS *Wampanoag*, an ideal ship to serve the traditional American continental naval philosophy of guerre de course — what naval historian Kenneth Hagan called a historical policy of "great virtue."[37] Yet Porter's construction of the innovative, high-speed ram *Alarm* reflected his openness to new technologies that he perceived as enhancing the navy's capabilities within the larger strategic paradigm.

Engineer-in-Chief Benjamin Franklin Isherwood designed *Wampanoag*'s record-breaking, geared-drive propulsion plant. The high-speed *Wampanoag* was part of the plan to build fast commerce raiders for use against Britain should that country enter the war for the Confederacy.[38]

Isherwood's biographer, Edward Sloan, has documented the vigorous campaign waged against engineering officers in the wake of the Civil War.[39] Isherwood had a broad spectrum of rivals and became Porter's, and the line's, whipping boy during their efforts to institutionalize line supremacy over engineers. When Isherwood claimed he was entitled to the rank of rear admiral as head of a bureau, Porter told a friend that "to punish him for his folly we intend not only to strip him and the engineers of all honors, but to make them the most inferior corps in the Navy." This effort continued after Porter; in 1876 line officers established a fund to lobby against naval engineers in Washington.[40]

It would be incorrect to link the line's rejection of *Wampanoag* to abandonment of guerre de course and a resurrection of Secretary of the Navy Benjamin Stoddert's 1798 call for ships of the line to serve a strategy of guerre d'escadre. When a crisis occurred with Spain in 1873 over the execution of Americans running guns to Cuban insurrectionists (the *Virginius* affair), the navy's deficiencies were painfully clear in the subsequent naval exercise. Porter looked beyond the navy's failure in squadron operations at Key West and reaffirmed commerce raiding — traditionally a "lone wolf" operation — as U.S. naval strategy in 1874.[41]

After Key West, line officers were on the horns of a dilemma. The navy had to be modernized, but what technologies should be incorporated into a "new," reinvigorated American navy? How could first-rate ships be designed and operated without empowering naval engineers, naval constructors, and their politically well-connected technical bureaus? Line officers were leery of untrammeled technological change and the subsequent upheaval that might bring to the status quo. Yet in suppressing naval engineers, line officers ran the risk of continued naval obsolescence in a time of rapid technological change within foreign navies. Line officers might retain a sense of cultural purity by disdaining certain technologies but they, and the U.S. Navy, could become as irrelevant and vulnerable in the international context as the Mamelukes in Egypt after 1516.[42]

While the U.S. Navy stagnated, other navies were constructing ships based upon the most advanced technology that modern industry and emerging, scientifically based naval architectural engineering could produce.[43] Aware of the growing gap between the outdated ships of the Porter era and those of major naval powers such as Britain and France, a consensus grew among American naval officers for technologically sophisticated ships so the navy could remain a viable force. The payoff of this consensus came in 1883, when Congress authorized the first four ships of the new steel navy (*Atlanta, Boston, Chicago,* and *Dolphin* — the "ABCD" ships).

The ABCDs marked the beginning of the American naval renaissance but not an abandonment of the traditional American naval strategy based upon commerce raiding. The ABCDs lagged behind foreign designs, but the "correct" technological path to an optimum navy was far from clear and was clouded further by the differing strategic visions of the U.S. Navy and the major naval powers.[44] The type of ship good for commerce raiding was not the best ship for guerre d'escadre as practiced by the dominant British Royal Navy.

The ABCD ships demonstrated that the naval profession was intent on mastering complex machines. This was important for the naval officer corps seeking to enhance its professional status in the United States during the 1880–90s. The establishment of the Naval War College, the Line Officers Association, and the U.S. Naval Institute with its *Proceedings* reflected this drive for recognition as a profession.[45]

Of more immediate concern for many officers was finding a way around the professional stagnation in a navy with too many officers and too few ships. As Peter Karsten observed, many line officers, stuck amid a seniority-based promotion process, linked a better professional future to naval expansion.[46] The Line Officers Association lobbied toward this end in Congress, in the private sector, and before the public.[47] Yet new ships without the latest technological developments diminished the naval profession and ill-served the nation. To expand the navy, the line had to rely on staff officers: steam engineers and the more tolerable naval constructors. While many senior officers were barnacle-encrusted reactionaries or merely neutral but ignorant regarding technology,[48] others, especially younger line officers, considered new technology crucial to building a world-class navy. The issue was how to garner the latest ship designs without fueling a naval engineering renaissance and a strengthening of the technical bureaus that already enjoyed solid congressional support.

Some line officers advocated creation of a line-officer naval general staff to gain professional hegemony over naval engineers, and thereby control the trajectory of the navy's technology. Many others, usually the younger line officers who operated the electric motors, lights, and complex hydraulic systems on the new ships, were increasingly at a loss technically. Some perceived the need for engineering graduate education to master these new technologies. These line officers were in the minority. Most still held romantic images of warships in marked contrast to naval engineers, who retained a general sense of ships as machines.[49]

In an 1897 reprise of Gideon Welles's call for every officer to be a steam engineer, a personnel board under Assistant Secretary of the Navy Theodore Roosevelt argued that every officer had to be an engineer. The result was the Personnel (or Amalgamation) Act of 1899, in which engineers, naval constructors, and other staff officers were absorbed into the line on an inferior basis. On the surface, the line-engineering controversy ended. However, rivalries between older engineers and line officers continued while younger line warriors railed at technocrats and bureaucrats who did not fight but still defined the technological basis of the profession.

Compounding the line dilemma was the institutional autonomy granted to engineering and other staff officers by Congress's creation of the bureau system in 1842. Naval administration was divided among bureaus whose chiefs answered only to the secretary of the navy, traditionally an individual appointed to repay a political favor and who had little knowledge of naval affairs. By the 1890s the line-controlled Bureau of Navigation held a preeminent place within the bureau hierarchy and assigned all officers to duty, including those in the Engineering Corps. The autonomy of the primary technical bureaus — Construction & Repair and Steam Engineering — was enhanced by their administration of industrial contracts and the resulting strong congressional alliances.

The new battleship strategic philosophy defined the social framework and technological paradigm of the navy after 1890. Yet engineering officers — including naval constructors — controlled the course of development of naval technology and defined the future of the service and profession. Independent line officers who traced their warrior tradition to John Paul Jones, Edward Preble, Stephen Decatur, and David Farragut were concerned that naval constructors and steam engineers, using their unique professional knowledge, could finally diminish the traditional superiority of the seagoing officer. This would relegate line officers to the status of mere operators of technologies designed by engineers who had no interest in maintaining the existing ethos and social fabric of the navy. The way to prevent the engineering tail from wagging the sea dog was to control rigidly the activities of naval engineers from the minute they entered the naval service, inculcating in them the profession's thought style and, ideally, a sense of their subservient role in producing the artifacts that served the professional status quo.

It was within this environment, between the Civil War and even into the years immediately following World War I, that the conflict between naval engineers and line officers played out for control of the technological basis of the naval profession. The struggle involved the issues of naval engineer training and attempts to establish officer homogeneity, first through curriculum controls at the Naval Academy during the 1880s, and later through the Amalgamation Act of 1899.

ANNAPOLIS AND A HOMOGENIZED OFFICER CORPS

The U.S. Naval Academy was founded in 1845, more than four decades after the Military Academy at West Point was established on the scientific-engineering model of the École Polytechnique. Before the establishment of the Academy curriculum in steam engineering for cadet engineers in 1861, naval officers received no training in science or engineering, other than that provided by their participation in the Coast Survey.[50]

After the Naval Academy's cadet engineer program was pronounced dead in 1868, the navy reestablished a two-year cadet engineer program in 1871, the year after Porter was promoted to succeed Farragut as admiral of the navy.[51] In February 1874 Congress abolished this latest engineering program and established a four-year course of study to parallel that provided to cadets of the line (cadet midshipmen). Funds were appropriated for up to twenty-five cadet engineers per year.[52]

The Naval Academy administration ensured that the cadet engineers conformed to the line ideal and the overarching navalist worldview. Most of the cadet engineers' courses during the first two years were taken with the cadet midshipmen. During their junior and senior years, while the line cadets studied seamanship, the engineers were enrolled in engineering courses and studied carpentry, blacksmithing, boilermaking, foundry work, and engine-building. During the summers, when the line cadets were serving at sea, the cadet engineers visited naval shipyards and private shipbuilding firms, establishing early professional relationships with private industry.[53]

By 1881, the postwar glut in the officer corps forced congressional action and provided line officers a venue to reduce the naval engineering community. Between 1878 and 1881, relaxed academic standards at the Academy resulted in a surplus of naval cadets. The graduation rate for cadet engineers in the Academy classes of 1879–82 averaged 57 percent against line cadets' graduation rate of 89 percent. For the classes of 1880–82, the disparity was even greater: line cadets graduated at a rate of 100 percent while only 45 percent of cadet engineers graduated.[54] Congressional appointments for new cadets were fixed by law and the lack of academic dismissals resulted in too many Academy graduates for a navy reduced severely in the years following the Civil War.

To remedy the situation, Congress, with input from line officer lobbyists, added a personnel section in the naval appropriations bill that became law in August 1882. This act fixed the number of officers in the navy at the 1882 level and limited the naval officer corps to one hundred engineers. The 1882 appropriations bill combined the cadet engineers and the cadet midshipmen into a homogenous undergraduate body of "naval cadets."[55] The Act stipulated that the graduates who would be commissioned were those with the highest class standing determined by examination. Those who fell below the cutoff were to be given their diplomas and one year's severance pay and released from the navy.[56] This sparked a rebellion among cadet engineers at Annapolis that eventually led to the U.S. Supreme Court.[57]

With the engineering curriculum canceled and the prospect of vacancies in the Engineer Corps slim, the cadet engineers appealed to Secretary of the Navy William Chandler to permit them to complete their engineering degrees:

> Owing to the present state of the Navy, almost all of us will be discharged within a few years. The first three years of our course at the Naval Academy have been devoted to the study of Steam Engineering to the complete exclusion of Seamanship, Navigation and Gunnery. This study, however, has been mainly preparatory, this [final] year being devoted mainly to the study of Engineering as a science. If, during the coming year, we take up new studies, Seamanship, etc., and merely review our former elementary course in Engineering—the course proposed—after graduation we shall be neither seamen nor engineers. . . . Though we shall be graduates of the best institution of its class in the world, yet we shall know less of practical, professional subjects than the graduates of any second or third rate college in the country.[58]

The cadets' evaluation of their education at Annapolis had merit. The 1878 Paris Universal Exposition had awarded its Diplôme de Medaille d'Or to the Naval Academy for its excellent engineering curriculum.

Secretary Chandler was sympathetic to the engineers' plight. The Academy superintendent, the joyless disciplinarian Captain F. M. Ramsay, was not and urged Chandler to disapprove the request of the cadet engineers.[59] Chandler acknowledged the impracticality of granting the full request, but asked Ramsay if those who would choose to leave the navy after completing their fourth, and final, year at the Academy could "pursue during that year studies as Engineers only."[60] Ramsay refused and Chandler had little latitude in applying the law without Ramsay's help. Chandler denied the engineers' request using the overcrowded officer corps as an excuse: "The statute seems to have been passed from a conviction that the Naval Service, in its lower grades, was becoming overloaded with graduates from the Naval Academy, and from a deliberate determination to prevent any further increase in the number of Officers in the Navy."[61]

During the 1882–83 academic year, Superintendent Ramsay took administrative action to ensure that the rebellious cadet engineers were unable to attain high enough class standing to receive commissions. Secretary Chandler believed he and Ramsay had agreed that the "late Cadet Engineers" would be examined and graded separately "on account of the impossibility of justly combining them with the late Cadet Midshipmen."[62] But there was no justice in Ramsay's intent. On 16 April 1883, a resolution of the Academy's Academic Board, acting under Ramsay's guidance, lumped the cadet engineers and cadet midshipmen together for commissioning testing. Chandler balked at approving the resolution since "it will place them [cadet engineers] very low down upon the combined list."[63]

Away from Washington and preoccupied with his ailing mother, Secretary Chandler proposed that three of the ten commissions thought to be available in 1883 should be in the rank of assistant engineer and the remaining seven should be ensigns (line officers).[64] A compromise was eventually reached with Ramsay and the class of 1883 graduated fifty-four naval cadets, twenty-one of whom were former cadet engineers. Of these twenty-one engineers, seven were retained after graduation and went on to achieve commissioned rank.[65] Those engineering cadets whose ranking was too low to receive a commission had to settle for "certificates of graduation, honorable discharges and one year's sea pay, to return to civil life, after the most thorough discipline and with the best education that this country can furnish, obtained without expense to themselves, or to their parents or relatives."[66]

Some line officers fell victim to the 1882 legislation, but the greatest attrition occurred among the engineers. In a wry twist of "every officer a steam engineer," the line had finally succeeded in imposing homogeneity on the future members of the officer corps during their important early years at the Naval Academy. The Naval Academy was now free to carry on with a more focused emphasis upon traditional naval topics, such as gunnery and navigation, with only minimal acknowledgment of the new scientific and engineering basis of the naval profession as practiced abroad.

While successful in eliminating the separate engineering program at the Naval Academy, members of the post–Porter line generally were aware of the stagnation and technical inferiority of the navy. European powers were developing new and better warships, and this contributed to the decision to send U.S. naval officers abroad to receive postgraduate training in naval architecture. Such a program was designed to ensure a certain level of technology acquisition. The challenge for the line was how to maintain control of its foreign-educated naval constructors and the technology they would create. The answer came as the result of an action taken by the British Admiralty in 1896.

ENGINEERING GRADUATE EDUCATION

Pursuit of graduate technical education was an important indication of the intellectual shift among younger American naval officers during the late 1800s. The issue came to the foreground in 1896 when some members on the British Board of Admiralty, sensitive to the importance of engineering to naval power and fearing technology transfer, were finally able to sustain a vote to exclude foreign students from the naval shipbuilding course at the Royal Naval College (RNC), Greenwich.[67] Before this, one or two of the top U.S. Naval Academy graduates entered the Construction Corps each year and received graduate training in naval architecture at Greenwich. When the RNC course was full, U.S. officers were sent to the less attractive courses at the École d'Application du Genie Maritime in Paris or to the University of Glasgow.[68] The preeminence of the Royal Navy made RNC, Greenwich, the much-preferred course, but the British prohibition forced the Navy Department to investigate alternate courses of study. Congressional sentiment existed to establish a graduate training course for naval constructors in the United States. Such an action was consonant with the burgeoning graduate educational system in late-nineteenth-century America and would enhance America's position as a naval power. The creation of an American naval architectural school also would satisfy those jingoes distrustful of European influence.

The first proposal for an American school of naval construction was submitted to the Naval Academy's newly established Committee on Post-Graduate Courses by Assistant Naval Constructor Richmond P. Hobson in February 1897.[69] Hobson had ranked first in his Annapolis class and entered the Construction Corps after his graduation in 1889. He proposed a three-year domestic course of study to replace the three-year course at Greenwich and the two-year course at the École du Genie Maritime. The curriculum would involve training in "shipbuilding and design and naval architecture," with auxiliary courses in "higher mathematics, applied mechanics, hydraulics [hydrodynamics], physics, chemistry, steam engineering, technology (machine tools, metallurgy, mechanisms), naval engineering administration, and modern (technical) languages [French and German]."[70] Since naval architecture required "object lessons more than the studies of any other profession," Hobson specified close ties between naval constructors and the private sector and, under his plan, students would spend their summers at private shipyards.[71]

While Hobson was proposing his graduate naval architectural course for the Naval Academy, the Bureau of Construction & Repair investigated educational options at private universities. In July 1897 the Bureau sent an inquiry to the Massachusetts Institute of Technology (MIT) concerning the possibility of establishing a three-year course in naval architecture for naval constructors.[72] A civilian naval architectural program already existed at MIT, having begun as a course option in mechanical engineering in 1889, but by 1893 it had become a separate department. The textbook was written by the department head, Professor Cecil Peabody, and was based largely upon French texts and the French practice of naval architecture,[73] a mathematically intensive, theoretical approach least preferred by the navy.

In September 1897 MIT forwarded a draft curriculum to Chief Constructor Philip Hichborn at the Bureau of Construction & Repair. Later that month, a line officer, Captain Dickens of the Bureau of Navigation, arrived at MIT to discuss the proposal. The naval architecture department was moving to a new building and the students were on vacation. Dickens saw only Professor Peabody and was not impressed with the facilities. Judging budding naval constructors fresh from Annapolis incapable of maturity and self-discipline, Dickens warned that "there is no method by which students can be kept up to their work by any means of discipline at the disposal of the Institute" and recommended that the navy establish the postgraduate naval architecture course at the Naval Academy.[74] There, students could be kept on a short leash.

Meanwhile, Assistant Naval Constructor Hobson was transferred from Newport News to the Naval Academy, where he submitted a more polished set of "provisional regulations" for his graduate course.[75] When the Academy superintendent, Captain Philip H. Cooper, doubted the need for a three-year course of study, Hobson responded that naval architecture was "an embodiment of the applications of modern science, covering practically all of the abstract sciences and most of the natural and concrete sciences," and the "preparation of the graduate of the Naval Academy for this advanced profession is only rudimentary at the best."[76] Hobson reinforced the importance of naval architecture, and indirectly, the importance of engineering knowledge to modern naval power: "the programme . . . is on a minimum basis consistent with efficiency; to reduce it in scope and in time would strike a blow at the future efficiency of the Construction Corps, which the world has recognized universally and the United States particularly the incontestable fact that upon the efficiency of the Construction Corps must depend the efficiency of naval materiel."[77] This was just the kind of technocentric argument that played on the deepest fears of line officers concerned about losing control of their profession.

A year elapsed before the navy took any action on the graduate course at Annapolis. Hobson went to war and obtained notoriety for his failed attempt to bottle the Spanish fleet in the harbor at Santiago de Cuba. In October 1898 Rear Admiral A. S. Crowninshield, acting for the secretary of the navy, directed Rear Admiral John Howell, senior member of the Naval Examining Board, to chair a committee to consider "the matter of providing for a thorough course of Naval Architecture, at the U.S. Naval Academy . . . [and to make] special recommendations in regard to the length of the course, corps of instructors, branches of study, and all questions relating to a full and successful line of instruction."[78] The Howell Board included Professor of Mathematics William Hendrickson, USN,[79] and Naval Constructor John G. Tawresay. The superintendent of the Naval Academy was directed to begin a provisional course of study based upon the Hobson proposal while the Howell Board conducted its deliberation.

In January 1899 the Howell Board recommended a three-year graduate course with emphasis upon theoretical instruction, but with practical experience to be gained through summertime work at the navy yards and private shipyards.[80] The first two years of study were to be devoted to theoretical and applied mathematics, mechanics, physics, and chemistry. The "professional subjects" — steam engineering, naval construction, and naval architecture — would be left for the third year, along with the study of French.[81] A naval constructor would have general charge over the courses in naval construction in the new "School of Naval Architecture" under the command of the Academy's line officer superintendent. Assisting the naval constructor would be a professor and an assistant professor of naval architecture. Three other professors, specializing in mathematics, mechanics, and marine engineering, were to be appointed from civilian life as chairs of their respective departments.[82] Three cadets, with proven abilities in mathematics, were to be selected each year by the Academy's Academic Board. During their senior year at the Academy, they would pursue a modified course of study that devoted more time to mathematics, mechanics, and steam engineering and less to traditional line subjects such as seamanship, ordnance, and navigation.[83]

The Howell Report underscored the importance of science-based engineering education in naval architecture to contemporary naval power. The Board feared a naval architect gap and argued that "national pride demand[s] that such a school, shall reach, as quickly as possible, a position second to none. Our country does not desire, and can not afford, to have its Naval Architects less thoroughly equipped than those of the leading maritime powers."[84]

Four days after receiving the Howell Report, Secretary of the Navy John D. Long asked Congress for $21,996 to establish the School of Naval Architecture at Annapolis. Besides facilities, the funds would be used to hire four professors and one assistant professor.[85] Applications were received from civilian naval architects for the faculty positions in February, and the graduate course in naval architecture appeared to be on track.[86] Secretary Long directed the Academy superintendent to put the "main features" of the Howell Plan into operation "as far as possible to do so without legislation."[87] The naval constructor in charge of the interim postgraduate course, Lawrence Spear, reported that the present "Hobson" course approached "as nearly as possible" the recommendations of the Howell Board.[88] Nevertheless, Spear complained that the niggardly allocation of facilities by Superintendent Frederick V. McNair had forced a combination of the two classes presently enrolled, and that the course of instruction would have to be curtailed to graduate the first class of naval constructors in 1900.[89]

Apparently, interference by McNair derailed the Howell Plan the following month. On 7 March 1899, Spear submitted a letter that resulted in the cancellation of the entire postgraduate naval architectural course at Annapolis.[90] What occurred among Spear, Superintendent McNair, and the Bureau of Construction & Repair during late February 1899 is unclear. Spear had repeatedly complained of a lack of support from, and interference by, the superintendent regarding facilities for the interim naval architectural course. Spear also had run-ins with the Academy's academic board that had responsibility for all aspects of undergraduate education.[91] The magnitude of the conflict resulted in the Bureau of Navigation canceling the interim "Hobson" course of study at the end of the year. Those advanced students whose work was satisfactory would be sent to Glasgow for one year's study to finish their instruction. The first-year students were sent to Paris for two years of study at the École du Genie Maritime. Two naval cadets would be selected from the Academy classes of 1899 and 1900 and sent, if possible, to the Royal Naval College, Greenwich. If the British prohibition on foreign students was still in effect, the Bureau of Navigation planned to send these students to Paris.[92]

Failure to implement the Howell Plan left the navy in the same position as 1896 when the British closed the RNC naval constructor course to foreigners. The navy again decided to investigate civilian institutions within the United States. Three months after the last textbooks were shipped from Annapolis to the students transferred to Paris, the Office of Naval Intelligence reestablished contact with MIT.[93] The MIT course was deemed satisfactory, and three Naval Academy graduates began their graduate studies at MIT in the fall of 1901.[94] The naval constructor course, designated Course XIII-A by MIT, lasted three years and resulted in a master of science degree. The program was monitored closely by the Bureau of Construction & Repair and the officer students were assigned to the naval yard at Charlestown. This apparently satisfied the concern over discipline and engineer independence Captain Dickens had raised in 1897.

The establishment of the navy's naval architecture course at MIT marked the beginning of combined navy–Institute work in naval architecture and warship design that continues to the present. No longer would American naval constructors have to be sent abroad to learn their profession. The "American School of Naval Architecture," advocated by Hobson in 1897 and embraced by the Howell Board in 1899, had come into existence at last.

Other fields within the navy required officers with technical skills, and many younger line officers needed engineering graduate training in such fields as electricity, metallurgy (for ordnance), and optics. In February 1902, seven months after the first naval graduate students arrived at MIT, the superintendent of the Naval Academy was directed to appoint a board to explore the establishment of a one-year postgraduate course in engineering at Annapolis.[95] This special board recommended courses in marine engineering, electrical engineering, and ordnance.[96] Each program would include the study of physics and chemistry as well as thermodynamics, pneumatics, hydraulics, and metallurgy.[97] Forwarding the Board's recommendations to the chief of the Bureau of Navigation in April 1902, Superintendent Richard Wainwright recommended that the first students be of the rank of lieutenant (junior grade) or below and advocated that *all* naval officers eventually take the one-year engineering course.[98]

Engineer-in-Chief George Melville, chief of the Bureau of Steam Engineering, saw this graduate program as an opportunity to enhance engineering within the navy, but also as a potential threat to the autonomous power of the technical bureaus. He moved quickly to exert control over the postgraduate program, recommending the immediate dispatch of two naval engineer specialists to Annapolis to develop the details of the projected course. Citing the "absolute difference of train of thought and study" between graduate and undergraduate study, Melville wanted to reduce academic dilution by establishing a graduate engineering faculty separate from that of the Naval Academy.[99] He wanted the graduate course to start slowly and initially be limited to trips to industrial establishments to view the manufacture of ordnance and engineering appliances and ships. In a tongue-in-cheek jibe at the up-and-down, post-1882 Academy curriculum, Melville justified these field trips as a much needed respite from the "arduous and exacting work to which the young officers have been subjected during the past five years [as Naval Academy students]," which had "unfitted them, in part, for continuous and severe mental labor."[100]

Led by Admiral of the Navy George Dewey, the senior line officers of the newly formed General Board opposed the graduate engineering program. The secretary of the navy established the General Board in March 1900 as a counterweight to the bureau system. Line officers had been lobbying for a naval general staff, but Long saw that as an erosion of civilian control and opted for the General Board.[101] Dewey, concerned with the contemporary officer shortage in the fleet, minimized the link between engineering education and naval power, viewing postgraduate training as a luxury that the navy could ill-afford.[102] Melville thought Dewey and the General Board's position short-sighted, but in an attempt to compromise, asked that only a few officers be detailed for study. Melville hoped to increase engineering consciousness among the officer corps gradually, and believed that "the establishment of a post-graduate course in Engineering at the Naval Academy might have a very important effect in making the [amalgamation] experiment a success."[103]

Dewey's obstinacy forced the referral of the Post Graduate School question to a committee consisting of the chiefs of Bureaus of Navigation, Steam Engineering, and Ordnance. In a show of bureau solidarity, the chief of the Bureau of Navigation sided with the two technical bureau chiefs and rejected the position of the rival General Board.[104] Secretary of the Navy William H. Moody approved the bureau chiefs' report that called for the annual assignment of four passed-midshipmen to postgraduate study in ordnance and four to marine engineering.[105]

The first graduate studies in ordnance engineering began in 1904, not with passed midshipmen, but with five commissioned officer students. The Bureau of Ordnance was unable to assign any instructors, so each new student determined his own course of study, which amounted to little more than on-the-job training at the bureau, the Naval Gun Factory, and the Naval Proving Ground. Over the next few years, the course of instruction was systematized, and by 1912 a total of thirty-five ordnance specialists had received some form of graduate training by the Bureau of Ordnance.[106]

Graduate studies in marine engineering under the Bureau of Steam Engineering began in 1905, with ten students assigned to on-the-job training at naval stations. As with the ordnance program, the initial results were poor and the bureau recognized the need for closer control and direction. This came in June 1909, when a graduate School of Marine Engineering was established at Annapolis. Ten officer volunteers were selected for the first class, with nine or ten students to be appointed each subsequent year. Each year, two graduates of the course would be selected for designation as specialists in marine engineering and assigned to the Bureau of Steam Engineering.[107]

The establishment of the in-house ordnance and marine engineering programs did not mean that the Navy Department had forsaken academia. Three months before the marine engineering course was created at Annapolis in 1909, the navy established a two-year course in marine engineering at MIT for officers who wanted to specialize in marine engineering design rather than engine operation and repair.[108] The navy reinforced its relationship with academia by broadening the Annapolis Post Graduate School curriculum in October 1912 to include courses in electrical engineering, radio engineering, and naval construction. University professors served as guest lecturers in various fields to "secure the best talent available."[109] Harvard University professors E. L. Dawes and H. E. Clifford received $1,000 and $1,500 respectively for a series of lectures in electrical engineering during the 1913 spring term. Joseph Ames, of Johns Hopkins, was paid $1,000 for lectures on physics, while A. L. Walker of Columbia University split a similar fee with a colleague to lecture in chemistry.[110] These lectures were transcribed by the Annapolis Post Graduate School staff with an eye toward future use by the one "engineering" instructor, two instructors in "electricity," one instructor in physics, and one humanities lecturer hired under the navy's fiscal year 1913–14 budget. The Post Graduate School also requested $3,000 to continue guest lectures by civilian academics.[111]

According to an assessment in the *Journal of the American Society of Naval Engineers* in 1916, the consensus within the naval officer corps favored a one-year course of study at the Annapolis Post Graduate School to be followed by a one-year course at "universities not under naval control, for the purpose of broadening their viewpoint by contact with civilians."[112] In January 1916 seventy-six officers were enrolled in technical postgraduate courses of study. The thirty-one students at Annapolis included sixteen marine engineers, four electrical engineers, one radio engineer, five ordnance men, and five naval constructors. Twenty naval students were enrolled at Columbia University: nine marine engineers, ten electrical engineers, and one "metallographist." Thirteen naval constructors were enrolled at MIT and three civil engineers at Rensselaer Polytechnic Institute. The remaining nine students were studying ordnance engineering at various steel plants, the Naval Gun Factory, and the Naval Proving Ground.[113]

While opposition to engineering training for line officers lingered in some older officers of the line, the Post Graduate School was accepted widely as a valuable asset to the naval establishment and naval profession.[114] The hostility toward engineering by senior officers raised in the old line was

being replaced by a technological consciousness among the younger officers of the post-1899 amalgamated line. The civilian-academic model was
in place at the graduate school at Annapolis, and a direct naval-academia
relationship had been established between the technical bureaus and civilian universities such as MIT, Columbia, and Rensselaer Polytechnic.

THE NEW "AMALGAMATED" LINE

In 1899 engineering officers were absorbed into the "new" line. This
amalgamation was the result of a study made by the Personnel Board under the chairmanship of Assistant Secretary of the Navy Theodore Roosevelt in 1897. Amalgamation was tacit recognition of the technological basis of contemporary naval power and the necessity of engineering training
for all naval officers. As Roosevelt had said, "On the modern war vessel,
every officer has to be an engineer whether he wants to or not."[115] Amalgamation was a recognition of the importance of technology, but also
served the purposes of the line by eliminating the independent corps of
naval engineers. Despite Roosevelt's pronouncement, the Amalgamation
Act did not require line officers to study engineering.

Naval engineers, conversely, were given two years in which to pass a test
on seamanship to qualify as deck watch officers. Failure meant loss of their
commissions.[116] Although forced to prove a proficiency in seamanship,
most naval engineers viewed amalgamation as a victory in their fight for
status with the old line. Naval engineers received naval rank titles and comparable pay. However, despite statutory equality, resistance to naval engineers and the debate over who should control naval technology continued
for fifteen more years. In 1909 the former chief of the Bureau of Steam Engineering, Engineer-in-Chief George Melville, summarized the naval conservatism that worked against true amalgamation:

> Mr. Roosevelt referred to this change [need for engineering knowledge] as an
> "evolution" and not a "revolution." This was true and yet it was an extremely
> radical change and it is taking time to have the change come about completely.
> Naval officers, from the very nature of the conditions which surround them, are
> intensely conservative, and men who have passed fifty and have been trained in
> a certain way find it very hard to adapt themselves to absolutely novel condi-
> tions.[117]

Melville compared the resistance to engineering to the

old fight in education between the classics and the sciences. The men with the classical education held the important positions and the advocates of scientific training for a long time had a difficult task to secure adequate recognition. It has been something like this in the Navy. The older officers, who are still the ruling faction and fill the higher positions, find it difficult to adopt enthusiastically a system which is entirely contrary to their training and most cherished traditions.[118]

This conservatism was manifested in ambivalence, or even distrust of new technology even after its reliability had been proven. Five years after the first American battleships were authorized, four of the gunboats built in 1895 were equipped with sails and steam propulsion.[119] In 1897, the Naval Academy received authorization to build a new training vessel, propelled by sail power alone. Melville found it "incredible that a sailing ship should be used for the training of men who are to serve on and command vessels entirely propelled by machinery and without any sails." He churlishly rejected the reasoning as "exactly the same as that given by the classicists for the study of the dead languages: that it gives a kind of training which is supplied by nothing else and which is of immense benefit whatever may be the future line of work."[120]

The internal navy controversy over amalgamation became public in 1905 after a steam explosion in USS *Bennington* killed several sailors and cast aspersions upon the qualifications of the young ensign serving as chief engineer. In a 1905 article in the *U.S. Naval Institute Proceedings* entitled "Is Amalgamation a Failure?," Lieutenant Commander Lloyd H. Chandler presented an extensive defense of the amalgamated line.[121] Chandler had matured in the post-1883 steel navy and spoke for those officers who accepted the pervasiveness of engineering-based technology within the modern navy. Chandler portrayed amalgamation as the result of "the ever-increasing realization of the fact that the major part of the duties of the line and of the engineer corps were purely mechanical and that equal engineering knowledge was required of both corps as to all sorts of machinery. . . . In other words people saw that the naval officer of the day must of necessity be a competent engineer."[122] Chandler felt the amalgamated officer corps "blazed the way toward the ideal condition," drawing other naval powers such as Britain along. The "blindness of caste [that ruled] . . . that a man cannot be a military officer and a mechanic at the same time," claimed Chandler, was finally routed.[123]

Many of Chandler's colleagues, and older line officers, did not agree and had no desire to become or be identified as mechanics. Opposition to the Amalgamation Bill had been strong as it moved through Congress, and few line officers took engineering duty seriously. The expansion of the navy after 1898 resulted in a shortage of officers, especially engineers, but the Amalgamation Act did not offer any redress by compelling line officers to achieve proficiency in engineering. Writing in 1904, the chief of the Bureau of Steam Engineering saw this as a fatal flaw: "So few officers of the line are taking up engineering seriously that the situation is becoming alarming."[124]

Regret over amalgamation was not limited to line officers. In 1906 Rear Admiral (former Chief Engineer) G. W. Baird condemned amalgamation for "working a great harm to the naval service."[125] He despaired of the naval profession's seeming repudiation of specialization so valued in Progressive-era America. Officers were becoming jacks of all trades and masters of none: "The amalgamated officer of to-day must 'qualify' in so many sciences, arts, professions and trades that he is not likely to be a specialist in any. While everyone else is specializing, the United States Navy alone is generalizing."[126] Baird characterized the absorption of naval engineers into the line as the culmination of a "struggle between individuals, cliques, and classes for supremacy." He feared the decline of the American navy and nation since "Every nation has been strong, prosperous, rich and great, just in proportion to its producing classes, i.e., its mechanics, farmers, artists, etc., and when these begin to wane, the nation begins to decay."[127] American naval decay had begun with the elimination of the productive engineering "middle class" by the line aristocracy.

The Navy Department had promised Congress that amalgamation would lead to the end of the "disagreeable contention" between naval engineers and the line. The service newspapers reported that the Navy Department had threatened to "call to account" any officer who opposed the Amalgamation Bill before Congress in early 1899. According to Baird, the ramming through of such an "unreasonable and so un-American" bill surprised many naval engineers.[128]

Amalgamation was a watershed in the social and institutional history of the navy. Old sea dogs joined with proud naval engineers to decry the mongrelization of their respective officer corps. Their complaints were eventually drowned out by the rise of the younger officers of the "new" line, trained for employment in both seamanship and engineering. Progressives may have been dismayed at the rejection of specialization. Nevertheless, amalgamation ended, at least superficially, the divisive social tensions that could no longer be tolerated within a service asking Congress to fund newer, larger, and more technologically complex ships.

The professional homogeneity promised by amalgamation was not achieved quickly. As late as 1913, the chief of the Bureau of Construction & Repair complained to the House Naval Affairs Committee about "dissatisfaction existing in the Staff of the Navy by reason of delayed promotion in the Staff as compared with officers of the Line of the Navy."[129] Four members of the Construction Corps resigned in the ten months preceding February 1913 — three with more than twenty years' service. The chief constructor blamed this loss of expertise to the private sector, in part, on the fact that naval constructors routinely were promoted to the rank of commander fifteen years later than their line counterparts.[130]

Amalgamation underscored the acknowledgment of the complex technical basis of naval power by the political hierarchy and naval officer corps of 1899. Despite nominal equality, engineering specialists were still relegated to secondary status within the profession. Senior line officers did little to encourage engineering training for former seamen officers of the old line. This lack of interest in engineering diminished with the ascension of the post-1883 generation into the naval hierarchy.

TECHNOLOGY AND NAVAL SOCIETY

The widespread intrusion of the steam and iron material culture into naval warfare between 1861 and 1865 threatened the social status quo of the American naval officer corps. As a result, the line regulated naval engineers. As the purveyors of the technology that threatened the prewar sailing ship ethos, naval engineers were regulated severely in the years following the Civil War. Suppression of engineers during the Porter era contributed to an American navy mainly consisting of obsolescent, but strategically consonant, sail-dependent warships at a time when major naval powers such as Britain and France were building complex iron and steel warships. As a result, American line officers grew increasingly embarrassed by the artifacts upon which their profession was based. Technological innovation was clearly necessary, but how could modern warships be built without increasing the power and influence of naval engineers and the technical bureaus?

By the end of the 1890s, and especially after the victory over Spain, American battleship construction continued at high speed and a new consensus was emerging in the line regarding the importance of technology to naval power. Using Theodore Roosevelt's pronouncement that all naval officers must be engineers, the line absorbed the separate Engineering Corps through the Amalgamation Act of 1899.

Factional disputes within the American naval officer corps took up a great deal of attention between 1861 and 1910. The common line-engineer acceptance of the navalist thought style and of the post-1890 battleship strategic paradigm did not ameliorate the antiengineer and antitechnical bureau campaigns of the line officers. These grew even more intense as the relationship between strategy and technology grew more clear-cut after 1898. The struggle to suppress naval engineers, through control of their education process and limitations on their social status, foreshadowed the continuing struggle to break engineering control over naval technology between 1900 and 1910. This conflict also foreshadowed the technologically based factional disputes that would rack the navy between the world wars.

Steam engineering officers were the lightning rods for general line displeasure within the changing technological framework of late-nineteenth-century naval warfare. Yet hostility to engineers varied. Naval constructors fared better because of the small size of the Construction Corps and their importance to naval power and the American naval renaissance of the 1880s. As ship designers, they did not typically go to sea or operate the noisy, leaky, dirty steam machinery that made life in a steam warship a shadow of what naval life had been in the "Old Navy" of wooden ships and iron men.[131] Starting in 1879, the top one or two graduates in each academy class, who so wanted, were sent to Europe for graduate education in naval architecture and were commissioned as staff officers in the Construction Corps.

However, naval constructors and their Bureau of Construction & Repair, which controlled the entire ship system, were potentially much more dangerous "engineer-sociologists" — to use Michel Callon's term — to the technological status quo than steam engineers.[132] Naval constructors defined the exemplar of the navy's technological paradigm. The Construction Corps was limited to forty officers by the amalgamation legislation of

1899. Construction and maintenance of the battleship fleet increased demand for naval constructors, and the corps was expanded by legislation to forty-six in 1902 and to seventy-five in 1903.[133] A faction of activist line officers before World War I believed that the growing Construction Corps also had to be brought to heel. That effort would continue through 1908.

CHAPTER 2

Competing for Control

Line Officers, Engineers, and the Technological Exemplar of the Battleship Paradigm

In 1890 Congress authorized the first technological exemplars of the new battleship-based strategic paradigm: the large-gunned, steel battleships, *Indiana, Oregon,* and *Massachusetts.* The battleship strategy was based in guerre d'escadre, the historic strategy of strong maritime powers such as Britain, in which line-of-battle ships engaged similar enemy fleets. To the chagrin of navalists, guerre d'escadre and the expensive fleet of ships of the line it required had not been the strategy of the U.S. Navy before 1890. Publication that year of Captain Alfred Thayer Mahan's *The Influence of Sea Power on History, 1660–1783* linked Britain's rise to great power status to its maritime dominance based in the practice of guerre d'escadre. Mahan's widely read book struck a chord in America that now had the wealth, worldwide commercial network, political cadre, and industrial base to challenge the world's premier naval powers. Mahan's conceptualization of the proper type of war at sea drew on land warfare theory and focused mainly on a single, Jominian-style, decisive engagement between battleship fleets.[1]

In many ways the *Indiana*-class ships were a reaction to Britain's battleship and modern cruiser building program instituted to serve Britain's adoption of a two-power naval standard in 1884.[2] Line officers used the British program to garner congressional, business, and popular support for construction of modern warships to escape the postbellum naval stagnation. Designated "coastal battleships," as were the three pre–Spanish-American War ships that followed, the three 1890 ships had a limited steam-

The Mahanian battleship-based strategy was rooted in the large gun and steel armor. Both are evident in the forward 13-inch gun turret mounted in Oregon, *one of the first three battleships authorized in 1890.* (Courtesy of Allan J. Drugan, from the album of Lewis H. Rockey. Naval Historical Center, NH 42969)

ing range. This reflected Congress's reluctance to move away from a defensive, continentalist naval strategy and the historical view that ships of the line were tools of empire.[3]

The naval war against the Spanish in 1898 seemed to justify the country's investment in battleships. In actions that would have made the frustrated navalists of the Early Republic cheer, Congress, between 1898 and 1905, authorized the construction of fourteen more battleships, enabling the United States to rise to the status of a world naval power. This naval buildup was quite expensive, consuming an average of 17 percent of the annual federal budget during Theodore Roosevelt's and William Howard Taft's administrations.[4]

The tremendous cost of a battleship navy was not without political and popular opposition. In 1905 Michigan businessman Fred Hobbs castigated Admiral of the Navy George Dewey over the battleship program while presenting a classic criticism of capital ships: "many of us *do* object to your extravagant proposals for the building of battleships. . . . For once put yourself in our place and study the question of expense from the standpoint of the producer and then see if it is wise to spend millions on a single ship that could be torpedoed and reduced to scrap in the twinkling of an eye."[5]

To counter opposition to battleship construction, Lieutenant Commander John Gibbons published a naval manifesto in 1903. Gibbons resurrected the idea of a naval general staff to provide line officers control over the technological basis of the profession and the "technical bureaus and industrial interests" in order to provide a "continuous building program." Gibbons called for a "campaign of education among the people, through the medium of naval leagues and press agitation, which have proved so successful in England and Germany." Gibbons thought such proselytizing would result in the allocation of a "proper share" of federal tax revenues to naval expansion over a period of four to eight years.[6]

Besides countering public opposition to battleship programs, some line officers also waged an intranavy struggle to wrest control of the technical basis of their profession from the engineering specialists within the technical bureaus, who, many line officers believed, failed to have an adequate understanding of strategy. Both line officers and naval engineers supported the battleship-based strategy of guerre d'escadre but differed over details of battleship technology. Each group believed it was best suited to control the "normal" refinement of the battleship technological paradigm.[7]

The line-engineer battle over who would control the technological basis of the battleship navy was joined over design details on the *Connecticut*-class battleships authorized in 1902. The furor over the *Connecticut* designs informed the engineers, and many members of the older line hierarchy, that the generation of more technologically savvy line officers raised in the post-1883 steel navy would no longer tolerate the engineers' and technical bureaus' monopoly over naval technology.

MAHANIAN NAVALISTS AND THE RISE OF THE BATTLESHIP

Captain Alfred Thayer Mahan is perceived widely as a spokesman for American imperialism and as the paradigmatic American navalist. Mahan was characterized by the naval historian Margaret Sprout as an "evangelist of sea power" who "precipitated and guided a long-pending revolution in American naval policy." A more recent, balanced account has been offered by Philip Crowl. A more muted appraisal of Mahan's influence is offered by Peter Karsten, who argued that commerce raiding was already a passé American strategic policy as early as 1865. Mark Shulman has offered the most recent and detailed analysis of the 1880s and the pursuit of an offensive-defensive navy.[8]

Karsten linked line pressure for an enlarged, modern navy to "career anxiety" over agonizingly slow promotions during the 1870–80s.[9] This anxiety led the "Young Turks" of the line to embark on an aggressive public relations campaign for an enlarged navy as an integral tool to achieve expanded markets overseas.[10]

Alex Roland described the strong identification that exists between a naval officer and his ship.[11] This was an important factor that drove younger, more energetic line officers to push for modern ships in the post-Porter era. As one young officer put it in 1886, a naval renaissance would provide "a few vessels that we will not be afraid or ashamed to show to foreign powers, and that will keep up the dignity of the great American Republic."[12] Dignity was tied to professional self-image and that was linked intricately to technology. Commodore Foxhall Parker elucidated a more concrete need for new ships during the 1870s: in case the navy encountered "a hostile force at sea." Parker longed for a time when the United States would "awaken from her lethargy . . . and once more put forth her strength upon the deep."[13]

Thanks to lobbying by naval officers, it became increasingly clear to Congress, and to some extent the American public, that expanding American markets could not be protected by antiquated Civil War ironclads suited only for coastal defense. The answer, according to navalists, was to build ships that would embody the most modern technologies and thus uphold the dignity of the nation and of the American naval profession. When America's rival for world trade, Great Britain, announced her plans to construct ten large-gunned, steel battleships and sixty cruisers between 1889 and 1894, Rear Admiral Stephen B. Luce, founder of the Naval War College, quickly called for the construction of an American "battleship" fleet to counter the Royal Navy.[14]

The American battleship navy originated during the administration of Secretary of the Navy Benjamin Franklin Tracy (1889–93).[15] Against the

British naval modernization program, the U.S. Navy could only offer the armored ABCD ships,[16] the first ships of the modern steel navy, still full-rigged with masts and sails, and the four modern heavy cruisers, *Texas*, *Maine*, *New York*, and *Olympia*, all of which were woefully inadequate against the battleships the British were constructing.

Tracy took up Admiral Luce's banner. In his first annual report as secretary of the navy (1889), Tracy called for the creation of a fighting force based upon "sea-going battleships." According to Tracy's biographer:

> The secretary rather bluntly proclaimed that the United States Navy had fallen behind not only Great Britain, France, Russia, and Germany, but even Turkey, China, and Sweden. The profitless program of building unarmored cruisers . . . could not provide defense for an unprotected American coastline of 13,000 miles with its twenty centers of population, wealth, and commercial activity. Tracy believed that a more powerful deterrent force was distinctly necessary.[17]

Tracy dismissed guerre de course as ineffective in deterring a naval attack: "The capture or destruction of two or three dozen or two or three score of merchant vessels is not going to prevent a fleet of ironclads from shelling our cities or exacting as the price of exemption a contribution that would pay for their lost merchantmen ten times over." Tracy's answer was "a fleet of battle-ships that will beat off the enemy's fleet on its approach." The United States needed "a navy that will exempt it from war, but the only navy that will accomplish this is a navy that can wage war." That, to Tracy, was a navy of armored battleships.[18]

The March 1889 crisis over a Pacific coaling station on Samoa involving the United States, Germany, and Britain aided Tracy's advocacy of battleships. A showdown was averted by a typhoon. When the storm had subsided, only Britain's HMS *Calliope* had been powerful enough to steam to the open sea and avoid disaster. The underpowered, obsolescent American ships were driven ashore and American naval power in the Pacific all but ceased to exist.[19]

Tracy made sure that the lesson of Samoa was not lost on Congress and the White House. President Benjamin Harrison had called for the construction of a "sufficient number" of modern powerful warships in his inaugural address. After reading Tracy's annual report, Harrison urged Congress to appropriate funds for the immediate construction of eight "battleships."[20]

Harrison's Republican Party controlled both the House of Representatives and the Senate, and naval enthusiasts in Congress offered bipartisan support for the eight battleships. Senator Eugene Hale of Maine submitted legislation for the Tracy–Harrison program in December 1889. His bill had an excellent chance of being enacted until a navy Policy Board report, published in January 1890, greedily called for the construction of ten long-range and twenty-five short-range battleships, twenty-four cruisers, and one hundred torpedo boats at an estimated construction cost of $281 million. If this program were completed, the U.S. Navy would rank second only to Britain's Royal Navy in tonnage.[21] Popular outrage against such large-scale naval expansion translated into stiff congressional opposition to Hale's bill.[22]

Senator William E. Chandler, Tracy's predecessor as secretary of the navy, spoke out forcefully against the battleship concept. When Chandler denounced any worldwide mission for the American navy, he ignored almost nine decades of distant naval operations in support of commerce, beginning with the operations against the Barbary Corsairs during the Early Republic. Chandler argued that the nation's legitimate defense needs could be served by heavily gunned and armored monitors that could protect the coast from hostile battleships economically.[23] Chandler's antibattleship stance elicited strong support from Midwestern isolationists and pacifist organizations.

Compromise legislation was introduced in the House of Representatives to build three "seagoing, coastline battleships" to the specifications of Tracy's original medium-range battleships. After a bitter fight against the Populist opposition, the legislation passed the House and a roughly similar bill passed in the Senate. In its final form, the Naval Bill of 1890 authorized $18 million for three battleships, each with four 13-inch guns, eight 8-inch guns, four torpedo tubes, and driven by reciprocating steam engines to a top speed of 16 knots.[24]

The navy's next battleship, authorized in 1892, marked an escalation in American pursuit of naval power. This ship, *Iowa*, came about in partial response to President Harrison's call for a modern fleet, not to protect the coast, but to protect American lives and property throughout the world. Tracy asked for appropriations for two *Iowa*-class battleships. Antinavy opposition resulted in the authorization of only one ship. Although a "coastline" battleship, *Iowa* was a giant step forward in American naval power since it was the first battleship not limited in range or size (tonnage) by statute. The ship carried 200 tons more coal than the 1890 battleships and that allowed *Iowa* to operate beyond the 1,000-mile limit previously stipulated by Congress.[25]

A failing economy aided the Democrats in capturing majorities in both houses of Congress and the White House in 1892. The Democrats' platform, under which Grover Cleveland was elected to his second presidency, called for moderate tariffs and fiscal responsibility, which included reduced naval expenditures. Cleveland's new secretary of the navy, Hilary A. Herbert, supported curbs on naval spending. A former Confederate army officer, Herbert had served for sixteen years as a representative from Alabama and had spent four terms on the House Naval Affairs Committee, one of these as chairman.

Despite his initial allegiance to the party platform, Herbert had a complete change of heart shortly after taking office. Crediting Alfred Thayer Mahan for his conversion, Herbert supported the Naval War College and its advocacy of the new battleship strategy, and opposed the administration's cutbacks in naval construction. Herbert used Mahan's histories to support his call for appropriations for five battleships from Congress.[26]

Secretary Herbert, like Tracy before him, became the point man for the battleship and the strategy of guerre d'escadre. Herbert chose to ignore the effectiveness of Confederate naval commerce raiders while remembering the failure of lightly armed Confederate blockade runners and commerce raiders when confronted with heavily armed Union ships.[27] He rejected the policy of building unarmored warships and called for the construction of at least one battleship in the next building program. His proposal embarrassed the Cleveland administration and came at a most inauspicious time for the domestic economy. As Walter Herrick observed, such a program contradicted the Democratic Party's tradition, and President Cleveland cautioned Congress that large-scale naval construction must await the return of prosperity.[28]

Cleveland was able to shelve Herbert's battleship plan, but his efforts to stabilize the economy split his own party and led to a Republican majority in the House of Representatives and Senate after the off-year election in 1894. Always friendly to the navy, the Republicans wanted to restart battleship production.

In 1895 Herbert convinced Cleveland to work with congressional Republicans, arguing that a naval construction bill would act as a catalyst on the depressed economy. This was similar to the Gladstone government's

The superposed gun turrets of the two 1895 Kearsarge-class battleships caused a series of design problems. When assigned to Kentucky in 1900, Lieutenant William Sims sent a lengthy criticism to the Bureau of Construction & Repair and later referred to the ships as the "worst crime in naval construction ever perpetrated by the white race." (Courtesy of the Filson Club, Louisville, KY. Gift of Mrs. Alexander M. Watson. Naval Historical Center, NH 92507)

increase in the 1884 naval estimates to stimulate the depressed British economy.[29] Abroad, the emergence of Japan as a Pacific power, and the Venezuela-British Guiana border dispute, which threatened a showdown with Britain over the Monroe Doctrine, reinforced the need for a stronger American navy.[30] Domestically, the arguments of Alfred Thayer Mahan were well entrenched in American culture. Considering these factors, Cleveland asked Congress for three, large (10,000-ton) battleships.[31]

The Republican-controlled House readily agreed to an appropriation of $12 million for three battleships. Populist opposition reduced the final authorization to two battleships of 11,520 tons, each carrying four 13-inch and four 8-inch guns, an armament more powerful than that of the preceding *Iowa.* Much the same formula was repeated in 1896, with Herbert asking for two battleships, the House voting for four, the Senate voting for two; a conference resulted in legislation approving the construction of three.[32]

The election of 1896 brought the Republicans — with their strong naval platform — the presidency and continued majorities in both the House of Representatives and the Senate. Despite the pro-navy components of his party's platform, President William McKinley was personally opposed to a further naval buildup. He appointed the former governor of Massachusetts, John D. Long, as secretary of the navy, a man whom fiery Assistant Secretary of the Navy Theodore Roosevelt thought "lukewarm if not actually hostile, to further expansion of the Navy."[33]

Political pressure forced McKinley to appoint the unpredictable Roosevelt to his administration. McKinley's apprehensions were soon justified when Roosevelt began to lobby for naval expansion and the active employment of the fleet in foreign policy.[34] Roosevelt allied himself with Henry Cabot Lodge and moved to annex Hawaii to thwart potential Japanese expansion. Later, with Secretary Long out of town, Roosevelt, preferring not to "drift into war butt end foremost, and go at it higgedly-piggedly fashion," readied the fleet for war after the *Maine* blew up in Havana in February 1898.[35]

In the wake of the *Maine* disaster, Long expanded the navy, buying two cruisers being built in Britain for Brazil, and purchasing or chartering merchant ships for use as naval auxiliaries. The war proved a boon to the navy as Congress authorized the construction of three new battleships within a fortnight of the Spanish declaration of war on 23 April, an indication of how synonymous the battleship had become with naval power in the United States.

The naval successes against Spain at Manila and Santiago de Cuba demonstrated the usefulness of America's new naval power. The potential for future conflict with Japan, Germany, and Britain underscored the importance of technological superiority, and strategic planning, to the application and maintenance of naval power. The correlation between technology and strategic and tactical superiority was obvious to most naval officers. What was needed, according to some members of the line, was a new institutional structure within the navy which reflected this. Many of these line officers viewed the creation of the General Board of the Navy in 1900 as a means to institutionalize the subordination of naval technology to the line. Line officers wanted to ensure that they—the strategists, tacticians, and operators of this technology—determined design specifications and approved warship plans, prerogatives previously reserved for the chiefs of the technical bureaus who dominated the navy's Board on Construction. The first conflict between the General Board and the technical bureaus involved the design of the six *Connecticut*-class battleships Congress authorized in July 1902.[36]

THE OPENING SKIRMISH FOR TECHNOLOGY CONTROL

The General Board provided the secretary of the navy with an advisory body of seagoing line officers outside the bureau system. The General Board was to "insure efficient preparation of the fleet in case of war and for the naval defense of the coast."[37] The Board had two categories of members: the full-time executive committee and ex officio members, senior officers holding specific posts, who attended monthly board meetings.[38] The first president of the General Board was Admiral of the Navy George Dewey, who served until his death in 1917. Besides Dewey, the ex officio members of the Board included the president of the Naval War College, the director of naval intelligence, and the chief of the Bureau of Navigation. Although a bureau head, the chief of the Bureau of Navigation was a line officer and was often at odds with the three main technical bureaus: Construction & Repair, Steam Engineering, and Ordnance.

The General Board was a watered-down version of the naval general staff proposed by a line officer, Captain Henry C. Taylor, in February 1900.[39] Taylor denied any attempt to interfere with the bureau system, claiming that his push for a general staff was motivated only by a desire to improve American naval war planning. To smooth any ruffled feathers in Congress and within the navy, Taylor sent a personal note to Colonel William Church, editor of the *Army and Navy Journal*, stating the raison d'être of the Board:

> The Board is to plan Naval Campaigns, to examine where our naval bases
> should be created in different parts of the world, to develop our Coast Defense,
> and stimulate the practical searching out of these and other facts which are
> purely Naval, Military, and connected with War. The Board *has nothing to do*
> *with material, manufacture, or any details of equipment, armament and con-*
> *structing* [emphasis added].

The General Board was the line's best hope to break the power of the
bureau system and to wrest control of the basis of the technological paradigm from the "engineer-sociologists" in the Washington bureaus. The
Board was to develop war and campaign plans, to provide advice regarding
logistics, to work with the War College on tactics, and to "consider the
number and types of ships proper to constitute the fleet."[40] Despite Taylor's
declaration, the General Board was headed for conflict with the technical
bureaus. Even had Taylor and Dewey wanted to focus exclusively on
"War," the interrelationship of technology, strategy, and tactics precluded
such a narrow focus by the General Board.

Secretary Tracy had established the Board on Construction in 1889 to
coordinate the bureaus' efforts to produce optimal warship designs. In 1901
the Board consisted of the chiefs of the Bureaus of Construction & Repair,
Steam Engineering, Ordnance, and Equipment, as well as the navy's chief
intelligence officer.[41] Although the Board on Construction had a rotating
presidency, the chief constructor, who headed the Bureau of Construction
& Repair, served as the de facto chairman of the board, placing the definition of warship technology firmly under naval engineering control.[42]

On 6 March 1901, Secretary Long directed the Board on Construction
to consider the designs for the *Connecticut*-class battleships to be submitted to Congress in December. After reviewing eight preliminary designs,
the Board on Construction split over which to recommend. The chiefs of
Ordnance, Construction & Repair, and Steam Engineering sided against
the chief intelligence officer and the chief of the Bureau of Equipment.
The majority report called for a main battery of four 12-inch guns in two
turrets with 10-inch armor; the minority favored the same turret arrangement but with less armor. A more serious difference involved the secondary
gun battery: the majority favored installing new, rapid-fire, 7-inch guns in
casements along the beam while the minority called for 6-inch and 8-inch
guns in six turrets, two of which would be superposed on the 12-inch gun
turrets.[43] Such a superposed arrangement (one turret directly on top of the

other and training in tandem) had been employed in the two battleships of the *Kearsarge* class, authorized in 1895.

The superposed turrets of *Kearsarge* and *Kentucky* had been the subject of controversy within the navy and sparked resentment by fleet officers. Lieutenant William S. Sims had gone straight from attaché duty in France to the newly completed *Kentucky* in 1900. Familiar with contemporary European battleships, Sims was horrified when he saw his new ship, remarking, "*Kentucky* is not a battleship at all. She is the worst crime in naval construction ever perpetrated by the white race." He later commented, "It is almost incredible that white men who have reached the present stage of civilization could have built a ship like that."[44]

The heavy 13-inch guns installed in *Kearsarge* and *Kentucky* had affected the overall design adversely. The weight of the guns required a redesigned recoil system that forced the guns to be mounted 14 inches back from the gun ports in the turret. This created an opening that exposed the guns' powder to premature ignition from an enemy shell or shrapnel, a problem Sims summed up as follows: "You could stand on the quarter deck and look into that turret, and you could spit right down into that powder magazine. Now that is a fact. You could stand in the bottom of the powder magazine, or rather, in the chamber out of the powder magazine, and looking up see a man smoking a cigarette right up on the bridge."[45]

Besides insufficient armor protection of powder rooms and magazines, the superposed turrets had a cascading, negative effect upon the overall design of *Kearsarge* and *Kentucky*. The Bureau of Ordnance was delighted with the superposed turrets since the main barbette armor for the 13-inch guns simultaneously protected the ammunition hoists for the 8-inch guns, simplifying their design. The Bureau of Construction & Repair had opposed the superposed turrets because they raised the center of gravity of the ship and reduced the ship's stability. Construction & Repair had wanted superfiring turrets, that is, separate 8-inch turrets mounted behind and above the 13-inch turrets. Ordnance thought such an arrangement would result in blast damage to the 13-inch turrets from the 8-inch guns. The 8-inch turrets, needed for defense against torpedo boats, could not be sacrificed and the Bureau of Ordnance stood firm for the superposed turret design. To offset the resulting higher center of gravity for the ship, Construction & Repair was forced to lower the height of the main deck, making *Kearsarge* and *Kentucky* "wet" in moderate seas and "criminal" designs to many seagoing officers, such as Sims.[46]

Defending the casement broadside guns in the *Connecticut*, the technical-bureau majority within the Board on Construction claimed that reports from line officers serving in the war with Spain favored the casement arrangement over that of stand-alone or superposed turrets for the auxiliary battery. The technical bureau chiefs belittled the worth of the 8-inch gun by pointing to its miserable record in 1898. At the Battle of Manila Bay, Dewey's flagship, *Olympia*, with 8-inch guns in turrets, fired only one-third the number of shells of the nonturreted 8-inch battery in USS *Boston*.[47] During the engagement at Santiago de Cuba, 319 8-inch shells were fired, with only 13 hits achieved (4.07 percent). At Manila Bay the hitting percentage of the 8-inch guns was 6.38 percent.[48] The new, quick-firing, longer-range 7-inch gun was viewed as superior (i.e., faster firing, but not necessarily more accurate) than the older 8-inch design. Also the use of 7-inch guns exclusively, as opposed to the mixed-caliber arrangement favored by the minority, had the added advantage of standardizing the secondary battery ammunition.[49]

The deadlocked Board on Construction requested guidance from the secretary of the navy. With his technical advisers at a stalemate, Secretary Long referred the question to Dewey and the General Board, providing the Board an entree into technological matters.[50] Dewey was opposed to superposed turrets. He knew nothing of the new 7-inch guns, but because of a sense of tradition and his own position as admiral of the navy, felt inclined to defend the majority position.[51] Dewey proposed a compromise design with an auxiliary battery of 8-inch guns with improved armor for the broadside gun casements.[52] Dewey opposed turrets since their multiple guns were pointed by a single individual and "the element of personnel error of one man, whether due to lack of skill, nervousness or temporary physical condition" could reduce the effectiveness of the battery in combat.[53]

With Dewey and the General Board invading its bailiwick, the Board on Construction conducted its own survey of over "eighty naval officers of prominence on the subject." The result was a compromise design featuring a mixed auxiliary battery of four 8-inch gun turrets mounted on each corner of the main deck and the newer 7-inch broadside guns in casements.[54] Secretary Long approved the Board on Construction report and Congress authorized construction in July 1902.

Line officers, through the General Board, had their first high-level involvement in the definition of battleship technology. But the General Board's recommendations counted for little when the Board on Construction presented a revised design which was accepted unanimously. Although the bureaus had agreed, a precedent had been set for General Board involvement in controlling the trajectory of the battleship technological paradigm. Among the eighty commentaries received regarding the *Connecticut* gun arrangement was one that called for a total reorganization of the process of battleship design — under line officers.

SIMS AND THE FRENCH MODEL OF WARSHIP DESIGN

Lieutenant William S. Sims, who had criticized the design of the battleship *Kentucky* so vehemently, warned the chief of the Bureau of Construction & Repair, Rear Admiral Francis T. Bowles, against design by an opinion poll of the line. Many were ignorant of developments in foreign naval technology: "[their opinions] are likely to be of little value and even dangerous (if given equal weight), through a lack of sufficient acquaintance with the progress made abroad. It is a singular and apparently unaccountable fact that the apathy of the great bulk of our officers regarding these matters is extreme, and that consequently they are largely unacquainted with even the main lines of the progress in naval construction abroad."[55]

Sims sent Bowles a fifty-nine-page paper on the Board on Construction and the American method of designing battleships. In it, Sims attacked the engineers who dominated the Board on Construction and argued that the officers who determined the technical characteristics of new ships should possess "nautical, tactical, and military experience to apply it without violation of the immutable military principles involved. It follows, therefore, that these officers should be selected from the purely military corps."[56] While naval constructors were "indispensable in arranging, perfecting, and completing the details of construction," they were unqualified to decide the general arrangements of equipment and weapons, as this required a "knowledge of matters not included in their profession."[57] Warship designs should be left to line officers who have "handled the ammunition, drilled the crews at the guns, and exercised them at target practice; maneuvered the vessel in fleet formation; cruised in foreign waters, and enjoyed frequent opportunities of comparing the details of our vessels with foreign nations."[58] Sims was describing himself.

As the American naval attaché to France, he had become familiar with contemporary European warship design and was embarrassed by the best technical artifact his navy could produce in 1901 — *Kearsarge* and *Kentucky* — with their superposed turrets, wet decks, and unsatisfactory armor protection.

Naval power was measured in qualitative, technological terms and Sims found American battleships lacking. This was quite a statement from a 42-year-old lieutenant with no appreciable engineering training. Sims considered line officers as idea men and the battleship as a concept to be generated by line officers with the design details left to naval engineers. Sims justified this by the fact that the line had to command and fight the ships "for whose success they must bear the sole responsibility."[59]

Sims's duty in Paris led him to recommend the adoption of the French system, in which naval constructors and engineers designed warships according to the dicta of the line hierarchy. Sims provided Bowles with an example of his nouveau regime, recounting his conversation with Emile Bertin, chief constructor of the French Navy. When Sims asked Bertin why the forward turret of *Henri IV* had been placed two decks, instead of one deck, higher than the after turret, Bertin had told him, "For certain tactical reasons; it was so recommended by the Superior Council, and the ship was therefore designed to carry the turret in that position."[60] French naval constructors obeyed the orders of seagoing officers, a situation Sims thought the U.S. Navy should emulate.

Sims's desire was clear: to remove any important aspects of battleship design from the technical bureaus. He denied that there was "any logical ground for giving the votes of a constructor and an engineer equal weight with those of experienced admirals in deciding questions that are purely nautical and military."[61] Sims overlooked the fact that contemporary admirals of the line typically had less comprehension of the important nuances of technology than younger line officers raised in the post-1883 "modern" period. Sims also ignored the ease with which a simple military decision — superposed turrets — had affected adversely the complex technological system of the *Kearsarge*-class battleships. The superposed turret of his hated *Kentucky* was popular among senior line officers as well as with the technical specialists in the Bureau of Ordnance who had created it. Its adoption had been largely a military decision that harmed the overall design of *Kentucky* and *Kearsarge* by requiring a lower main deck to reduce the excessive topside weight. This had resulted in a ship with less reserve buoyancy, poor sea keeping capabilities, and reduced fighting abilities in rough weather.[62]

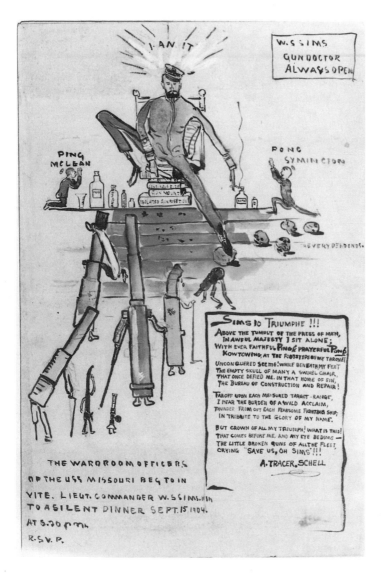

Sims's attack on naval engineers brought him fame among line officers. In September 1904 the wardroom officers in the battleship Missouri *made Sims their guest of honor, praising him in verse: "Unconquered see me! While beneath my feet the empty skull of many a swivel chair that once defied me, in that home of sin, the Bureau of Construction and Repair!" (Collection of Adm. W.S. Sims, USN. Naval Historical Center, NH 89489-KN)*

Sims concluded his critique by suggesting that the secretary of the navy specify a greater role for the line officers of the General Board in the design of new battleships.[63] He also recommended that the tactical qualities of the various preliminary designs be subjected to analysis by the line officer staff at the Naval War College.

Chief Constructor Bowles had been appreciative of Sims's suggestions in the past and had even gone so far as to send him warship plans for comments on a regular basis.[64] When Bowles received Sims's magnum opus calling for the restructuring of the Board on Construction and subordination of naval constructors to the line, Bowles's cordiality ended:

> As I have written you before I welcome your criticisms of our vessels and receive them wholly in the spirit in which they are made. I do not always agree in the matter, nor approve wholly in the matter of making them. I am only led to this remark from the fact that I feel that your last paper, above referred to, would injure your usefulness to the service in a valuable capacity, from attempting to deal in a carping spirit upon matters upon which you may not have the best information, and which will only result in bad feeling.[65]

Bowles soon found that Sims's "carping" was quiet compared to the turmoil generated when the Board on Construction eliminated the torpedo tubes in the *Connecticut*-class battleships.

The Torpedo Tube Debate

In November 1902 Lieutenant Commander A. P. Niblack presented a paper entitled "The Tactics of the Gun" at the 10th General Meeting of the Society of Naval Architects and Marine Engineers.[66] Niblack characterized the battleship gun as only one weapon of offense, its mission to neutralize the guns and personnel of an enemy battleship. Complete destruction could only be wrought underwater: "It is the function of the ram and the torpedo to penetrate the under-water body of a ship, and breaking down its water-tight subdivisions, or striking it in a vital point such as a magazine or boiler, to destroy it . . . it is only by luck or by indirection that a modern battleship can sink another by gun-fire alone."[67]

Niblack's argument for the torpedo came at the same time the Board on

Construction eliminated torpedoes in the *Connecticuts*. In an attempt to "keep the pot boiling," anonymous line officers sent Niblack's paper to Secretary Long, who inquired why torpedo tubes were no longer being installed in American battleships. The Board responded that seagoing officers preferred the submerged torpedo tube over a trainable, "overwater" tube. However, the Board went on to characterize the military advantages of *all* torpedoes as "remote to say the least." Underwater tubes were prone to flooding and their inclusion in a ship called for some serious strategic and tactical sacrifices:

> The torpedo outfit of the [battleship] MAINE [authorized 1898], which has but two underwater tubes, weighs 27 tons, about equal to the weight of two 6-inch guns and mounts, and it is safe to say that most commanding officers would much prefer two such guns to two underwater tubes. The space assigned for the two underwater tubes of the MAINE is, approximately, 10,000 cubic feet, which is almost equal to that assigned for [the ship's total] provisions [food].[68]

The Board on Construction had inferred, from complaints by commanders at sea, that the torpedo was a nuisance and inappropriate.

Apparently, Niblack's enthusiasm for the torpedo was not held by several battleship captains who had applied for the removal of their above-water torpedo tubes.[69] Even assuming that the underwater tubes functioned perfectly, the Board felt "it is a grave question whether it is advisable to sacrifice any important feature on a battleship or armored cruiser for the installation of a secondary weapon of remote value, and which requires tactics differing entirely from those best adapted to the gun, the primary weapon of such vessels."[70] The torpedo, as Roland and O'Connell have both observed, was a weapon beyond the pale.[71]

The Board on Construction was not unanimous in excluding torpedoes from the new battleships. Rear Admiral Royal B. Bradford, the line officer chief of the Bureau of Equipment, defended the torpedo. Bradford thought the torpedo question a tactical one that should be referred to "a considerable number of tactical line officers."[72] Such a dangerous proposal sparked a pointed response from the technical bureau chiefs:

> Who are the tactical line officers of the Navy whose opinions would be of greater value than those of the members of the Board on Construction, whose atten-

tion is constantly drawn to all matters of pertaining to the designs, military features and other details of ships of war? It can be confidently asserted that no other officers of the Navy have a better knowledge of the intricacies of and the difficulties attending the use of torpedo discharge tubes than the members of this Board.[73]

Secretary Long upheld the Board on Construction majority position abolishing battleship torpedo tubes, but the question was far from settled within the line.

Commander J. B. Murdock used the torpedo issue to attack the naval constructors' control over battleship design in the pages of the *U.S. Naval Institute Proceedings*.[74] Murdock argued that only line officers, with their appreciation of tactical requirements, should rule on weapons technologies like the torpedo. Murdock pointed out that American prejudice against torpedoes would not make U.S. battleships torpedo-proof:

> we have never given as much weight to torpedoes as other nations have done, and have largely ignored the possible results of their use against us. In short, we are building battleships with neither torpedo tubes nor [antitorpedo] nets, apparently laying down the law that battleships need not use torpedoes, and have nothing to fear from those of an enemy.[75]

Murdock elucidated the effect national strategic and tactical policies, and naval culture, had upon the "normal" refinement and trajectory of the battleship technological paradigm. The U.S. Navy had always placed emphasis upon gun power and structural strength — and in armor as that came into vogue — in its warships. Murdock appreciated that naval and military establishments tend to design their weapons systems to defeat ships similar to their own and to fit their own strategic and tactical thought styles. This mirror approach dismisses counterweapons: if we have not thought to employ torpedoes, our enemy will not. This was as true of the battleship gun as it is of most of today's weapons.[76]

Murdock's article resulted in a clamor by prominent line officers for torpedo tubes in U.S. battleships.[77] In July 1903 Lieutenant Frank K. Hill, of the Bureau of Ordnance, filed a minority opinion with the Board on Construction calling for restoration of the torpedo tubes. Hill cited Niblack's paper and Murdock's article and claimed that foreign battleships were equipped with torpedoes with ranges up to 3,000 yards. The chief intelligence officer, Captain Charles D. Sigsbee (a line officer), had voted with the majority for removal of the battleship torpedo tubes. He reaffirmed, to the Navy Department, the absence of any intelligence data that could confirm Hill's claims. The Board on Construction attributed Hill's enthusiasm to his work with torpedoes and claimed that his duties did not "extend to questions of policy or tactics, but to material only" — the same "crime" for which the technical bureaus were being indicted.[78]

Running into a dead end at the Board on Construction, torpedo supporters appealed to the General Board. Apparently, Captain Sigsbee was not aware of all the intelligence available on foreign torpedoes. Younger members of his staff at the Office of Naval Intelligence informed the General Board that foreign torpedoes had the capability to strike at ranges up to 3,000 yards. Admiral Dewey warned the secretary of the navy that

> Since gun-fire, in order to result in a decisive action, must be delivered at a range not greatly exceeding 3,000 yards, it follows that the tactics of fleet actions will hereafter be influenced by the presence or absence of torpedoes. . . . Tactical war games played at the Naval War College between fleets with and fleets without torpedoes have been won by the former whenever the result of the game has been decisive; in all cases, the fleet armed with torpedoes holds the command of the situation in tactical maneuvers.[79]

Faced with such a strong statement from Dewey and the General Board, the Board on Construction was forced to reopen the torpedo question. As a result of further study, and "in view of present foreign practice . . . and recent and prospective improvements in the service torpedo of the U.S. Navy, the Board recommends for battleships . . . as many torpedo tubes as can be possibly installed."[80]

Sigsbee's ignorance of the current intelligence regarding foreign torpedoes, along with the Board on Construction's abrupt course change, decreased the credibility of the Board on Construction with the office of the secretary of the navy. Conversely, the General Board's prestige and ability to influence technological decisions increased. The navy retrofitted torpedo tubes in the first five *Connecticut*-class battleships after the circumnavigating cruise of the "Great White Fleet"[81] in 1907–8. The sixth ship, *New Hampshire*, was delivered with her tubes installed.[82]

FAILED LEGALIZATION OF THE GENERAL BOARD

The *Connecticut* controversy enhanced the prestige of the General Board and underscored the line officers' position that they should control the warship design process. Line officer control would ensure technological conformity with the strategic paradigm and assure that the navy was on the proper technological trajectory — a "further articulation" of a technological paradigm, "partly influenced by the selection environment."[83]

In April 1904 Secretary of the Navy William H. Moody called for statutory recognition of the General Board to empower it vis-à-vis the bureaus and the Board on Construction. The General Board had been created by an executive order of the secretary of the navy whereas the bureau organization was fixed by an 1842 law.[84] Moody told the House Naval Affairs Committee that statutory recognition of the General Board was a move that would not "supplant the present organization of the Navy Department . . . [but was] one which supplements that organization."[85] As might be expected, Moody's proposal generated almost universal opposition by the bureau chiefs, who foresaw a reduction in their own power and autonomy should Congress place the General Board on a par with the bureaus.

In a move that struck at the very basis of the technical bureaus' power, Secretary Moody also proposed that the General Board be authorized to determine what types of ships the navy should build. Moody's statement that "the technical bureaus should say how those ships should be designed and built" sounded a great deal like Sims's proposed reorganization on the French model. Only the details of the actual technology would be left to the engineers.[86]

In his testimony in support of his plan, Moody blithely dismissed any "conflict between the General Board and the construction department." He did admit that the General Board "could not settle the detailed characteristics of the warship types. They have not sufficient knowledge to do that."[87]

According to Admiral Dewey, the clear linkage among technology, strategy, and tactics revealed during the torpedo dispute had expanded the duties of the General Board to include determining "the manner and types of ships proper to constitute the fleet."[88] When pinned down before the House Naval Affairs Committee on the Board's responsibilities, Admiral Dewey stated that the General Board only recommended the "numbers and types" of ships, implying that the details of battleships, for example, were left entirely to the Board on Construction.[89] But the General Board's consideration of the "manner" of ships had grown to include more than numbers and types. There was little doubt among the bureaus, and their friends in Congress, that the General Board desired a stronger voice in the technological policy of the navy.

The chief of the Bureau of Construction & Repair, Rear Admiral Washington L. Capps, told the House Naval Affairs Committee that he had no objection to the General Board as long as its duties were "clearly defined and restricted." Capps was quite specific as to where control over naval technology should be vested: "Definite responsibility in all matters affecting the design and construction of ships of war, including hull, machinery, armament, and fittings, should be lodged with those whose technical and military training, and already legalized official duties, make such responsibility direct and unavoidable. This responsibility is now definitely assumed by the four technical bureaus of the Navy Department, the chiefs of which constitute the Board on Construction." Capps was willing to leave the "disposition of the completed ship" and "matters affecting the military efficiency of the fleet" to the General Board.[90]

Rear Admiral Charles W. Rae, chief of the Bureau of Steam Engineering, characterized the General Board as a "wedge" being used to break the power of the technical bureaus and centralize the control of the navy in a board of line officers. Rae belittled the competence of the General Board, pointing out its stubborn call for the construction of protected cruisers — a ship protected solely by an armored deck at the waterline — that the Board on Construction judged (and was) technically obsolete.[91]

Despite Secretary Moody's strong support for congressional recognition of the General Board, the legislation failed to leave the committee, torpedoed by Moody's subordinate, veteran Assistant Secretary of the Navy Charles H. Darling. Darling testified forcefully against any increase in the General Board's power. In Darling's opinion, "a vast amount of the work in connection with the administration of the Navy is civilian work pure and simple."[92] There was no need to expand the General Board's military influence in battleship design since "The building of a battleship is purely civilian work. . . . It is entirely a matter of business administration," work that fell under the province of the Washington bureaus.[93]

Although thwarted in its attempt to attain status equal to the bureaus, the General Board continued to expand its involvement in decisions regarding U.S. naval technology. In October 1905, in accordance with its charter, the General Board recommended the types of ships to be included in the building program for Congress. However, the Board also specified the number and caliber of main guns, the freeboard (the distance between waterline and main deck) — something well within the province of naval architects — and the armor protection for the proposed battleships. The General Board requested three 18,000-ton battleships (2,000 tons larger than the *Connecticuts*). The Board also requested three scout cruisers, four destroyers, four torpedo boats, four submarines, and four gunboats at a total cost of $35.96 million.[94]

Sensitive to the General Board's intrusion into its domain, the Board on Construction recommended three smaller battleships, three scout cruisers, and two gunboats for a total of $28.7 million. In case some members of Congress had not yet comprehended the centrality of the battleship to American naval power, the Board on Construction recommended that Congress cut the battleships last as they were the "primary and most important" type of vessel. The Board on Construction rejected the General Board's 18,000-ton battleships since a "most effective vessel of this type can be provided on a displacement which will not involve excessive cost or size."[95]

Angered by the Board on Construction's opposition to bigger battleships, Dewey proposed replacing the Board on Construction with a permanent Board on Designs. Dewey feigned concern that the members of the Board on Construction were being overworked, as their "responsible duties of research and administration" as bureau chiefs left them "scant time" to devote to the Board on Construction. Dewey's proposed Board on Designs would include two civilians "having expert knowledge and experience" and five officers, one of whom should be a naval constructor. None of the Board members would be chief of a bureau. The sole duties of this Board on Designs would be "to examine and pass upon all designs submitted to the Navy Department for vessels to be built."[96] Such a line officer-controlled Board on Design was necessary due to the "glaring faults that are being repeated, and which lessen in a great measure the fighting qualities of ships of the battle line."[97]

Not surprisingly, other line officers rallied behind Dewey's proposal. Rear Admiral Charles H. Stockton, president of the Board of Inspection and Survey, recommended that the General Board alone generate the building program sent to Congress. Once Congress had authorized the numbers of ships, Stockton thought that a Board on Design, consisting of

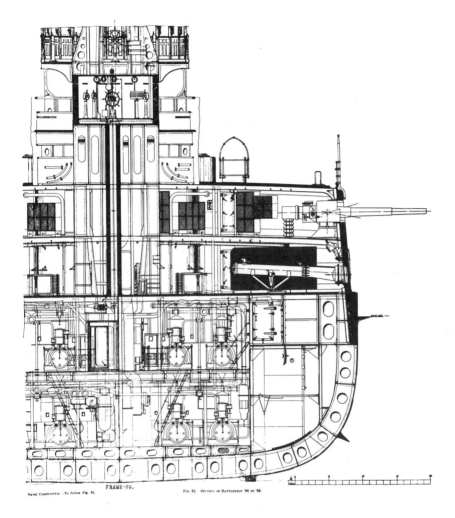

FRAME-33.

Naval Construction -To follow Fig. 81.

FIG. 92. SECTION OF BATTLESHIP '96 TO '99.

This 1895 plan of the midship section of the Kearsarge-class shows a gun of the secondary battery projecting from the armored casement typical of pre-dreadnought battleships. An above-water torpedo tube is located on the next deck down. Both weapons were at the center of the 1901 controversy between the Board on Construction and the General Board. Placement of the vertical armor belt (shaded) along the waterline generated subsequent line officer attacks on the Bureau of Construction & Repair. The ships' thicker vertical armor was intended to defeat flat-trajectory shells in the close-range battleship duels anticipated prior to the Russo-Japanese War. (Copied from Naval Construction, R. Robinson, 4th ed., Annapolis, 1917, 160–61. Naval Historical Center, NH 76632) the Board on Construction and "other sea officers of high rank and experience," should put the "technical designers in touch with those who use and fight the vessels." Stockton's rationale was strikingly similar to Dewey's: "The fact that our vessels of recent construction and design lack equality in speed and military qualities with their contemporaries in the British service emphasizes the need of such a Board to relieve and assist the members of the Board on Construction whose time is mainly absorbed by their duties as administrative heads of the Bureaus of the Navy Department."[98]

Unfortunately for Dewey and his line colleagues, the reorganization they desired could not overcome the strong historical links between the bureaus and Congress. Dewey had grown more vocal in his attacks upon the Board on Construction and the bureau chiefs, but as the navy's senior officer he was well aware of the danger of publicly claiming technical bureau incompetence so soon after the *Bennington* disaster and the public furor over amalgamation. Any expression of doubt over the quality of American ships would provide wonderful ammunition for congressional opponents of the Roosevelt naval buildup.

THE SITUATION IN 1905

The deadlock in the Board on Construction over the secondary gun batteries in the *Connecticuts* provided the line, through the increasingly powerful General Board, its first opportunity to express opinions at the highest level on the technological basis of the profession. To forestall this challenge to their control over the battleship, the exemplar of the profession's technological paradigm, the members of the Board on Construction conducted a survey of "expert" seagoing officers (many of dubious technical sophistication), setting a new precedent for bureau-line cooperation.

The exclusion of torpedo tubes from the *Connecticut* battleships provided a second opportunity for critics of the technical bureaus. Faced with widespread criticism and condemnation by Admiral Dewey, the Board on Construction changed its position and accepted torpedoes in the *Connecticut* class. The Board on Construction was weakened and the secretary of the navy convinced of the necessity for high-level line involvement in the design process.

When Secretary Moody tried to grant the General Board increased power through a legislative reorganization of the Navy Department, the technical bureaus mobilized their allies in Congress to prevent any change in the institutional status quo. Senior officers of the line were unwilling to force the issue publicly as a new controversy with the technical bureaus would come too soon after the amalgamation controversy and jeopardize battleship appropriations. Two events in 1905 — the Battle of Tsushima and construction of the British battleship *Dreadnought* — provided the line with new opportunities to charge the technical bureaus with malfeasance for designing and producing inferior battleships. Line officers' efforts to attain hegemony over the battleship technological paradigm increasingly occurred in an environment in which senior officers of the line sided with the bureaus against younger insurgents advocating ever-larger battleships.

In a Kuhnian sense, both bureau engineers and seagoing line officers accepted the basic strategic paradigm — guerre d'escadre based on a battleship fleet. They differed on who should define the battleship and control the trajectory of the battleship technological paradigm. After 1905 there was a relatively sharp split between advocates of bigger and smaller battleships. However, this professional division no longer ran along the engineer-line officer fault line.[99]

CHAPTER 3

Refining the Technological Ideal

The Simsian Uproar, Engineer Bashing, and the All-Big-Gun Battleship

Much like late medieval astronomers arguing over the significance of a new comet to the Ptolemaic universe, U.S. naval officers differed over what conclusions should be drawn from the Russo-Japanese War. The battleship paradigm, like the pre-Copernican Ptolemaic cosmogony, was intact but contained certain puzzles that required refinement, such as the size and type of battleship which best exemplified the paradigm.[1] The resulting discussion yielded further ammunition for a continuing line-officer assault on the technical bureaus.

In their struggle with the technical bureaus over definition of the battleship, some line officers, centered around William S. Sims, saw President Theodore Roosevelt as a potential ally. Roosevelt viewed naval expansion as an integral part of America's transformation into a world power and the navy as a positive instrument of national policy: "If we build and maintain an adequate navy and let it be understood that . . . we are perfectly ready and willing to fight for our rights, then the chances of war will become infinitesimal."[2] To Roosevelt, an "adequate" navy required warships that reflected the "greatness of our people," that is, modern battleships. The ships of the *Connecticut* class, despite the controversy over their gun battery and torpedo tubes, were on a par with the best foreign designs and mirrored the international battleship technological paradigm. Yet the Simsian rebels continued their attacks on the technical bureaus' production of "inferior" battleships and endangered the Roosevelt naval buildup.

Just as the cannon had threatened the castle, torpedo boats such as Ericsson *imperiled the battleship. As with the castle, defensive efforts focused on creating a "defense-in-depth" using cruisers and torpedo boat destroyers to keep the threat away.* (Naval Historical Center, NH 63744)

GUNS, ARMOR, AND SPEED

The shifting geostrategic realities of the early twentieth century placed great stress on the U.S. Navy. The historical, singular rivalry with Britain was replaced with concerns over the rise of Germany and the Anglo-Japanese Alliance of 1902 and a potential two-ocean war. The Mahanian-Jominian dictum of concentration of forces forbade splitting the battle fleet between the Atlantic and the Pacific. The Panama Canal was one answer in the works, but its width later would place design limits on U.S. battleships.[3] The acquisition of the Philippines from Spain had extended greatly the area of operations of the U.S. Navy and complicated strategic planning. It was quite difficult, as William Braisted observed, to "defend the interests of the United States in two oceans with a one-ocean navy."[4] Despite navalist efforts, appropriations for a two-ocean navy would have to wait until the summer of 1940 and the fall of France.

Line officers and engineers in the technical bureaus had grown up in a service in which the geographic expanse of the Pacific dominated operations and was one of the most basic design criteria for U.S. warships.[5] As in Vice Admiral Porter's day, steam-propelled ships required fueling bases and ships that could carry sufficient fuel to steam long distances without replenishment. Acquisition of Guam from Spain helped, but the Philippines drew an even stronger U.S. naval presence to the Far East. Complicating the design problem was the finite volume and displacement of a ship's hull that had to be apportioned among guns, armor, propulsion machinery, fuel, and provisions. Simply put, battleships for Pacific operations had to be much larger than the early "coastal" battleships of the 1890s.

In an article in the *U.S. Naval Institute Proceedings* in 1900, the professional journal of the naval officer corps, Captain Asa Walker called for the construction of larger battleships to protect American imperial responsibilities in the Far East.[6] To achieve a desirable level of "all-around efficiency," Walker argued that armament and armor should take precedence over speed. Walker wanted battleships with moderate speed (not to exceed 17 knots) and coal capacity to provide for 7,000 miles of steaming. His emphasis upon invulnerable armor designs and maximum gun power was in line with the brown-water Civil War experience that emphasized armor strength and gun power and harkened back to the same features in the 44-gun Humphrey superfrigates of the 1790s. Far East operations, on the other hand, required warships with the capability for long-range, economical steaming, which could only be had at the expense of speed, reduced armor, or fewer or smaller guns.[7]

In a March 1902 article, Lieutenant Matt Signor proposed a battleship with more emphasis upon a heavy main battery, in keeping with the "traditional desire" of the American naval officer to possess "long range guns."[8] However, the most unabashed call for large American battleships came from Lieutenant Homer Poundstone in January 1903.[9] Poundstone believed that the battleship was "one of the cases where 'The biggest is the best' and . . . 'The best is the cheapest.'" Larger ships were necessary to optimize the major technological subsystems of the battleship: guns, armor, and the propulsion machinery. To Poundstone, cost was incidental. What was important was the expansion of the navy and the construction of large numbers of invulnerable battleships: "if we should build and maintain a fleet of even as many as 50 *Connecticuts*, the expenditure involved would not constitute an extravagant insurance on the material wealth and for the continued peace of our country."[10] Echoing President Roosevelt's desire for the United States to "take the position to which it is entitled among the nations of the earth," Poundstone's writings epitomized the navalism firmly rooted in the battleship-based strategic paradigm.[11]

Naval officers established the Naval Institute in 1876 to promote the free exchange of ideas. The *U.S. Naval Institute Proceedings* provided the environment for an honest discussion of naval issues. The Institute's editorial policies deemphasized, as much as possible, an author's naval rank or position. The articles written by Poundstone and his colleagues were serious proposals and considered as such by Naval Institute members.

The psychology underlying battleship design, and the tactical employment of battleships, reflected a fortress mentality. Battleships truly had become castles of steel.[12] Their survival lay in their armor. Battleship design was a continuing competition between improved naval guns, with armor-piercing shells, and steel armor dependent upon metallurgical and manufacturing processes as well as the geometry of armor installations.

The effect of naval ordnance on ironclad warship design after the Civil War was analogous to the introduction of bronze cannon and iron cannon balls to medieval fortifications. Naval constructors became modern versions of the fifteenth-century Siennese military engineer Francesco di Giorgio Martini[13] in their experimentation with new armor designs to counter the effect of improved naval guns. Armor systems thought to be impervious to penetration by an enemy's shells *could* foster a sense of invulnerability and lead to an aggressive course of action in battle. However, in most cases, battleship commanders had to deal with an enemy whose strengths and weaknesses were unknown.[14] This could, and often did, lead to unpleasant surprises as with British battlecruisers at Jutland in 1916 and the battlecruiser *Hood* in 1941.[15] In a battleship engagement, the armor protection either worked or failed and naval officers, quite understandably, wanted to sail in the best design possible. Poundstone's advocacy of the biggest and best battleship was understandable in the all-or-nothing environment of battleship combat. For some line officers, the fact that naval constructors and other bureau engineers rarely rode ships in combat underscored the engineers' more "academic" interest in the battleship.

Poundstone's call for fifty battleships of the *Connecticut* class might seem excessive, but it was in line with the General Board's own call for the construction of forty-eight battleships — critics claimed one for each state.[16] Yet underneath the numbers, Poundstone was arguing for the construction of fewer, large battleships rather than many smaller ones built at the same cost. In an argument similar to that later used by Admiral Jackie Fisher in Britain, Poundstone claimed that a larger number of small battleships could not defeat one large battleship.[17]

Poundstone's article stimulated the antibureau faction within the line. Manifesto-writing Lieutenant Commander John Gibbons lamented that the eloquent Poundstone had not been on hand to defend the large battleship, supported by navalists in the Senate, against attack by the House of Representatives and their friends in the Board on Construction.[18] Defending the technical bureaus, Assistant Naval Constructor T. G. Roberts challenged Poundstone's bigger-is-better logic, asking, "If the largest is best, why not make it [the next battleship] 25,000 tons?" Roberts would only have to wait seven years for the *Arkansas* and *Wyoming* in 1910.[19] Naval Constructor Horatio G. Gillmor sought to educate line officer critics on the delicate interrelationship among speed, horsepower requirements, weapons, and armor: "by a reduction of two knots from the speed requirements, there can be produced a very powerful battleship upon a moderate displacement, presenting a correspondingly smaller target to an enemy."[20] To Gibbons, Poundstone, and other big-battleship advocates, the nattering details of design trade-offs were unimportant. The smaller battleships described by Gillmor did not interest them.

The growing agitation for large battleships by naval officers sparked a condemnation from Representative James Tawney, chairman of the House Appropriations Committee, who bemoaned the fact that "like children competing for the most expensive and glittering toys, we must compete with the nations of the world in the construction of the largest and most expensive battleships in order to satisfy our national pride and vanity."[21] While Tawney had reason to complain, the push for larger battleships involved more than vanity. The rapid evolution of various technologies promoted quick obsolescence in warships.

In 1904, on the eve of the Russo-Japanese War, there was a general recognition of the need for larger guns and battleships capable of long-range operations. The technical bureau officers demonstrated support for the big-gun battleship with the design of *Michigan* and *South Carolina* that same year. The geography of the Pacific Basin, and the growing threat from Japan, had resulted in an American battleship design that was in many ways quite comparable to Fisher's *Dreadnought*. However, to many senior officers, including the publicly acclaimed creator of the battleship strategy, Alfred Thayer Mahan, such a ship was an aberration and violated the lessons of the Russo-Japanese War.

THE EVENTS OF 1904–1905

In August 1904 the first engagement involving modern battleships took place between the Tsarist and Imperial Japanese navies in Pechili Gulf.[22] This battle, and the more important battleship engagement at Tsushima in 1905, were analyzed and discussed by the admirals and naval constructors of every major naval power. The large-bore naval gun, mixed-bore guns for close-in fighting, the torpedo, and the mine each found champions.

On the other side of the world from Japan, Admiral Jackie Fisher was appointed first sea lord and launched his "Scheme" for revitalizing the British navy. Besides the elimination of obsolete ships and the modernization of the shore establishment, Fisher convinced the Board of the Admiralty to restructure the Royal Navy to include just five types of fighting ships: battleships, armored cruisers, oceangoing destroyers, coastal-service destroyers, and submarines.[23]

Fisher sought to maintain Royal Navy dominance by redefining the technological basis of naval power and basing it on an all-large-caliber armament. According to Fisher's Committee on Designs, "Both theory and the actual experience of [the Russo-Japanese] war dictate a uniform armament of the largest gun, combined with speed exceeding that of the enemy, so as to be able to force an action."[24] The Committee cited the Battle of Pechili Gulf as proving the uselessness of the smaller-caliber guns — the secondary armament that had split the Board on Construction on the *Connecticut* design.[25]

No longer would naval battles be fought at close ranges. The Japanese battleship *Mikasa* had been struck by a Russian 12-inch shell at a range of 13,000 meters (more than 7 nautical miles).[26] Fisher's push for a battleship sporting an all-large-caliber gun battery was bolstered by secret intelligence that the war had convinced both the Russians and the Japanese to build their future battleships as all-big-gun designs.[27]

In addition to an all-big-gun battery, Fisher specified high speed as a requisite feature in his new type of battleship. Fisher's design was to steam at 20 knots at a time when the new American battleships of the *Mississippi* class, authorized in 1903, were barely reaching 17 knots, the top speed of Isherwood's *Wampanoag* in 1869. Fisher's reasoning was reflected in the Committee on Design's statement on battleship speed and its importance within the expanded naval battlefield:

> There is no question that the first desideratum in every type of fighting vessel is a greater speed than that possessed by a similar class of the enemy's ships. It is the "weather gauge" of the olden days. Strategy demands it, so as to get the deciding factor (the battle fleet) to any desired spot as quickly as possible. Tactics demand it to afford choice of range at which the action is to be fought. This will naturally be a long range, so that gunnery skill can be used to the best advantage. Apart from the size of the ship's turning circle, and the dread of the torpedo, close ranges level individuality of marksmanship, and therefore are to the advantage of the least trained guns' crews.[28]

To achieve higher speeds, Fisher convinced the Admiralty to reject traditional steam reciprocating engines and to install Parsons marine steam turbines in the *Dreadnought*-type battleships.

The gun arrangement of Fisher's *Dreadnought* reflected the primary British tactical environment — the stormy North Sea. Unlike the American *Kearsarge*-class battleships, whose smaller-caliber broadside guns were often unusable in moderate seas due to the low main deck, *Dreadnought's* main gun battery was on the upper deck, so that "the guns can be fought in all times and in all weathers."[29]

Fisher's dreadnought battleships were also designed to counter the effects of the torpedo and the submerged mine. Both the Russian battleship *Petropavlosk* and the Japanese battleship *Hatsuse* had sunk in less than two minutes after hitting submerged mines. "Inviolate" watertight bulkheads in *Dreadnought* would prevent progressive flooding after damage. The ammunition magazines were moved to the ship's centerline and elevated to place them as far as possible from the effects of a mine or torpedo detonation. For additional protection, Fisher pushed, unsuccessfully, to armorplate the hull sides and bottom, an impractical addition to the ship's displacement.

Through his emphasis upon accurate, long-range gunfire and high speed, Fisher was attempting to expand the naval battlefield and make existing battleships, especially those of Germany, obsolete. As Jon Sumida has demonstrated, *Dreadnought* was an interim step toward Fisher's ideal "armoured vessels" that would be a "fusion" of the battleship and armored cruiser.[30] Fisher was planning for battleship actions at a range of 15,000 yards shortly after Admiral Dewey and the General Board were demanding battleship torpedoes and defining the maximum effective range of battleship gunfire at 3,000 yards.[31] In fairness, Dewey's pronouncement had come on the eve of the Russo-Japanese War. American naval officers, perhaps more than their British counterparts, differed on just what conclusions to draw from the war. They also differed about what to make of Fisher's *Dreadnought.*

DEBATING THE LESSONS OF THE RUSSO-JAPANESE WAR

In the Naval Institute's prize essay for 1905, Commander Bradley Fiske endorsed the construction of very large, high-speed battleships, similar to those of Homer Poundstone.[32] Fiske was attacked for his advocacy of increased size and greater speed.[33] In his rebuttal, Fiske claimed that a truly "compromiseless" battleship, and one consonant with American strategic commitments in the Pacific, could be had at a displacement of 20,000 tons.[34] Fiske reported consulting with naval constructors who told him that such a large ship was possible from an engineering standpoint.

Fiske was calling for a ship much different than Fisher's *Dreadnought.* Despite the success of the long-range guns in the battleship actions between the Russians and Japanese, Fiske was still thinking of battleship duels at close range. Larger than Fisher's *Dreadnought*, Fiske's proposed compromiseless battleship had a top speed of 18 knots with *torpedoes* to keep an enemy 4,000 yards away, and carry such armor as to be "invulnerable, *beyond torpedo range*, to any guns built, or building [original emphasis]."[35] Fiske's proposed battleship was merely an enlarged *Connecticut*, not an all-big-gun battleship like *Dreadnought* or the U.S. version, USS *Michigan*, whose design work started in 1904. Fiske believed that the majority of line officers desired such large ships and that "Congress will give us some, provided we all agree, and *keep stroke* in advocating them. . . . But a very formidable antagonist in the way of our getting such ships exists — one whose opinion has tremendous weight in Congress, probably more weight than all the rest of the navy combined — Captain Mahan [original emphasis]."[36]

Mahan threw his weight around in an article in *Collier's Weekly* in June 1905. Mahan interpreted the sea battles of the Russo-Japanese War as endorsing the small battleship since "Japanese success [at Tsushima] has been the triumph of greater numbers, skillfully combined, over superior individual ship power, too concentrated for flexibility of movement."[37]

Fiske accused the "preceptor in Naval Strategy" of "fallacious" reasoning. Fiske acknowledged that flexibility, the property of a fleet consisting of many, small battleships, had been of great use to the Japanese, and allowed them to surround and prevent the Russians from escaping, but only "*after* the Russians had been whipped [original emphasis]" by long-range gunfire.[38] Although still visualizing battleship fighting at close range, Fiske believed that war experience dictated the requirement for larger guns, which in turn required larger battleships. According to Fiske, America's future as a major naval power depended upon it. Flexibility, so important to Mahan, could only be gained "by sacrificing power of concentration" — another Mahanian dictum. According to Fiske, when "great resistance" in the form of an enemy battleship fleet "is to be overcome concentration is required." However, when confronting a "feeble" enemy force, the "principal effort is to keep an enemy from escaping [and] flexibility [many small ships]" would be necessary. The issue was whether to be a first-rate power, building "our fleet so that it shall be able to overcome the resistance of a powerful enemy, or build it so we shall be able to prevent a weak enemy from escaping."[39]

Fiske acknowledged that large ships would not provide strategic or tactical flexibility if so large as to allow the construction of only one or two. However, Fiske was not interested in a fleet of one or two, but envisioned a fleet of at least eight "compromiseless" battleships as flexible, and easier to manage, than a large fleet of smaller ships. The 20,000-ton battleship was what Fiske wanted, and according to him, what America needed to achieve true naval power.[40]

In June 1906 Mahan responded with his own analysis of the "Battle of the Japan Sea" in the *U.S. Naval Institute Proceedings*.[41] He criticized Fiske's emphasis on speed, since a fleet was limited to the speed of its slowest battleship and the construction of faster battleships was pointless unless enough were constructed to retire all the older ships. Mahan was prescient in describing the destabilizing effect of technological innovation on modern arms races: "the standards of size and consequent speed depend upon the ship your neighbor is laying down. We are at the beginning of a series

William Sims's failed attack on the Bureau of Construction & Repair during 1907–8 focused on the placement of the horizontal armor belt on U.S. battleships. This 1898 view of the launching of Kentucky *shows the lighter band of teak planking over which the armor belt will be installed.* (Courtesy of the Filson Club, Louisville, KY. Gift of Salem H. Ford. Naval Historical Center, NH 92506)

to which there is no logical end, except the power of naval architects to increase size."[42] Mahan, quite correctly, was unable to fathom why Fiske believed that the size of future battleships would stop at 20,000 tons.

Mahan defended the mixed-bore gun batteries of the pre-dreadnought battleships of the Russian and Japanese fleets (and *Connecticut*) and attacked Fisher's all-big-gun *Dreadnought* and its supporters in the United States. Mahan maintained that the Russian defeat was due to the Japanese numerical superiority in smaller-bore broadside guns since the Russian fleet had almost double the number of large-bore guns than the Japanese. This should have given the Russians the long-range advantage, as *Dreadnought*'s supporters claimed, yet the Russians reported being "blinded" by the volume of shells from the Japanese secondary battery."[43]

Mahan attributed the Russian defeat to their beginning the action in a fleet formation in which the battleships "could not quickly develop the full power of the broadside."[44] In an interesting sailing metaphor, Mahan compared the ships' funnels with masts and sails. He analyzed the destruction the secondary batteries of 6-inch and 8-inch guns wrought upon the ships' funnels in both fleets. Perforations in the funnel, Mahan claimed, reduced a ship's speed, forcing the battle line to slow to maintain fleet integrity while protecting the damaged ship. Besides slowing enemy ships, the smaller guns also served a more important, antipersonnel mission which was often the cause of defeat during the age of sail: "It has long been my own opinion that the so-called secondary battery is really entitled to the name primary, because its effect is exerted mainly upon the personnel rather than the material of the vessel."[45] Curiously, Mahan ignored the effect that a 12-inch shell, fired from a big gun and delivered from long range, could have on *both* materiel and personnel.

Mahan recognized the slippery slope of ever larger battleships and the threat posed to the status quo by *Dreadnought*, but except for decrying "this wilful premature antiquating of good vessels" as a "growing and wanton evil," could offer no solution:

> Practically, the navies of the world have now committed themselves to solving their problem by progressive increases in size, which affects national expenditure in two principal ways: first, increase of cost by bigger ships, and, second, by prematurely relegating to the dump vessels good in themselves, but unable to keep up with the last one built. To-day's *Dreadnought* has no immunity from the common lot of all battleships. In a fleet, to-day, her speed will be that of her slower sisters [pre-dreadnought battleships]; more *Dreadnought*s must be built to keep up with her; and upon them in turn, according to the prevalent law of progress, she will be a drag, for her successors will excel her.[46]

While urging retrenchment based on economy, Mahan was driven, as was Porter, by a strong attachment to the thought style typical of officers of his generation. The dreadnought battleship threatened to make Mahan's old navy obsolete and invalidate the historical framework he had constructed in his interpretation of naval power. If a modern dreadnought had a useful technical life of but a few years before being superseded, the long-term identification of the officer corps with its artifacts, the very mortar of the naval profession, was in danger and perhaps, in some insensate way, Mahan perceived a future technology-based presumptive anomaly that would overthrow the strategic paradigm of battleship-based guerre d'escadre.

Mahan's opposition to the new dreadnought battleship was attacked by Lieutenant Commander William S. Sims, appointed inspector of target practice by President Roosevelt.[47] Taking care to treat the great Mahan respectfully, Sims cited "recently available" information — a publication of charts and chronology of the Battle in the Sea of Japan provided by a Russian officer-observer.[48]

For Sims, the efficacy of the large-caliber gun had been proven in the first minutes of the battle when two Russian battleships were pounded out of the line by Japanese long-range gunfire. The battleship *Osliabia* was sunk and in *Suvaroff*, the Russian admiral was wounded. Despite their superiority in large guns, the Russians preferred a surprisingly short 1,800-yard battle range. The superior speed of the Japanese ships allowed them to keep the Russians at an unfamiliar distance. The Russian ships were also overloaded with coal and stores so that their waterline armor belt, located along the side of the hull, was submerged and ineffective — a charge Sims would soon level against the Board on Construction regarding U.S. battleship designs.[49]

The lesson of the Russo-Japanese War, according to Sims, was that all things being equal, higher speed conferred tactical advantage. The commander of a faster fleet *always* had the ability to: "(1) Refuse or accept battle. (2) Choose his own range. (3) Control the rate of change of range. (4) Control the compass bearing, thus taking advantage of the weather conditions that favor his own gun-fire."[50] If armor and guns were untouchable, the higher speed advocated by Sims could only come from more powerful propulsion plants, which would require much longer and larger battleships.[51]

Sims pointed out that the new all-big-gun battleships like *Dreadnought* were impervious to anything but fire from the heaviest guns and had made battleships like the *Connecticut* obsolete. In the new dreadnoughts, the turrets were completely enclosed in 8-inch armor plate, making the gunners invulnerable to fire from the smaller guns of Mahan's favored secondary battery. Given the need to build battleships capable of meeting those of other nations on equal terms, Sims thought it "unwise to equip our new ships with a large number of small guns that are incapable of inflicting material damage upon the all-big-gun, one-caliber ships of our enemies, or upon the personnel manning their guns."[52] Sims accused Mahan of confusing "volume of fire" with "volume of hitting," or as Sims preferred, "rapidity of hitting, which is the only true standard of efficiency for all kinds of gun-fire."[53]

Ignoring the issue of technological obsolescence, Sims argued that ten 20,000-ton, all-big-gun battleships would be more economical to acquire and to maintain ($100 million) than twenty small battleships ($120 million to $130 million). Sims surprisingly claimed that the larger ship would require half the officers and men as its smaller, more labor-intensive counterpart. Sims may have been thinking in terms of gun crews and characteristically ignoring engineers as well as deck personnel required to maintain the ship. The larger battleships would cost less yet would be "greatly superior in tactical qualities, effective hitting capacity, speed, protection, and inherent ability, to concentrate its gun-fire, and have a sufficient sum left over to build one 20,000-ton battleship each year."

Sims's argument presaged later U.S. Navy efforts to reduce crew size. The 1970s *Spruance*-class destroyers were highly automated ships nominally incorporating the latest technology and providing a solution to the navy's postconscription manpower concerns. This approach, however, did not prove to be a panacea. The smaller crew, on a larger ship, meant routine maintenance such as chipping rust and painting was hard to accomplish. The high percentage of trained technical specialists in the crews of these ships were not pleased by the prospect of manual labor, exacerbating retention of these highly skilled personnel.[54] During the mid-1990s, the so-called arsenal ship was touted for its minimal crew and reduced crew size continues to be an issue in the twenty-first-century navy. It is likely that Sims's ideas would have produced similar problems a century ago.

In his biography of Sims, Elting Morison has portrayed his father-in-law as only interested in constructing the most efficient and powerful battleship possible, and downplayed any inference that Sims was a big-navy advocate.[55] While Sims did not write with the same fervor as Gibbons and Poundstone, he was not exactly a voice for moderation in battleship construction. Perhaps Park Benjamin, a contemporary writer who graduated from the Naval Academy in 1867, said it best:

The fleet of big overtopping ships which Lieutenant-Commander Sims treats, and no doubt would like us to treat, with easy familiarity, will, on his own argument, be open to defeat, for precisely similar reasons, by half their number of still larger ships; and these, in turn, by a still smaller number of vessels bigger yet; and finally, what is to prevent (*pace* the naval constructors) the reduction (*ad absurdum* if you like) of the last pair, like "two single gentlemen rolled into one," to an isolated and massive identity? Think of that! Perhaps but one ship may be needed to scare off the effete navies of Europe from the Atlantic.[56]

Sims's all-big-gun battleship was in line with the "traditional desire" of the American naval officer to possess "long range guns," elucidated by Lieutenant Signor in 1902. Mahan's arguments were rejected by his colleagues and overtaken by the all-big-gun design of the battleship *Michigan*, produced by the technical bureaus and funded by Congress in 1905. *Michigan* was soon followed by true dreadnought battleship designs for the U.S. Navy featuring steam turbine propulsion.[57]

Dreadnought battleships replaced the mixed-gun-bore pre-dreadnoughts and, like nuclear weapons today, became the quantifiable and defining measure of national power for maritime nations. Even Brazil and Argentina ordered dreadnoughts, not for any valid defense reason, but solely as a matter of national pride. By 1909, the cry in Britain was for dreadnoughts: "We want eight and we won't wait!"[58] Still, as Mahan had predicted, the dreadnought battleship was but another step in a technological arms race and, before the outbreak of World War I, the powerful dreadnought was superseded by the larger superdreadnought with more effective guns.

Big battleship advocates, such as Sims, sincerely believed that large dreadnoughts were economical, and therefore in a progressive sense, more efficient alternatives to their smaller predecessors.[59] It was a happy time for navalists and naval officers since the future, under Theodore Roosevelt, promised more and ever-larger battleships. Unfortunately for Sims and his colleagues, control of battleship technology still rested with the technical bureaus. Having quashed criticism of the all-big-gun ship, and having defeated the great Mahan, Sims set his sights on the hated technical bureaus and the engineer-dominated Board on Construction. Sims publicly criticized American dreadnought design in one more attempt to achieve control over the technological basis of the profession.

MUTINY IN THE LINE: THE SENATE INVESTIGATION OF 1907–1908

William Sims had never been bashful in presenting his views on the shortcomings of the service to the Navy Department. Besides calling for the subordination of the technical bureaus to the line in 1901, Sims had successfully appealed to President Roosevelt to force a change in naval gunnery technique in 1902.[60] Ordered to Washington to serve as the inspector of target practice, Sims observed the fight over the *Connecticut* at first hand. He was intolerant of the naval bureaucracy and sought the "blood" of those guilty of "indirection and shiftiness . . . in high places."[61] Caring more for "the vital interests of our great service" than "saving face," Sims did not share Dewey's fear of a public fight for control of the profession.

In 1907 Sims indicted American battleship designs in an article in *McClure's Weekly*, penned by Henry Reuterdahl, a civilian writer on naval affairs. The article savaged the design of American battleships, placing the blame squarely upon the technical bureaus and the Board on Construction.[62] Sims hoped to create enough controversy to force a congressional hearing, during which the malfeasance of the technical bureaus would be exposed. Sims's ultimate goals were the reorganization of the Navy Department under a line-officer general staff, establishment of a system of promotion based upon ability, and an end to bureau control over battleship technology.

The Reuterdahl article was published one month after President Roosevelt proposed a greatly expanded battleship construction program to counter HMS *Dreadnought*. The *McClure's* article created just the effect Admiral Dewey had feared in 1904—useful ammunition for congressional opponents of naval expenditures. Coming hard on the heels of the economic panic of 1907, the article placed Roosevelt's dreadnought navy in jeopardy.[63]

Recognizing Sims's handiwork in the *McClure's* article, the powers within the Navy Department lost no time in accusing Sims, now serving as presidential naval aide, and Lieutenant Commander Frank Hill, who had challenged the Board on Construction over battleship torpedo tubes, of complicity with Reuterdahl.[64] Hill proclaimed his innocence, but did acknowledge his long-standing opposition to the method of placement of the waterline armor belt on American battleships, a key criticism in the Reuterdahl article.[65] To protect his naval program, Roosevelt ordered a temporary halt in the Navy Department inquiry directed at Sims.[66] Although Roosevelt had succeeded in keeping his naval aide under wraps, the technical bureaus appealed to Senator Eugene Hale, chairman of the Naval Affairs Committee, for vindication.

Hale, a Republican from Maine, opposed larger "imperialist" battleships, and had little tolerance for their advocates.[67] Previously a pro-navy congressman, Hale allowed his support for naval expansion to erode in line with his strong opposition to the war with Spain. Elting Morison attributed Hale's aversion to large battleships to the claim, undoubtedly voiced by Sims, that the shipyards in Hale's home state could not accommodate large battleships and were losing choice construction and repair contracts—a charge belied by the construction of the large (14,900-ton) battleship *Georgia* at Bath Iron Works in 1903. According to Robert Albion, Roosevelt was often forced to bypass Hale and utilize Senator Albert Belveridge to push many of the naval appropriations bills through the Senate.[68] Hale had waged a stiff fight with the House Naval Affairs Committee in 1903 over the three 16,000-ton *Connecticut*-class battleships voted by the House. He forced a compromise in which the three *Connecticuts* were joined by two smaller 12,000-ton battleships of the *Mississippi* class. Hale thought

> [the smaller battleship] does not fit the fancy of a naval officer. She has not the room on her for staterooms and for comforts and for conveniences and for all the intricate machinery that a 16,000 ton ship has. But she has the same number of 12-inch guns and the same number of turrets and the same efficient force, so far as the great guns go. She is smaller, more efficiently handled, and when she goes to the bottom, as any battleship will from the effect of a torpedo . . . instead of seven or eight million dollars going to the bottom, it is only $5,000,000.[69]

In preparation for the Senate hearings, Hale garnered statements from Rear Admiral Washington Capps, chief of the Bureau of Construction & Repair, and Rear Admiral George Converse, a retired line officer serving as president of the Board on Construction.[70] Assured by Capps and Converse that the charges against American battleship designs were groundless, Hale optimistically opened the hearings designed to vindicate the technical bureaus and expose the big battleship insurgents as irresponsible spendthrifts.

Capps had recently assumed the duties of chief of the Bureau of Construction & Repair and was defending his predecessor's design for the *Delaware*-class dreadnoughts currently under construction. Capps had graduated from the Naval Academy in 1884, four years after Sims, but was entitled to the temporary rank of rear admiral as chief of a bureau.[71] Newspapers supporting Sims sneered at "Rear Admiral" Capps, but his lengthy experience as a naval architect qualified him to head his bureau.

In his statement to the committee, Capps refuted the charges made by Sims and his compatriots.[72] Capps presented comprehensive data on contemporary warship design practices that supported the bureaus' contention that American warship design philosophy was comparable to foreign practice. In contrast to the claim made in the Reuterdahl article, the data also indicated that American armor belt design was consonant with the design practices of the British and Japanese navies.[73] Capps reported that the Japanese *Aki*-class battleship "indicates clearly that the Japanese, with all their experience derived from the battle of the Sea of Japan, have confirmed in 1906, so far as concerns freeboard, water-line armor protection, height of guns, etc., the American design of two years previous, that is the *Connecticut-Vermont* class."[74]

Capps pointed out that Sims's ideal battleship, HMS *Dreadnought*, had only one-third of her coal on board when the design waterline was calculated. When fully loaded, *Dreadnought* was almost 2 feet deeper in the water than when designed, submerging her armor belt even farther than the worst American ship. When Sims suggested that Capps's data was in error and that a little legwork by Capps could "dig out" the correct information, Capps's reply was peremptory: "It is not a case of digging it out; it is simply a case where those who have to deal with these things and are responsible know, and those who carelessly criticize do not know."[75]

Technically, Sims was correct in accusing the Bureau of Construction & Repair of arbitrarily establishing the waterline at which the armor belt was installed, but it was a decision designed to foster strategic consonance of the battleship force. In 1896, the Walker Board had recommended fixing the design waterline based upon a two-thirds load of coal, stores, and the like.[76] But given the strategic and tactical necessity for battle fleet integrity, an important Mahanian concept, the Board on Construction had decided that the varying rates of fuel consumption of different classes of battleships had to be taken into account. The "normal" load (upon which the design waterline was based) should include enough coal to provide a common radius of steaming, allowing the fleet to operate together. As a result, some ships had their belt armor location based upon a coal load of 67 percent while others might have only 45 percent of their coal allowance on board. In any case, the actual draft variance was small between ships. In no case did any American ship ever exceed the design waterline draft by more than 11 inches. That occurred with one ship. The next highest overdraft at full load was 7 inches.[77]

Arguing that it was better to have the belt too high rather than submerged, Sims, as "a practical naval officer in contradiction to a scientific one,"[78] scoffed at the need for underwater protection from gunfire as a naval shell "always ricochets with extreme suddenness off the water. It never goes down. . . . The water is entirely impenetrable."[79] Sims, the gunnery expert, was scientifically naive and out of touch. Capps informed the committee that army experiments, done in cooperation with the navy, had demonstrated that it was possible for a shell to penetrate the water and strike a ship's side.[80]

Sims was ill-suited to counter Capps's technical evidence and expertise and fared poorly before the Senate Committee. Sims shifted his attack to his claim that line officers were excluded from the battleship design process. But Sims's dislike of the technical bureaus blinded him to the increasing cooperation between seagoing officers and engineering specialists that had produced the *Michigan* design in 1904.

The designs for the *North Dakota*-class dreadnoughts, which Sims condemned, had been subjected to extensive review within the navy. In 1907 the secretary of the navy's General Order No. 49 solicited comments on the proposed battleship design from *all* officers of the navy. The *North Dakota* design was selected after the consideration of twenty preliminary plans, including designs prepared by naval architects from the private sector.[81] The Board on Construction selected the best plans and forwarded them to a special board for review. This board consisted of the assistant secretary of the navy, Truman Newberry; three line officers from the General Board, Rear Admiral Merrell and Captains Wainwright and Rodgers; the chief of the Bureau of Construction & Repair; the chief of the Bureau of Ordnance; and the chief of the Bureau of Steam Engineering. The final plans for the two *North Dakota*-class battleships were approved by this line-engineer special board, forwarded to Secretary of the Navy Charles Bonaparte, and approved by Congress in the naval appropriation bill of 2 March 1907.[82]

Sims had forwarded a response to General Order 49 in which he charged that the armor belt on the *North Dakota* design was 30 inches too low.[83] Sims's complaint was referred to the Board on Construction, which rejected his assertion and recommended no change in the design. Sims filed a rebuttal, and in an extraordinary expression of tolerance, the secretary referred the matter to two line officers, Rear Admiral Willard Brownson, chief of the Bureau of Navigation, and Rear Admiral Robley D. "Fighting Bob" Evans, commander of the Atlantic Fleet, a former battleship commander and popular hero from the Battle of Santiago de Cuba. Brownson supported the Board on Construction. Evans, who admitted that he had believed American armor belts were too low, changed his mind, stating that he had been "misinformed" as to the facts.[84]

Secretary Metcalf then went one step further and submitted the armor design of the *North Dakota* class to other line officers. With one exception, they informed the secretary that "these ships, namely, the *Delaware* and *North Dakota*, were amply protected; in fact, better protected than the ships of any other service."[85] With Sims's complaint rejected by a large number of seagoing officers, Metcalf testified that

> [the Navy Department's] highly trained and legally appointed official advisers [Board on Construction and General Board] should have the dominating influence in all matters connected with the design of naval vessels, it appears to me that in this question of the location of the water-line armor belt, the Department has gone out of its way to obtain opinions from the highest possible authorities among the seagoing branch of the Navy, and that there is really no ground for further contention with respect to the matter of the water-line belt armor on the *Delaware* and *North Dakota* or on the *South Carolina* and *Michigan*, whose armor belts are very similar in location and character.[86]

When the armor belt issue surfaced in the *McClure's* article, Secretary Metcalf had forwarded a report to Senator Hale written by Rear Admiral Robley Evans during the cruise of the Great White Fleet:

> it would appear that better protection might have been afforded had these belts been originally placed between 6 inches and 1 foot higher; this on the theory that the Commanding Officer would admit sufficient water before an action to sink the belt to within 18 inches above the waterline; but even this is open to

The alleged deficiencies in the dreadnought North Dakota *led to the Sims faction's failed attempt at Newport to increase the guns mounted in* Utah *(seen here) and* Delaware. *Commander Albert Key criticized the placement of turret three, immediately behind the second stack.* (Naval Historical Center, NH 44256)

question, for it has been noted that even when heavy laden and in the smooth to moderate seas which have thus far characterized this cruise, the ships frequently expose their entire belt and the bottom plating beneath it. It must be remembered that even a 5 or 6-inch shell (of which there would be a great number), could inflict a severe and dangerous injury if it struck below the belt, while otherwise the waterline, even with the belt entirely submerged, is, on account of the casemate, armor and coal, immune to all except the heaviest projectiles.[87]

Secretary Metcalf informed Senator Hale that Captain Roy Ingersoll, late chief of staff to Admiral Evans, had informed Metcalf that "the undoubted consensus of opinion of officers of the Atlantic Fleet was that the armor belt of battleships of the fleet should not be raised; and that this conclusion was based upon careful observation during the recent cruise . . . and involved a distinct and definite change of view on the part of some officers who had held a very different view before the cruise began."[88]

Sims received the congressional hearing he wanted, but the resulting testimony characterized American dreadnoughts as comparable to foreign designs. Sims's complaints revealed that the line now played a strong role in defining the battleship technological paradigm. Sims's inability to support his charges against the armor belt, his inaccurate knowledge of foreign practices, his naivete regarding the density of water, his attempts to talk about past issues, and his open disdain for Capps and the Board on Construction did not serve his case.[89] The increasing influence of the General Board, the significant line officer representation on the special battleship design review board of 1907, the fact that the current president of the Board on Construction, Rear Admiral Converse, was a line officer, and the statement of the fleet commander, all undermined Sims's claims.

Elting Morison has claimed that Sims and his colleagues "proved their case, but lost the decision."[90] However, they never did prove their case. They erroneously challenged the expertise of the Bureau of Construction & Repair on a technical point of debatable importance. When Rear Admiral Capps could place American designs within the mainstream practices of foreign navies (in a Kuhnian "normal" sense), he demonstrated that American battleship designs were the product of a conservative, well-founded appreciation of a complex technical problem.

Some segments of the press sympathetic to Sims continued to charge a bureau cover-up, and Secretary Metcalf urged Hale to release the reports of Admiral Evans and Naval Constructor Robinson, compiled during the cruise of the Great White Fleet, for publication.[91] When combined with the Board on Construction's comments, Metcalf thought that these reports should ensure that there remains "no doubt whatever in the minds of impartial critics that the battleships of the United States Navy are equal, if not superior, to battleships of foreign navies of corresponding date of design."[92]

Sims and his colleagues failed to discredit the technical bureaus. For the navy, the publicity was extensive, but not as damaging as Admiral Dewey might have feared. Yet Sims was not finished. He dismissed the senior line officers who endorsed the North Dakota design as ignorant, and appealed to President Roosevelt to set things right.

THE 1908 NEWPORT CONFERENCE

Thwarted in Congress, Sims and his compatriot, Commander L. Albert Key, turned to President Roosevelt in the hope that, through him, they finally could break the bureau system and establish a general staff. The focus of their attack was the design for the new dreadnought *North Dakota*, but this time the point man was Albert Key.[93]

Key was due to take command of the new cruiser *Salem* under construction at the Fore River Shipyard in Massachusetts where *North Dakota* was being built. After inspecting *North Dakota*, Key claimed that she could not prevail against foreign dreadnought battleships. Key described the installation of 5-inch armor to protect the secondary battery of anti-torpedo-boat guns as a waste, since it would not stand up to 12-inch shells from an enemy dreadnought. Key cynically observed that it would be better to mount the secondary guns "in the open than behind armor."[94] Key also criticized the location of the number three 12-inch turret between the boiler and engine rooms. The powder magazine for this turret was, according to Key, "completely enveloped, at both ends and sides, by the main steam pipes." Only a 12-inch air gap and a cooling water collar — around the pipes where they penetrated the bulkheads — kept the heat from the pipes from being conducted to the bulkheads of the magazine.[95] Ignoring the recent Senate hearing, Key also complained about the location of *North Dakota*'s armor belt.

According to Key, the chief culprits for the *North Dakota*'s abysmal design were Washington Capps, the Bureau of Construction & Repair, and the Board on Construction. Key reported that an "expert designer" had informed him that an increase of 25 feet in the length of *North Dakota* would allow turret number three to be shifted aft of the engine room and situated to fire over number four turret.[96] In all probability, Key's expert designer was in the employ of the Fore River Shipbuilding Company, which would profit handsomely from such a major modification to *North Dakota*.

Key's letter was referred to the Board on Construction by Acting Secretary of the Navy Truman Newberry on 17 July. Key had also forwarded a copy of his letter to Sims, who used his position as naval aide to the president to send the letter on to Roosevelt. Sims asked Roosevelt for an internal navy hearing to review Key's complaints. Sims wanted a conference to be drawn from officers of the line, specifically, from the General Board and the Naval War College. After reviewing Key's charges, the conference could provide the president with their opinion on the "defects of design" of the new dreadnoughts, whether these defects could be remedied, and a list of all the "military characteristics" required of battleships to be built in the future.[97]

Roosevelt agreed to Sims's request, and a conference was convened at Newport in the summer of 1908. Unfortunately for Sims, the conference was not packed with his supporters. After initial discussions, the conference members agreed that the position of number three turret in *North Dakota* was faulty and that the armor protection for the 5-inch guns offered no protection from large-caliber shells. The location of the armor belt was not criticized. The faults found in the *North Dakota* were considered "minor" and relocation of number three turret was judged too expensive. The original location of the turret had been chosen to avoid "excessive girder stress" in the ship's hull, a valid and significant naval architectural consideration, but just the type of engineering "detail" that infuriated Sims.[98]

While it was too late to change *North Dakota*, the two follow-on ships, *Utah* and *Delaware*, had been funded but not started. Sims and Key tried to use the admitted defects in *North Dakota* as an excuse to modify *Utah* and *Delaware* into even larger all-big-gun battleships. They advocated replacing the five 12-inch turrets of *North Dakota* with four 14-inch gun turrets in *Utah* and *Delaware*.

The possibility of an even more powerful dreadnought delighted Roosevelt, but the up-gunning of *Utah* and *Delaware* would require a quantum leap in gun technology. The majority of the conference members were not sure that the new guns could be completed by the time that the ships would be ready. Sims received assurances from Bethlehem Steel that they could produce the guns in time, but redesigning the ships to accept the new gun battery would postpone delivery for fifteen months. Fearing the growing "dreadnought-gap" with Britain, Roosevelt accepted the existing 12-inch gun battleships to prevent any delay in the delivery of two more dreadnoughts to the Fleet.[99]

Key, Roosevelt's naval aide before Sims, complained to Roosevelt about the "true inwardness of the Newport Conference." Returning to the issue of the naval general staff, judged by Sims and Key as the best way to bring the technical bureaus to heel, Key warned Roosevelt not to accept Acting Secretary of the Navy Truman Newberry's proposed Navy Department reorganization — that would empower the Bureau of Construction & Repair — at face value: "Mr. Newberry, in his attempt at a reorganization of the Navy Department, is merely increasing the main evils under which we now suffer. He is doubling at the expense of sea-going officers, the power and prestige of a shore-staying staff officer, whose main purpose is to preserve and increase his own and his corps' power, prestige and patronage."[100] Interestingly, Key's attack on Newberry came in the wake of Admiral Dewey's endorsement of Newberry for advancement to secretary.[101]

THE LEGACY OF THE 1907–1908 INSURGENCY

The first wartime use of modern battleships confirmed the trend toward larger guns typified by Fisher's *Dreadnought* and the United States's *Michigan*. Pre-dreadnought battleships, such as the ships of the *Connecticut* class with their mixed gun batteries, were rendered obsolete by the big-gun power of the new ships. Almost overnight, the naval battlefield had expanded from a few thousand yards to almost 7 miles.

William Sims was a central figure in the improvement of naval gunnery that led to the adoption of the all-big-gun battleship in the United States. In discrediting Mahan, Sims became the foremost spokesman for the "big" battleship. His importance in the move toward the big battleship should not be understated, but in looking at the history of naval technology of this period, one must be aware of Sims's biographer's confusion of conservative engineering practice with reactionary suppression of new technologies.

The Russo-Japanese War, and the construction of dreadnought-style battleships, provided Sims and his allies with a new springboard for their efforts to establish line control over all facets of the navy. This included the technical bureaus with their significant power over the profession's technological paradigm. Sims truly was out for blood and unconcerned over the possible damage to the navy, and to naval expansion, from his public attack upon the Washington naval establishment. Sims did not reckon with the hostility of the chairman of the Senate Naval Affairs Committee. In basing his attack upon the armor belt, Sims demonstrated his ignorance of contemporary warship design practices, and Chief Constructor Capps easily discredited him.

In the wake of the amalgamation of 1899 and with the beginning of postgraduate technical education for officers, a cautious cooperation had developed between the technical bureaus and the old line hierarchy. While many senior members of the line may have merely "rubber-stamped" the work of the technical bureaus and the Board on Construction, as claimed by Sims, the historical record does not justify an evaluation of Sims as the primary cause for any increase in line control over naval technology. If anything, Sims and his colleagues ignored the new postamalgamation working relationship between the technical bureaus and the line. As a result, the "mutiny" of 1908 was an insider-outsider conflict in which the Sims faction was fighting a naval profession that had evolved beyond Sims's frozen perceptions of it. Sims was a generation removed from Mahan and a generation older and out of touch with the younger officers in the postamalgamation line. These officers pursued graduate engineering education because they accepted, as Sims did not, that every officer in the dreadnought navy had to be a line officer and an engineer.

Resentment of the Sims faction lasted for many years. Albert Key, who left the naval service to sell insurance, reported running into Capps and Naval Constructor David Taylor, who, as late as 1917, "had not forgotten the Newport Conference days, [and] gave me the 'glassy' eye."[102]

In spite of the "Simsian" uproar, the officers of the amalgamated line, both engineering specialists and seagoing officers, were well aware of the need for cooperation due to the rapid changes in naval technology. The improving capabilities embodied in the naval technologies of Britain and Germany in the Atlantic forced the design and construction of increasingly larger battleships. Large size became an important determinant for the trajectory of the battleship technological paradigm. But above all, it was the geographic realities of the vast Pacific, and the potential conflict with Japan, that now governed American battleship design.

CHAPTER 4

Technological Trajectory

Geostrategic Design Criteria, Turboelectric Propulsion, and Naval-Industrial Relations

The naval profession's strategic paradigm of guerre d'escadre governed the battleship technological paradigm and the "normal" refinement of its technological exemplar — the battleship. However, in the United States and other leading industrial powers, the new technologies used in modern dreadnoughts were most often developments of the private sector. The issue for the American naval profession was how to control the technological trajectory — the development and incorporation of technologies — related to the battleship paradigm. This was a difficult undertaking in an environment of rich capital ship contracts, evolving naval-industrial relations, and the ever-present pressure of political-industrial alliances. Selection of the proper propulsion machinery for U.S. dreadnought battleships during the Pacific century was representative of the paradigmatic filtering through which new technologies had to pass.[1]

The U.S. Navy began commerce protection operations in the Pacific Ocean soon after the War of 1812. The East India Squadron was established formally in 1842 after years of operations in the region. The navy's early, inefficient steam warships were ill-suited for these distant operations and this constituted the primary technological tension in nineteenth-century steam propulsion for the U.S. Navy.

The extent of the Pacific necessitated territorial acquisitions for coaling stations or compromises in warship design.[2] The ill-fated USS *Maine,* commissioned in 1895 as a "second-class battleship," was originally designed to carry sails, in addition to steam engines, to extend her cruising radius.[3] The acquisition of the Philippines in 1898 and the emergence of Japan as a rival Pacific naval power underscored the geopolitical factors governing the design of U.S. naval propulsion technology.

Without a worldwide coaling network like Britain possessed, the United States required warships (as it would later require strategic bombers) capable of long-range operations. Captain C. W. Dyson, of the Bureau of Steam Engineering, described the unique situation facing the U.S. Navy:

> The former [European] nations were providing for operations in confined waters such as the North Sea and the Mediterranean, where they were never far from their bases. . . . Under the conditions foreseen high speed was more desirable than cruising radius, and the latter was sacrificed.
>
> Turning now to the conditions which confront us, we see . . . the entire line from Seattle to Panama to the southward, and from Seattle to Honolulu, thence to Guam and on to the Philippines and still further to Samoa, requiring our attention. The areas to be covered are great and the distances to bases and from base to base in some cases are magnificent. Fuel economy is of the highest value, even predominates over speed, as the refueling problem becomes a serious one and the greater our bunker capacities and the fuel economy of our machinery, by so much is the seriousness of the problem reduced.[4]

By 1909, American battleships were being designed with a steaming radius of 10,000 nautical miles, almost double that of the longest-ranged battleship, USS *Oregon*, which had fought at Santiago de Cuba in 1898. This increased steaming radius came only at the price of larger coal bunkers in bigger, more expensive ships.

The first twenty-four U.S. battleships, commissioned between 1895 and 1908, were powered by large, relatively economical, reciprocating steam engines. The development of the compact marine steam turbine, and its introduction in HMS *Dreadnought* in 1905, greatly reduced the internal hull volume required for propulsion machinery.[5] However, due to its direct connection to the propeller, the marine turbine brought a decrease in propulsion efficiency and steaming radius.[6] Although the U.S. Navy could not afford any decrease in propulsion economy, the turbine was attractive because its compact size offered potential improvements in the

One of the reciprocating steam engines destined for the battleship Wisconsin, *1898. The height of these engines required large engine rooms which adversely affected the ship's armor design, center of gravity, and stability after damage.* (Courtesy of the San Francisco Maritime Museum. Naval Historical Center, NH 75106)

fighting efficiency of a battleship, for example, the installation of an improved armor design precluded by the size of traditional reciprocating steam engines.

The ideal solution was to equip battleships with turbines, along with a speed reduction device that would allow the propeller to operate at a slower, more efficient speed. Designed by the General Electric Company (GE) and using an electric motor to power the propeller, the turboelectric drive was one of several systems developed to correct the reverse salient created by the introduction of the marine steam turbine.[7] The other systems were a mechanical reduction gear, developed in the United States and Britain, and a "hydraulic transformer," a type of fluid gear box, developed in Germany. Each was adopted by one of the major naval powers according to their strategic requirements.

DEVELOPING AMERICAN NAVAL TECHNOLOGY

The accretion of increasingly complex artifacts by modern steel navies cried out for a prescribed method for technology development. Yet in the United States, as in Britain, the navy left this task to private industry. The substantial budget of the Royal Navy ensured the steady growth of the British naval-industrial complex in the last decades of the nineteenth century while the fluctuating level of American naval appropriations during the Harrison, Cleveland, and McKinley administrations made U.S. naval-industrial relations a "hot and cold" affair.[8] While progressive officers within the navy clamored for modern battleships, the oppression of naval engineers by the seagoing officers of the line had precluded the establishment of any significant naval technical research facility (except for naval ordnance). The Experimental Model Basin, established at the Washington Navy Yard in 1899, limited its work to hydrodynamics. The Bureau of Steam Engineering's Experimental Station at Annapolis did not develop new technologies; it merely evaluated items offered for sale by the private sector.[9] By default, new technologies, for the most part, originated from private-sector inventors and entrepreneurs, or later, as products of the industrial research revolution.

Some naval officers realized the need for an institutional structure to address the navy's technological needs. In 1902, the navy's engineer-in-chief, George Melville, called for a research establishment modeled upon the German naval facilities at the Charlottenberg and Dresden Technical Colleges.[10] Melville's research institution was precluded by the line's desire to break, rather than to enhance, the power of the technical bureaus typified by the push to create a naval general staff. Bradley Fiske, an inventor as well as a line officer, proposed a department of invention for the navy in 1907 and lobbied Thomas Edison for his services in 1911.[11]

During his testimony on American battleship design before the Senate Naval Affairs Committee in 1908, Fiske described the navy's difficulties in developing new technologies, attributing the problem to the failure of "officers in high position to realize the duality of the naval profession; to realize that a navy consists of both personnel and material; the two of equal importance, and each useless without the other." Almost a decade after the amalgamation of the engineering corps into the line in 1899, senior officers, according to Fiske, still failed to "correlate the military and the engineering arts" and neither engineering specialists nor line officers possessed the perception to see the "relations that ought to exist between the two arts."[12] Fiske presented the emerging private-sector industrial research facilities as the solution:

> The remedy is easy to find, because it has been found already by the large industrial concerns. These are themselves large organizations; and competition between them is so keen that a concern which falls behind the times goes into the hands of a receiver very soon. Fortunately for our Navy foreign navies have been as lax as we and will continue to be so until one navy wakes up. Then they will all have to bestir themselves or get into a condition so obviously inferior that fighting would be a useless sacrifice of life and limb.[13]

Ironically, Fiske's plea for research according to the industrial model would be forestalled by Secretary of the Navy Josephus Daniels's 1915 call for a department of invention, which resulted in the ineffective and peripheral Naval Consulting Board.[14] As a result, the early-twentieth-century navy continued to rely upon the private sector for new technologies.

THE MARINE STEAM TURBINE AND WARSHIP PROPULSION

In Britain in 1894, Charles Parsons formed the Marine Steam Turbine Company, Ltd., and attempted to interest the Royal Navy in his reaction turbine for propelling warships. Parsons finally succeeded in bringing his invention to the attention of the Admiralty three years later when his *Turbinia* steamed at 31 knots down the battle line at Queen Victoria's Diamond Jubilee Naval Review.[15]

Although initially condemned in most quarters as unsuitable for naval use due to its high rotative speed and experimental nature, Parsons's turbine offered the potential to increase warship speed.[16] A faster battleship possessed obvious strategic and tactical advantages. If the turbine could be installed in destroyers, the navy would also have a means to increase the defensive zone around its capital ships concerning the high-speed torpedo boat. Unlike Isherwood's *Wampanoag*, Parsons's innovative technology could protect the naval status quo, based upon the supremacy of the battleship, from the challenge of the torpedo.[17]

In 1901, the Admiralty agreed to install Parsons turbines in two torpedo-boat destroyers, *Cobra* and *Viper*. Although lost at sea before operating experience could be obtained, both ships had successfully completed their trials at speeds much higher than destroyers equipped with reciprocating engines, demonstrating the turbine-equipped destroyer's potential tactical superiority over existing ships. Surprisingly, however, no additional turbine orders were placed by the Royal Navy until 1905.[18]

While Parsons was attempting to establish a turbine market in Britain, the General Electric Company, in the United States, was developing the Curtis turbine for use in electrical power generation. In 1900, Charles G. Curtis had approached General Electric with his concept for a more efficient version of the reaction-type steam turbine. Curtis's design was based upon impulse theory and offered the potential for increased operating economy and, perhaps more important, a patentable technology. Curtis entered an agreement with General Electric that stipulated the development of a working prototype within three years. Failure to do so would require General Electric to relinquish all rights to his design. If development was successful, Curtis would retain the rights to all nonelectrical uses of his turbine, a factor that became important during the 1917 controversy over electric battleship propulsion.

Development of the GE-Curtis turbine proceeded slowly, forcing GE Vice President for Engineering E. W. Rice Jr. to bring in one of his most successful engineers, William Le Roy Emmet. After reviewing the GE-Curtis developmental effort, Emmet reported that the Curtis turbine had tremendous potential for use in large, central-power generating stations and should be pursued. As a reward for his vision, Emmet was tasked with completing the Curtis turbine for General Electric.[19]

Emmet, an 1881 graduate of the line officer curriculum at the Naval Academy, was a self-taught electrical engineer. Under the provisions of the August 1882 appropriations act, Emmet's poor class standing had left him without a commission after his graduation cruise was completed in 1883. Fascinated by electricity, he used his aggressiveness to win a job installing streetcar systems for fellow Naval Academy alumnus Frank J. Sprague. From there, Emmet went to work for the Edison General Electric Company in Chicago, where he was introduced to Thomas Edison's systems approach to electrical power generation.[20] After the merger with Thomson-Houston, Emmet moved to Schenectady as an employee of the new General Electric Company, where his first major accomplishment was winning a large share of the Niagara Falls alternating current generating station contract from rival Westinghouse in 1894.[21]

Emmet and Curtis developed a large-capacity turbine that was more efficient than the Parsons reaction turbine and which threatened Parsons's
position within the U.S. turbine market. A 5,000 kw GE-Curtis impulse
turbine was installed in Consolidated Edison's Fisk Street Station in
Chicago in October 1903.[22] General Electric extended its challenge to Parsons by concluding a development agreement and worldwide market division of the GE-Curtis turbine with Allgemeine Elektrizitäts Gesellschaft
(AEG) the same year.[23] In Britain, Parsons remained one of two principal
manufacturers of equipment for electrical generating plants until World
War I.[24] Nevertheless, while Parsons shared the British domestic-power
market, he had established a virtual monopoly over the marine turbine
market in Britain, continental Europe, and Japan.[25]

By 1905, many navies were considering the marine turbine as a replacement for the reciprocating steam engine in warships. Responding to
British naval interest in turbines, the U.S. Navy signed contracts for the
construction of three cruisers in 1905, one to be equipped with Parsons turbines, one with Curtis turbines, and one with traditional reciprocating
steam engines to compare their relative economies.[26] Based upon these
1905 cruiser trials, the U.S. Navy decided to accept the steam turbine as the
prime mover in its capital ships.[27] Attracted by the compact size of the turbine, and undoubtedly copying Fisher's turbine-propelled *Dreadnought*,
the navy selected U.S.-built Curtis turbines for its first true dreadnought
battleship, USS *North Dakota*, in 1907. Parsons turbines were rejected
since they required an engine room 24 feet longer than the Curtis installation.[28] Unfortunately for Curtis, deterioration of the turbine steam nozzles adversely affected the performance of his turbines and resulted in the
utilization of Parsons turbines in the next four U.S. battleships.[29] The Parsons turbines did not deteriorate during use, but their relative inefficiency
reduced the battleships' steaming range. This led to a reversion to the more
economical, and strategically consonant, reciprocating engines in the two
New York–class dreadnoughts of the 1910 Program.

ATTACKING THE REVERSE SALIENT:
INCREASING TURBINE EFFICIENCY

Widespread acceptance of the marine turbine depended upon the development of a means to optimize the existing turbine-propeller system.

Until this could be accomplished, the turbine would be limited to ships requiring high-speed operation — warships and passenger liners. However, in the U.S. Navy, where speed took second place to propulsion economy, high speed was not a primary design criterion. In America, the turbine's position was precarious, as demonstrated by its exclusion from the battleships of the 1910 Program.

Hydrodynamic limitations on propeller design shifted the focus for improved propulsive efficiency onto the turbine. To regain the U.S. naval market in 1909–10, the turbine manufacturers increased the diameter of their turbines, reducing their rotative speed while maintaining the same level of power output.[30] Although reduced to several hundred rotations per minute, the turbine still operated at too high a speed for optimum propeller efficiency. Overall propulsive efficiency, which translated into fuel economy and long-range operations, still remained less than for a reciprocating steam-powered ship, but was offset, to a point, by the turbine's compact size that allowed for increased coal bunkerage and improvements in other ship systems.

To solve the problem of low propulsive efficiency and propeller cavitation, marine engineers pursued three paths to develop a speed-reducing device to operate between the high-speed turbine and an optimized, low-speed propeller. The first system was the mechanical reduction gear, pursued in Britain and the United States. The second system was the turboelectric drive, an idea investigated in Britain, Germany, and the United States. The third system was the hydraulic speed reduction, the Föttinger Transformer, developed in Germany.[31]

In 1904, a consortium had been formed by Charles Parsons, George Westinghouse, former U.S. Navy Engineer-in-Chief George Melville, and John MacAlpine, a marine engineer. Their goal was the development of a suitable mechanical reduction gear.[32] Metallurgical and machining difficulties precluded producing geometrically accurate gear teeth able to withstand the tremendous pressure and torque required for use with steam turbines. Melville and MacAlpine thought they could bypass this problem by designing a reduction gear that incorporated a "floating frame." Refinement of the design was retarded by the adverse economic climate of 1907 that forced George Westinghouse to reduce his funding for the project.

Drawing upon British expertise in metalworking, Parsons independently pursued a fixed-frame, high-tolerance mechanical reduction gear assembly. This type of gearing required accurate machining technology and it was not until 1910, almost thirteen years after *Turbinia*'s success, that the Parsons Marine Steam Turbine Company was finally able to develop and test a fixed-frame, experimental reduction gear in the 1,000 hp propulsion plant of SS *Vespasian*.[33]

Around the same time, a Professor Föttinger produced a hydraulic speed reduction device in Germany. Föttinger had originally developed an electromagnetic transmission system that provided an overall power transmission efficiency of 87 percent (compared with 60 percent in direct-drive turbines), but was abandoned due to its complexity and expense. Föttinger had more success with his next design, the hydraulic transformer, in which the turbine shaft rotated a waterwheel that forced water through a set of guide blades onto a larger waterwheel attached to the propeller shaft. This Föttinger Transformer, coupled to an AEG-Curtis turbine, was tested at the Vulcan Shipbuilding Company power plant for fourteen months before being installed in a specially built test boat in 1910.[34]

While Westinghouse, Parsons, and Föttinger worked on mechanical and hydraulic methods of speed reduction, William Le Roy Emmet, motivated by an idea advanced by the electrical inventor Reginald Fessenden, proposed the turboelectric ship propulsion system to eliminate the inefficiency of the directly connected marine steam turbine.[35] Emmet had been quite successful in the development and marketing of the GE-Curtis turbogenerator system within the American electrical power industry. However, the electrical industry could only absorb a limited number of turbogenerators. The U.S. Navy, on the other hand, was a growing organization and a wealthy potential customer and Emmet sought to extend his entrepreneurial success, for General Electric, to this new and lucrative market.[36] To break into the battleship propulsion market, General Electric had to compete against the Melville–MacAlpine gear of the Westinghouse Machine Company, and overcome the influence of the private shipyards that were Parsons licensees and advocates of the directly connected turbine drive.

A HARD SELL: THE U.S. NAVY AND THE TURBOELECTRIC DRIVE

The navy had previously rejected turboelectric propulsion when Fessenden proposed its use in 1908.[37] The Board on Construction thought the electric drive too experimental to risk its installation in a battleship.[38] Fessenden received permission to approach the "engineers of any manufacturing companies" provided he made it clear that there was no navy interest or sponsorship involved. Westinghouse and Allis-Chalmers rejected him outright, while Emmet at General Electric investigated Fessenden's proposal thoroughly.[39]

Emmet immediately saw the merit in Fessenden's idea. Pointing out the differences between General Electric's proven technology and the paper theories of Fessenden, Emmet wrote the secretary of the navy that the GE plan for electric warship propulsion "involves no feature which is not already an accomplished fact."[40] The new secretary of the navy, George von L. Meyer, directed the Bureau of Steam Engineering to investigate the GE proposal for battleship propulsion.[41]

The navy's inspector of equipment at GE-Schenectady, Lieutenant Commander J. McNamee, enthusiastically outlined the strategic value of the turboelectric system to the Navy Department: "[the electric drive] so far exceeded the economy of the Parsons system, that there can be no doubt that upon the completion of these vessels [battleships *Wyoming* and *Arkansas*], involving an expenditure of 20 million ($20,000,000) dollars, they will be so deficient in steaming radius, compared to what they might have been [with an electric propulsion system], as to cause the gravest concern." McNamee also raised the specter of loss of the electric drive to the rival Royal Navy, arguing that adoption of the electric drive in British dreadnoughts would render the latest American battleships "in a strategic sense, perilously near the category of lame ducks."[42]

Officers in the Bureau of Steam Engineering attributed Commander McNamee's enthusiasm to his close working relationship with General Electric and made no effort to speed up their evaluation. After four months of waiting, a discouraged Emmet, used to the quicker pace of the business world, expressed his frustration to his Naval Academy classmate, Congressman John W. Weeks: "Such improvements as I offer will never be realized unless somebody pushes them. Humanity is so constituted that nine out of ten men devote much of their time opposing or discouraging such things."[43] His patience exhausted, Emmet gave the system a push. He persuaded his brother-in-law, who regularly socialized with Secretary Meyer at the Harvard Club, to write Meyer portraying Emmet as a respected General Electric engineer, not a mere "promoter."[44] This may have had some effect, as the chief of the Bureau of Steam Engineering soon reported fa-

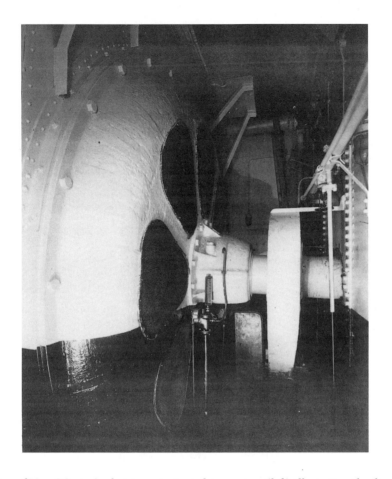

One of New Mexico's *electric main propulsion motors (left) illustrating the direct connection to one of the ships four propeller shafts (right).* (Courtesy of General Electric Co., ca. 1920. Naval Historical Center, NH 75106)

vorably, but cautiously, on the potential of the electric drive to Secretary Meyer.

Engineer-in-Chief Hutch Cone was a progressively minded officer whose support for William Sims in 1908 reflected the blurring of old allegiances in the amalgamated officer corps.[45] Despite his "revolutionary" past, Cone was a cautious engineer who was unwilling to accept Emmet's claims at face value. Cone recommended that the turboelectric drive be tested in some "unimportant vessel" such as the cruiser *Baltimore* or in one of the new colliers to be built the following year.[46] Emmet feared these ships would not provide a fair evaluation of his system, which was designed for a large battleship.

Since the Board on Construction had recently rejected Fessenden's turboelectric drive, Secretary Meyer, as a favor to Emmet's brother-in-law, used the conditional endorsement of the engineer-in-chief to bypass the board and refer the question of turboelectric propulsion to the line officers on the General Board. The General Board, demonstrating its new influence over naval technology, recommended installing the General Electric system in one of three new colliers.[47]

The collier construction program of 1911 was analogous to the cruiser propulsion trials of 1905, pitting Emmet's GE turboelectric drive against a new Westinghouse-backed system with the Melville–MacAlpine reduction gear. A third collier was equipped with a diesel submarine engine from the Electric Boat Company.[48] The entire financial risk of the experiment was carried by the competing companies: "in the case of failure of the experimental installation, the builders of this machinery [General Electric and Westinghouse] would receive no payment, and the entire cost of its removal, and the preparations of the foundations to receive reciprocating engines would be borne by them."[49]

Emmet, drawing upon his experience in developing electrical systems, fabricated and thoroughly tested the complete electric drive, including all nonelectrical auxiliary machinery, at GE-Schenectady.[50] Growing more concerned that the collier's small-horsepower requirements would not properly demonstrate the large-horsepower efficiency of the turboelectric drive, Emmet again offered the turboelectric drive to the navy for use in new battleship construction.[51] When the secretary's office directed the bureaus to reconsider the turboelectric drive for battleship use, Engineer-in-Chief Cone stood firm in his opinion that the suitability of the turboelectric drive could only be judged after the completion of the collier tests.[52]

Emmet decided to pressure the navy by embarking upon a publicity campaign to sell the GE turboelectric drive to merchant marine operators and to the public. Speaking before local engineering societies, he made much of the turboelectric propulsion system selected for the collier *Jupiter*. Emphasizing the light weight of the electric drive, Emmet claimed, with some exaggeration, that a comparable 30,000 hp turboelectric drive would weigh only one-tenth that of the Parsons direct-drive machinery installed in the new passenger liner, *Lusitania*. He also stated that the turboelectric drive could save a ship owner at least $20,000 per year on fuel. Not one to let sleeping dogs lie, Emmet blasted the navy, claiming that a turboelectrically propelled battleship of the *Wyoming* class could steam two-thirds of the way around the world on a single coaling. Emmet won many adherents and the partisan *Electrical World* trumpeted "that a great field for electrical service is close at hand."[53]

Despite Emmet's publicity blitz, the navy announced that the new superdreadnought *Pennsylvania* would be equipped with Parsons direct-drive turbines and a small-horsepower Parsons cruising turbine fitted with a reduction gear. Emmet protested vigorously to the new secretary of the navy, Josephus Daniels, characterizing the continuation of direct-drive turbines as a serious mistake in U.S. battleship design. Through Emmet, General Electric even offered to guarantee a $350,000 electric drive installation in *Pennsylvania*.[54] Assistant Secretary of the Navy Franklin Roosevelt rejected the General Electric offer, supporting the position of the Bureau of Steam Engineering. Its new chief, Rear Admiral Robert Griffin, believed, like his predecessor, Cone, that the electric drive must first prove itself in a collier before it could be considered for use in a battleship. Griffin wryly remarked that although there was talk in the press about the widespread use of the electric drive, so far the U.S. Navy appeared to be its only customer.[55] Bristling at Griffin's criticism, Emmet condemned the bureau engineers: "The methods proposed are entirely outside the experience of naval officers and I do not expect their technical indorsement." Ignoring the pecuniary basis of General Electric–navy relations, Emmet wrote to Roosevelt, "I have no financial interest in this matter and am simply seeking to be useful and to extend the uses of electricity."[56]

The report of the *Jupiter* trials, finally published in May 1914, justified Emmet's claims for the electric drive. *Jupiter*'s chief engineer, Lieutenant S. M. Robinson, reported that the turboelectric drive was easily operated by relatively unskilled sailors, provided for accurate speed control, and, most important, exceeded General Electric's guaranteed economy predictions by a phenomenal 18 percent.[57]

Emmet's thoroughness in land-testing the complete turboelectric drive system had paid off. The rival Westinghouse Melville–MacAlpine reduction gear system in the collier USS *Neptune* had failed its trial. Ironically, the problem was not centered in the reduction gear, but in a new turbine designed by the Westinghouse Machine Company and in a pneumatic turbine-control system designed by George Westinghouse to allow direct engine control from the bridge.[58] The third collier, *Maumee*, did not perform anywhere near expectations and the diesel engine was rejected.

In 1915 the U.S. Navy was faced with a choice between the proven success and economy of the turboelectric drive, the trouble-plagued Westinghouse Melville–MacAlpine reduction gear system, or a continuation of the inefficient direct-drive turbine augmented by a small reduction gear-equipped cruising turbine, first introduced in 1912. To the officers of the Bureau of Steam Engineering, the geopolitical realities of the Pacific component of U.S. naval strategic doctrine made the choice clear. In April 1915 Secretary Daniels announced the selection of the General Electric system to power the new superdreadnought *California*.[59]

THE 1917 ELECTRIC DRIVE CONTROVERSY

The shipbuilding industry, which manufactured the direct-drive turbine propulsion systems under license, did not protest the selection of the electric drive for *California* as two sister ships, *Idaho* and *Mississippi*,[60] were to be fitted with the shipyard-built, 1912-style Parsons and Curtis geared-cruising-turbine systems. Attacks on turboelectric propulsion only began in earnest when it was announced that all sixteen superdreadnoughts of the $588 million 1916 Program would be equipped with turboelectric drives. This stripped the shipbuilding industry of the profits that could have been derived from the manufacture of direct-drive propulsion components for this new generation of American capital ships.

With turboelectric propulsion, the capital ships of the 1916 Program possessed superior fuel economy and the small size of the propulsion components allowed for greater internal subdivision of the hull. This enhanced the survivability of the propulsion system in particular and the ship overall in the face of the new nemesis of the European War, the submarine-launched torpedo. In addition, the navy maintained that turboelectric propulsion was less costly to acquire and that it had shaved $200,000 from the cost of *California*, a figure disputed by the shipbuilding industry.[61] Even at this early stage, a necessary selling point for a new military technology was its acquisition cost. A blind eye was turned to operational ("life-cycle") cost — a continuing legacy of the annual congressional budgetary process.

In keeping with American design practice, which made speed subsidiary to large guns and maximum armor, the battleships of the 1916 Program still had a respectable top speed of 21 knots. This speed required a turbine de-livering a little more than 7,000 hp per propeller shaft, which was within the capability of the turbines used in domestic electric power plants. As a result, the decision to install the electric drive in the battleships was read-ily defensible from a technical standpoint. The six battlecruisers, on the other hand, were designed to reach speeds of 33 knots requiring a total of 180,000 hp, or 45,000 hp per propeller shaft, almost a sevenfold power in-crease over the battleships. The mechanical lobby focused their attack on the battlecruiser design.

The private shipyards indirectly protested the selection of the electric drive by submitting inflated construction bids for the 1916 Program battle-ships, citing the uncertainty of working with the unfamiliar turboelectric drive.[62] Less charitable assessments within the Navy Department attrib-uted the high bids to the loss of revenue caused by the elimination of the shipyard-built, direct-drive turbines. Conferences between the shipyards and Secretary Daniels finally resulted in battleship contract awards in De-cember 1916. Newport News Shipbuilding Company agreed to build two battleships, using GE turboelectric machinery, for $22.4 million. A similar agreement was reached with New York Shipbuilding to build two battle-ships using electric drives constructed by the Westinghouse Electric Man-ufacturing Company.[63]

The risk inherent in the large power requirements of the battlecruisers was used to justify even higher bids for the first four ships to be contracted. The navy had estimated building costs for each of the battlecruisers at $20.4 million, twice that of a single battleship. The 1916 Naval Appropriations Act authorized the award of cost plus fixed-percentage-profit contracts and the bids received from the four private shipyards were all on this basis, with profits ranging between 10 percent and 15 percent, but on top of a higher base cost than the navy estimate.[64]

While the Bureau of Construction & Repair reviewed the battlecruiser bids, its chief, Rear Admiral David Taylor, confirmed to the House of Rep-resentatives that half the battlecruisers' boilers would be located above the main armored deck. This apparent defect was added to the list of com-plaints about the electric drive. The exposed boilers and the large power output required of the electric drive were disparaged in the *New York Times* by Schuyler Wheeler, an electrical engineer with links to the private ship-yards. Three days later the paper printed a list of electrical engineering ex-perts, including Michael Pupin, opposed to the electric drive.[65]

The most strident criticism of the turboelectric drive appeared in *Scientific American*. Editorials by J. Bernard Walker carried titles such as "A Grave Defect in Our Battlecruiser Designs," and the letters to the editor provided a forum for the expression of "disinterested" engineering opinion on the electric drive. Walker even enlisted the president of the Naval War College, Rear Admiral William S. Sims, to fight the electric drive in his capacity as an *ex officio* member of the General Board.[66] The 10 February 1917 edition of *Scientific American* contained a letter from Charles Curtis condemning the turboelectric drive while praising the patriotic efforts of private shipyards to prevent the navy from making a grave technical error. Curtis called for the creation of an impartial engineering board to judge the situation. Rear Admiral Griffin rejected a review of the electric drive proposed by members of the civilian Naval Consulting Board since the evaluators would have been Benjamin Lamme of Westinghouse and William Le Roy Emmet, both of whose opinions were hardly impartial.[67]

The mechanical interests attempted to involve both the House and Senate Naval Affairs Committees in the electric drive dispute. The House Committee stood clear of the controversy, as "it was an entirely technical subject with which the bureau experts of the Navy Department had exclusively to do." Senator "Pitchfork" Ben Tillman, chairman of the Senate Naval Affairs Committee, on the other hand, was somewhat receptive to the complaints against the electric drive while navalists, such as Senator Lodge, wanted to provide the navy a forum for its defense. As a result, J. Bernard Walker and Rear Admirals Taylor and Griffin were called to testify before the Senate Committee during a special Saturday session in February 1917.[68]

The hearing focused on the placement of some of the battlecruiser boilers above the main armored deck. The testimony revealed that this was not due to a lack of volume within the armored citadel caused by the electric drive, per se, but by the number of boilers required for the high speed requirements of the battlecruiser, the number of boilers being the same whether the ship was driven electrically or mechanically. The navy design bureaus had decided to separate the boilers vertically to protect one-half from the effects of torpedoes exploding against the hull.[69] As the Senate hearing progressed, both admirals challenged Walker to name any civilian engineer who had publicly gone on record against electric propulsion yet had no financial interest in the situation. When Walker was unable to respond, the Senate Committee came to the same conclusion as the

House — that there was insufficient reason to challenge the technical expertise of the bureaus.[70]

After successfully defending the navy's selection of its capital ship propulsion technology before Congress, Rear Admiral Griffin took the unprecedented step of presenting the Navy Department's case to the public in the pages of *Scientific American*. Quoting from previous correspondence with Senator Tillman and Congressman Padgett, the chairman of the House Naval Affairs Committee, Griffin recounted Schuyler Wheeler's earlier visit to Secretary of the Navy Daniels in which Wheeler's limited large-scale electrical experience was revealed, painting him, in Griffin's opinion, as little more than a shill for the shipyard interests.[71]

The attempt of Walker, Wheeler, and Curtis to enlist congressional aid was the high point of their fight. Public statements in favor of the electric drive by such notable electrical experts as Nikola Tesla and Frank Sprague did much to sway public opinion in favor of the navy.[72] Even shipping industry publications such as *Marine Engineering* and *Shipping Illustrated* pointed out the financial motivation behind the opposition to the electric drive. The latter, who held "no brief for the electric drive," attributed the "incompetent criticism" to the shipyards' loss of business as patent licensees of Curtis and Parsons turbines. When the electric drive moved from an experimental collier built in a government yard to battleships built under contract by private firms, "the shoe began to pinch."[73]

Much like Emmet some four years before, Charles Curtis claimed that his motive was disinterested: "my efforts to set the Navy Department right in this important engineering matter are not due to any royalty or commercial considerations." But Curtis was ignoring the fact that he retained the royalty rights to the GE-Curtis turbine for all *nonelectrical* uses. Curtis disingenuously stated that he believed that any future navy geared-turbine installations would exclusively feature Parsons turbines.[74] Parsons or not, the selection of geared turbines for any of the superdreadnoughts of the 1916 Program would have prevented a turboelectric monopoly and an eclipse of the International Curtis Marine Turbine Company within the lucrative American capital ship market.

Lacking congressional and public support, the mechanical lobby called an end to the electric drive fight. In the spring, contracts were signed for the first four battlecruisers of the 1916 Program. Fore River Shipbuilding contracted with General Electric, and Newport News with Westinghouse, for turboelectric drives at an average price of $2.8 million per ship.[75] A

comparative study conducted by the Bureau of Steam Engineering revealed that at power levels below 29 percent of full power, that is, at cruising speeds, the turboelectric drive was approximately 20 percent more efficient than direct-drive Parsons or Curtis turbines and at least 6 percent to
8 percent more efficient than a mechanical reduction gear driven by a Curtis turbine.[76]

The vigorous campaign by the opponents of the turboelectric drive was
typical of the political maneuvering involving any large government contract. The 1980s struggle between Northrop (F-20) and General Dynamics
(F-16) over air force day-fighter contracts is a more recent example. The
importance of the electric drive was its contribution to a paradigm for subsequent naval-industrial relations that made strategic compatibility a pre-
requisite for consideration, let alone adoption, of any technology offered
by the private sector.

TECHNOLOGY SELECTION AND NATIONAL STRATEGIC PHILOSOPHIES

The U.S. Navy was not alone in imposing a strategic litmus test on its
warship propulsion technology. Despite its worldwide responsibilities, the
strategic focus of the British navy in the years before 1914 was the North Sea
and conflict with the German navy. The high speed provided by direct-
drive turbines in its dreadnought battleships was deemed essential for successful sorties against the Germans. This does not mean that the British
were not interested in improving the fuel efficiency of turbine propulsion,
but operations in the North Sea made fuel economy less important a design criterion than for the Americans.

The Germans had the same strategic focus as the British, but without a
global empire to patrol could design ships primarily for North Sea operations. Admiral von Tirpitz had specified that German capital ships must be
resistant to battle damage, and German naval architects were successful,
surprising the British with the German ships' ability to withstand punishment at the Battle of Jutland in 1916. British naval constructors attributed
the German ships' strength to the local focus of German naval strategy.[77]
They referred to German dreadnoughts as "coastal ironclads," and there is
a kernel of truth to this pejorative evaluation. Like the British, the Germans

required high battle speed, which could be provided by direct-drive turbines. Their capital ship designs were not driven by strategic requirements for steaming range as were the British, and to a much greater extent, the Americans. The wartime reorientation of German naval construction to the building of submarines resulted in a delay in the installation of the Föttinger transformer in a German capital ship. Three of the seven German battlecruisers authorized during the war, those of the *Ersatz Yorck* class of the 1916–17 Program, were to be equipped with Föttinger's transformer, but the ships were never completed.[78]

The Japanese navy expressed an interest in turboelectric propulsion after the war. In 1920 the newspaper *Tokyo Nichi-Nichi* reported that the Imperial Japanese Navy was appropriating several million yen to study the electric drive.[79] The following year, the Japanese navy ordered and took delivery of the auxiliary oiler *Kamoi*, equipped with a GE turboelectric drive, from the New York Shipbuilding Company.[80]

Although the Japanese had embarked upon indigenous capital ship construction with the superdreadnoughts *Fuso* and *Yamashiro* in the 1911 and 1913 Programs, they had a strong tradition of purchasing their naval technology abroad. This allowed the Japanese the flexibility to consider the technological selections of other countries. The fact that the technology being examined in 1920 was American rather than British reflected the change in post–World War I power relationships in the Pacific. Any Japanese desire to pursue turboelectric propulsion for their navy was cut short by the Five-Power Treaty ratified in 1922 with its ten-year construction halt on new capital ships.

Both the German and Italian navies, encouraged by the American naval experience with turboelectric propulsion, seriously considered the electric drive during the 1930s. A turboelectric drive was originally selected for the *Bismarck*-class battleships, but Siemens-Schuckert Werke would not accept the contract because of a fear that it could not meet certain technical requirements. Its withdrawal led to the reversion to a reduction-gear system, patterned after those installed in the *Scharnhorst*-class battlecruisers, favored by the conservative faction of naval constructors for its ease of repair.[81] The Italians also considered the turboelectric drive in the capital ship design studies performed in 1932 and 1933, but rejected it in favor of a more efficient reduction-gear system for the large 35,000-ton "treaty" battleships of the *Vittorio Veneto* class ordered in 1935.[82]

THE STRATEGIC IMPERATIVE

The selection in 1905 of the marine steam turbine to power HMS *Dreadnought* engendered an immediate interest in this form of propulsion among the other naval powers. The turbine was popular with naval officers since it was quieter and more tidy than reciprocating steam propulsion and it allowed a reduction in the volume of a warship devoted to an opaque machinery system seen as less worthy than guns and armor. The industrial, organizational, strategic, and tactical advantages of the turbine were enumerated by the Admiralty Committee on Designs in 1905 and applicable to the United States and Britain:

> While recognising that the steam turbine system of propulsion has at present some disadvantages, yet, owing to the great savings in weight and reduction in number of working parts, smooth working, capability of being started on short notice, saving in coal consumption and in engine-room complement, and to the increased protection which it is possible to obtain by the introduction of this system, the Committee have decided to recommend that it be adopted in this design [HMS *Dreadnought*].[83]

The "disadvantage" of the steam turbine system, the lack of efficiency that resulted from the direct-coupling of the turbine to the propeller, had little, if any, impact given the British strategic focus on the German navy and the North Sea — a strategic policy that required the higher speed, larger guns, and improved armor protection the turbine could provide.

The inherent deficiencies in a directly coupled turbine system were realized from the beginning. In 1904, a German naval architect expressed a desire that "all the ships of the enemies of Germany would have turbines."[84] Yet within three years, the German director of naval construction had announced that all future warships of the Imperial Navy would be powered by turbines.[85] The United States, Japan, and France all jumped on the turbine bandwagon. However, the poor economy of the turbine caused the U.S. Navy to reconsider its selection of the turbine in 1909–10. The engineering and entrepreneurial efforts of William Emmet induced the navy to accept turboelectric technology because it enhanced the strategic consonance of U.S. battleships.

All the naval powers agreed with Jackie Fisher's precept that ship design should be dictated by strategy. The use of direct-drive turbines by marine engineers in Germany and Britain does not mean that they were not interested in improving the efficiency of turbine propulsion, as the efforts of Parsons and Föttinger demonstrate. Yet given their strategic focus, neither had as pressing a requirement for improving the efficiency of direct-drive turbine propulsion as did the United States.

Like so many modern military technologies, the supremacy of the turboelectric drive was short. Both General Electric and Westinghouse pursued reduction gear technology, and the navy installed it on smaller warships such as destroyers. Reduction gears became more efficient, eventually transmitting 2 percent to 3 percent more power than the electric drive. By 1931, machining and metallurgical technologies could produce compact lightweight reduction gears with which the electric drive could no longer compete. The first electrically propelled battleship, USS *New Mexico* (ex-*California*), had her electric motors replaced with a General Electric mechanical reduction gear in 1931.[86]

When U.S. battleship construction recommenced as a part of the industrial recovery program of the New Deal, American capital ship design philosophy continued to favor large guns and the maximum armor possible. The Pacific strategy still required long-range steaming, but faster Japanese capital ships meant that high speed was now a requirement as well.[87] The Bureau of Engineering prepared fifteen propulsion system proposals, ten of which featured mechanical reduction gears, while the remainder were turboelectric drives. The electric drive was discarded since it required at least 10 percent more weight and volume than reduction gear systems of equal power and efficiency.[88] The battleship displacement limit mandated by the London Naval Treaty of 1930, and the subsequent Anglo-French-U.S. accords of January 1937 and March 1938, made volume and weight extremely critical design factors. As a result, U.S. battleships could ill afford the larger, heavier turboelectric drive, and it fell victim to treaty limitations and to the new emphasis on speed within the U.S. Navy's Pacific-oriented strategy.[89]

The turboelectric drive reflected the strength of the technological trajectory associated with the battleship technological paradigm to affect the industrial sector of U.S. society. Industrialists who were successful in establishing naval markets for their products had to be consonant with the overarching battleship paradigm. Competition among different *battleship* technologies was fairly traditional and the use of political influence, complaints about the bidding process, or method of evaluation was typical.

Successful naval-industrial relations could only be forged by companies offering technological refinements — not challenges — to the battleship. Naval officers were co-optive, less receptive, or downright hostile toward technologies that threatened to overthrow the technological paradigm and thereby threatened the profession's overarching battleship strategy.

CHAPTER 5

Anomalous Technologies of the Great War

Airplanes, Submarines, and the Professional Status Quo

Unlike the turboelectric drive, advances in airplanes and submarines during the World War presented patchy, potential anomalies that threatened the battleship technological paradigm. The navy's exploration of these technologies was in a Kuhnian "normal" sense. However, the anomalous nature of these technologies was enhanced by their use in an innovative wartime environment. As a result, submariners and aviators became more clearly delineated as new technologically defined thought collectives (defined by Ludwig Fleck as a community "mutually exchanging ideas or maintaining intellectual interaction") within the naval profession.[1] The dramatic sinking of three British cruisers by a single submarine at the outset of the war garnered respect for the submarine. The airplane, on the other hand, was still unrefined when the war ended, but many presumed it had the potential for future development and the ability to replace the battleship in fulfilling the guerre d'escadre strategy.

To flourish after the armistice, aviators and submariners had to develop a strategic thought style (Fleck's "stock of knowledge and level of culture") which avoided roles as counterweapons to the battleship.[2] Most aviation and submarine officers had been raised in and belonged to, the overarching, "conventional" battleship thought collective. They had little or no problem supporting a strategic continuum in which aircraft and sub-marines played subordinate strategic and tactical roles to the battleship.

Senior officers considered the battleship an evolving technology, able to weather any challenge from the air or the depths. They accepted aircraft and submarines, but saw their missions as limited to those that contributed to the continued supremacy of the battleship. Submariners, who in reality operated surface ships that occasionally submerged, were more conserva-tive in estimating their anomalous nature. Naval aviators, flying alone or in small crews, were more independent, operating outside the factory-like at-mosphere of the battleship with its thousand-man crew. As a result, avia-tion attracted individualists, some of whom were not at all shy in defining a broad, aggressive threat to the supremacy of the battleship-based techno-logical paradigm. This laid the groundwork for the presumptive anomaly based in naval aviation that would mature by the late 1930s.

Besides the creation of new aviation and submarine specialists, wartime conflict over the future of American battleship design divided the line. Many officers favored emulating Great Britain in the construction of battlecruisers, ships that sacrificed heavy armor for high speed. Battle-cruisers — like *Wampanoag* in 1869 — were the perfect ships for guerre de course. However, battlecruisers went against the traditional American technological paradigm, which favored long-range capital ships carrying the maximum armor and largest guns possible. The battlecruiser question was complicated further by Japanese construction of four, powerful, 30-knot battlecruisers of the *Amagi* class (1918–19 budget). These ships were said to carry large guns that could outrange and demolish U.S. battleships. In the end, the status quo prevailed because of the more complete post-war knowledge of the dramatic failure of three British battlecruisers at Jut-land in 1916 as well as the capital ship ceiling imposed by the Five-Power Treaty.[3]

Between 1914 and 1920 the naval officer corps divided into communities defined by the technologies they operated, a division, with its attendant rivalries, that continues to the present. These emerging factions — the avia-tors, the submariners, and the internally conflicted battleship sailors — illus-trate the social process that governed the introduction of new technologies into the navy. The same process, complicated by stronger naval-industrial-congressional liaisons, governs technology selection and influences strate-gic options today.

Submarine S-22, seen in 1929, was typical of inter-war submarines built to support the battle line. Ironically, larger "fleet" submarines developed during the late 1930s spent little time supporting the fleet during World War II. Instead, they waged an effective campaign of guerre de course against the Japanese. (Courtesy of RADM Ridley McLean, 1935. Naval Historical Center, NH 46339)

THE BATTLESHIP-BATTLECRUISER CONTROVERSY

The initial threat to the American battleship design philosophy of slower, longer-ranged "Pacific" ships with maximum armor and powerful guns came from Jackie Fisher's "battlecruiser," which followed hard on the heels of HMS *Dreadnought*. The three battlecruisers of the *Invincible* class, authorized as part of the British 1905 Program, carried 12-inch guns like *Dreadnought,* but were lightly armored to carry the propulsion systems needed to attain 25 knots.[4] The battlecruiser's primary defense was its speed; or as Fisher was fond of saying, "Speed is Armor."

As Jon Sumida has shown, the battlecruiser was an interim ship created on the way to Fisher's ideal fast armored ship to maintain British sea power.[5] Fisher was fond of pointing out the need for such a high-speed ship by citing a French minister of marine, who in a classic guerre de course argument ridiculed the ability of battleships alone to protect a nation's seaborne trade: "The battle fleet parades the seas in solemn pomp, but, inasmuch as it has insufficient speed, it cannot protect the traders that fly its flag from the depredations of an enemy's corsairs and can only win victories on condition that its enemy is complaisant enough to 'come and be killed.' "[6] Fast British battlecruisers could go, catch, and kill commerce raiders and lesser warships as they did to Admiral Graf von Spee's force off the Falkland Islands in December 1914.

Advocates of the battlecruiser also claimed a tactical advantage in guerre d'escadre for a battle fleet that included this new type of ship. Battlecruisers could destroy lightly armed enemy scout cruisers before they could find the battle fleet. With its scouts destroyed, the enemy fleet would have no intelligence on the battle fleet's location, strength, and disposition. Using the intelligence gathered by its own battlecruisers, the fleet could then maneuver freely for tactical advantage against its blind enemy.

This argument typified the concept of intra-artifact combat that permeates much of modern military thinking. It also ignored the dangers of evolutionary technologies Mahan had decried. The advantages of having a fleet with battlecruisers was greatest if one's enemy had none. This parallels Guilio Douhet's strategic bombing argument that lost some of its efficacy if one's opponent also had a strategic air force. To prevail against an enemy fleet incorporating battlecruisers, one needed either technologically superior battlecruisers, leading to a broadened naval race, or, if battlecruiser technology was relatively equal, greater numbers. Either option would be tremendously expensive. British pursuit of the battlecruiser threatened to pull the U.S. Navy down a technological path that ran counter to the American tradition.

In December 1914, almost a decade after the first British battlecruiser was laid down, the General Board recommended following Britain and constructing battlecruisers as adjuncts to the battleship. Nevertheless, the Board believed it prudent to complete the construction of a sufficient number of dreadnoughts for the battle fleet before allocating any construction assets to battlecruisers.[7]

The call for battlecruisers soon spread. In January 1915 former Secretary of the Navy George von L. Meyer, in a speech at the New York Republican Club, characterized the lack of battlecruisers as a fatal flaw in the U.S.

Navy: "We don't own a battleship that could keep up with the armored cruiser *Bluecher* [sic, *Blücher*], of the German Navy ... and yet the *Bluecher* was lost in the recent North Sea action because she was too slow. Japan could wipe out our whole Pacific commerce. Japan has battlecruisers and we have none."[8]

The lack of fast capital ships to protect American trade was underscored by a report in the *Army and Navy Journal* that the Japanese battlecruisers of the *Kongo* class possessed 14-inch guns compared with the 12-inch guns on U.S. battleships.[9] Reflecting a Mahanian focus on the guerre d'escadre strategy, the *Army and Navy Journal* ignored the commerce raiding potential of the Japanese battlecruisers and focused on the 6-knot speed advantage of the *Kongos*. Their high speed would allow them to maneuver and destroy an American fleet from beyond U.S. gun range.

The *Journal's* concern about gun-bore disparity was based upon the common misconception that larger-diameter gun bores translated directly into longer ranges, that is, a 16-inch gun could shoot farther and penetrate thicker armor than a 14-inch shell. All things being equal, this was true. In fact, the shell weight, muzzle velocity, and armor-piercing capabilities of the U.S. 12-inch/50-caliber, Mark 7 gun entering service in 1912 were comparable to the British-built 14-inch/45-caliber guns installed in the *Kongo*-class battlecruisers.[10] Yet even with gun power relatively equal, the faster Japanese ships could control when, where, and if a battle would occur, a quality Bradley Fiske included in his "compromiseless" battleship in 1905.

Battlecruiser proponents, including both line and engineering officers, campaigned for the battlecruiser in the pages of the *Proceedings* of the Naval Institute. Assistant Naval Constructor B. S. Bullard decried America's singular stagnation in battlecruiser construction, calling for the immediate adoption of this type of ship which had "proved its worth in actual battle."[11] Commander Yates Stirling maintained that the decision to build only battleships was sound, if all the preliminary operations leading to the final battle between battleships were ignored.[12] The dreadnought battleship, according to Stirling, had become little more than a defensive fortification, and he proclaimed that "The battle-cruiser is the mistress of the sea and he who commands the most powerful fleet of battle-cruisers commands the sea."[13]

Expounding upon its 1914 endorsement of the battlecruiser, the General Board concluded by July 1915 that the battlecruiser "gap" represented a serious weakness in American sea power. When Secretary of the Navy Josephus Daniels requested advice on the navy building program, the General Board told him that by 1925 the navy "should be equal to the most powerful maintained by any other nation" and "gradually increased to this point by such a rate of development year by year as may be permitted by the facilities of the country." As part of this gradual buildup to naval supremacy, the Board, in 1915, recommended the addition of "four battlecruisers, four dreadnoughts, six scouts, 30 coast submarines, [and] seven fleet submarines."[14]

The General Board proposal for battlecruiser construction was endorsed by the Bureau of Construction & Repair, which was compiling lessons to be learned from the war in Europe. In attempting to assess "desirable types of vessels," the Bureau reported that the British felt that the battlecruiser had "thoroughly established" its usefulness in the Dogger Bank action (1915) and by destroying von Spee's squadron off the Falklands.[15]

The 1916 Naval Appropriations Act authorized six battlecruisers "of the highest practicable speed" at a cost, excluding armor and armament, of under $16.5 million each. The Act also authorized ten battleships "carrying as heavy armor and as powerful armament as any vessels of their class" at a cost, excluding armor and armament, not to exceed $11.5 million each.[16] At 31,000 tons, the battleships of the 1916 Program were very large ships, underscoring the editorial opinion in *Scientific American* that "Size, thick armor and the big gun, as combined in the modern dreadnought, constitute the supreme controlling factor in naval warfare."[17] The magazine's editorial staff agreed with Poundstone's philosophy that bigger was better, claiming that there was "nothing of the freakish or phenomenal in the extraordinary growth in size of the battleship and the gun it carried"; the large dreadnought embodied engineering and economical "efficiency."[18]

The upward spiral in battleship size and expense, so long predicted, caused concern even among stalwart navalists in Congress. Attempting to set some ceiling on battleship size, Congress attached a rider to the 1916 Appropriations Act requiring Secretary Daniels to report on the "largest battleship which can be undertaken in the United States." In response, Daniels informed Congress that "the technical and military advisers of the department have prepared and considered a number of alternative designs" before arriving at an 80,000-ton, 975-foot battleship, sporting fifteen 18-inch guns.[19] The navy thought it would require the construction of five such ships at a cost of $50 million each. It appeared to some that Park Benjamin's satirical prediction concerning the "big ships" would come true.[20]

Within naval circles, the battleship emerged from the May 1916 Battle of Jutland stronger than before, but the worth of the battlecruiser was in doubt. Combat negated naval sentiment that the first hits would mean victory. Both German and British battleships had taken tremendous punishment from large-caliber naval guns at Jutland and endured. The battleship *Warspite*, for example, with her steering gear disabled, survived fire from six German battleships.[21] This seemingly vindicated the traditional view elucidated by Chief Constructor Washington Capps in 1908 that the battleship could absorb tremendous punishment from big guns.[22]

The value of the battlecruiser, on the other hand, became suspect in U.S. circles once more details of the destruction of the British battlecruisers *Indefatigable*, *Queen Mary*, and *Invincible* (3,319 dead) became known after America became a cobelligerent.[23] The thin armor and poor flash protection of Fisher's fast battlecruisers could not keep out the German shells or prevent ruinous secondary explosions, and the ships' high speed had provided no protection.[24] In a righteous defense of his battlecruiser designs, Fisher angrily noted on Jellicoe's letter to him after Jutland that battlecruisers were "never meant to get in enemy's range!"[25] Because of Jutland, the British redesigned the new battlecruiser HMS *Hood* to include more armor, but not enough to save her from a catastrophic explosion (1,415 dead) during her May 1941 engagement with the battleship *Bismarck* and heavy cruiser *Prinz Eugen*.[26]

Once enlightened about Jutland, the U.S. naval hierarchy began to rethink its own commitment to the battlecruiser. For the U.S. Navy, guerre de course remained marginal and the battlecruiser had no ability to counter the submarine, the dominant commerce raider of the war. In 1918 William Sims, now the admiral commanding U.S. naval forces in Europe, was asked to review the designs of the battlecruisers and battleships of the 1916 Program. Sims knew that the British had abandoned singular reliance on Fisher's "speed is armor" idea and were increasing the protection on *Hood*. Sims's London-based Planning Section recommended increased armor protection at the expense of high speed for the U.S. battlecruisers not yet laid down. The General Board, perhaps acting out of a need to defend its stated position, and also because of a certain lingering hostility toward Sims, rejected any change in the battlecruiser design. Sims called on the navy's senior officer, the chief of naval operations (CNO), Admiral William S. Benson, to support the increased protection recommended by his Planning Section.[27]

The Sims–Benson battlecruiser dialog came in the wake of a serious rift over wartime construction priorities. The anglophilic Sims advocated suspension of the 1916 capital ships to build destroyers and merchant hulls to save Britain from the German submarine offensive. Benson, distrustful of the British and with an eye on the postwar naval balance, pushed to continue the 1916 Program.[28] Benson favored proceeding with the six weakly armored 1916 battlecruisers if delays would be incurred by a redesign of their armor protection.[29] Sims glibly estimated the maximum delay at six months and argued that the redesign would provide conformity with the new, more heavily armored second wave of battlecruisers currently being proposed by the Department. With the total cost of the six battlecruisers running around $150 million, Sims legitimately argued that "in view of the enormous sums involved, it appears to me indefensible to actually start construction of 12 capital ships of what we actually consider to be obsolete types. . . . Everyone seems to be agreed that these enormous vessels, with only five inches of armor, would be a grievous error."[30]

With Sims and the General Board at odds, and Benson wavering, Secretary of the Navy Josephus Daniels suspended construction on the six battlecruisers early in 1919, pending a visit by him and his three principal technical advisers (the chiefs of the Bureau of Ordnance, Bureau of Construction & Repair, and the Bureau of Steam Engineering) to the navy departments of Britain, France, and Italy to discuss their conclusions drawn from the war.[31]

The American battlecruisers were never completed. The dispute over the value of the battlecruiser was ultimately resolved by the Five-Power Treaty, which limited the total number of capital ships the United States could possess. The navy decided to retain its traditional heavily gunned, long-range, slow ships, as they were more valuable for operations in the Pacific. Two of the 1916 battlecruisers did enter service, but as the large aircraft carriers *Lexington* and *Saratoga* allowed under the treaty as a result of U.S. lobbying.

The battlecruiser had caused a slight rift among battleship officers. Aggressive battlecruiser advocates, such as Stirling, may have seen the battlecruiser replacing the dreadnought and eventually acting as a true fast battleship as Fisher had in Britain. Yet for most, there was never any doubt that the battlecruiser would only serve as an adjunct within the battleship technological paradigm.

The submarine, on the other hand, had the potential to challenge the battleship's preeminence. Yet by the end of the war, the submarine would be incorporated into the normal practice of the battleship paradigm.

THE SUBMARINE

The sinking of the three British cruisers, *Aboukir, Cressy,* and *Hogue,* by the German submarine *U-9* at the outset of the war brought the submarine instant credibility.[32] The demonstrated efficacy of the submarine forced British battleships into a peripatetic life until fleet anchorages were secured by antisubmarine nets and other defenses.[33] On the open seas, battleships had to travel at high speed. They also had to follow zigzag courses which reduced their speed of advance and increased fuel consumption by as much as 30 percent.

Technological countermeasures directed at the submarine foreshadowed postwar antiaviation arguments in which the battleship was presented as an adaptable, superior technology. The U.S. Navy, like foreign navies, initially relied upon steel antitorpedo nets, suspended from booms swung outboard on each side of the ships, to protect battleships at anchor. In October 1914 the navy ordered torpedo nets manufactured by the John A. Roebling's Sons Company for the new superdreadnoughts *Nevada* and *Oklahoma* just as the Newport Torpedo Station proved that a torpedo could penetrate such nets.[34] It is unclear if the unwelcome Torpedo Station report was taken seriously for long, if at all.

In Britain, the German submarine success meant that capital ship sorties into the North Sea had to contend with the submarine as never before. During the Dogger Bank engagement in 1915, Vice Admiral David Beatty broke off his pursuit of the outnumbered German ships because he believed German submarines were in the area. Fisher, the first sea lord, warned Beatty that "to go out in daylight with a lot of large ships and the German submarines so ably handled is a very serious risk."[35]

In a rush to learn lessons from the European War, the House Naval Affairs Committee, in December 1914, looked into whether the battleship could be eliminated since its effectiveness appeared to have been impaired by the war in Europe. In his testimony before the committee, Rear Admiral Frank F. Fletcher, commander of the Atlantic Fleet, minimized the "submarine with its torpedoes" as "a weapon of opportunity" that is "formidable and destructive." However, according to Fletcher "a skillful enemy need not permit the opportunity to occur."[36]

For Fletcher the torpedo was a passing threat that could be countered by improving battleship design. A new weapon, according to Fletcher, always "appeals to our imagination, and we are apt to say that it changes the history of warfare."[37] In a classic military dismissal of a counterweapon, Fletcher stated that the torpedo, and weapons of its kind, can only "delay or obstruct the movements of the main force of battleships" until "the final clash is decided when the battleships come together."[38] The battleship remained supreme since "the development of any offensive weapon is naturally followed by the development of the defense, which does not appear until later. The submarine is the latest weapon and the defense against it is not yet fully developed, but in the natural progress of all such development the submarine [will become an] auxiliary weapon. . . . It will be able to delay, obstruct, and hamper the movements of the main force, but it will not win battles."[39]

Fletcher claimed that the navy could keep submarines one hundred miles from its battleships using cruisers and destroyers, and perhaps, even dirigibles. The submarine's slow speed restricted its ability to attack fast battleship formations, but Fletcher severely underestimated the modern submarine's capabilities. When pressed, Fletcher was unsure how submerged submarines could be prevented from closing with battleships. With development of acoustic detection devices still in the future, submerged submarines could only be found visually. As to fighting submarines, Fletcher admitted that the methods so far proposed — using torpedoes, grappling hooks, or rapid-fire guns — had not proven "absolutely positive as yet."[40]

Fletcher agreed with both the General Board and Navy Department position that called for the abandonment of the smaller "coastal defense" type of submarine, designed to destroy attacking battleship fleets, in favor of a new "fleet" type, designed to operate with the battleships and scout for, and harass, an enemy fleet.[41] In a perfect example of the battleship thought style and its rejection of the pre-1890 continentalist naval thought style, Fletcher responded to previous testimony advocating one hundred antibattleship coastal defense submarines. For the same money ($50 million), Fletcher said he would rather have four dreadnoughts.[42]

Fletcher's defense of the battleship technological paradigm was echoed by Rear Admiral Charles Badger of the General Board, who reaffirmed to the committee that "nothing has occurred to supplant the battleship with any other ship of war or impair the usefulness of the battleship."[43] In September 1912 the General Board had called for the construction of one hundred "fleet" submarines to "insure the strategic freedom of the main fleet in a theater of war."[44] The General Board retained this prewar view of the submarine as an auxiliary weapon.

The General Board position conflicted with that of the Bureau of Construction & Repair. In its study of the first year of the war, the Bureau crossed into the traditional strategic domain of the line. Bureau officers resurrected a heretical, continentalist, antibattleship strategy by calling for "submarines, as fast as possible, and of as great a radius [of action] as possible, and therefore of considerable displacement, and in as great a number as may be obtained." These submarines would serve as a *weapon of defense against a power of superior sea-strength* [emphasis added]." The naval constructors stopped short of championing the submarine at the expense of the battleship, cautioning that the submarine's "importance should not be unduly emphasized."[45]

The Bureau's clarion call for submarines, and the inherent, defensive continental strategy based in such a counterweapon, challenged the dominant goal in the naval officer corps to achieve paramount status in a battleship-defined naval world. It would be easy to attribute the Bureau position to engineers out of touch with the reality of the high seas (as someone like Sims would). But the massive 1916 capital ship program was a year away in a murky political future, and in 1915 the Wilson administration was eschewing naval expenditures and wistfully placing its naval preparedness in the hands of Thomas Edison and the Naval Consulting Board.[46] In such an environment, submarines were an economically attractive counter to an attacking battleship fleet. Yet as Robert O'Connell has observed, counterweapons have never been popular due to the uncertainty they bring to the established order.[47] Nor was there any reason for the leading philosophers of the battleship thought collective to pursue anything but an increase in parallel weapons development, that is, more dreadnoughts.

The submarine's anomalous counterbattleship mission crossed into the mainstream of the naval profession in a daring article in the U.S. *Naval Institute Proceedings* in January 1915. Ensign Valentine N. Bieg made a valiant attempt at career suicide by outlining an antibattleship mission for new American "fleet" submarines, whose 5,000-mile radius allowed them to interfere with "the safe navigation of hostile battleships." Deterrence of war, Bieg argued, no longer rested upon the dreadnought but upon the submarine.[48]

Publication of Bieg's article typified the remarkable permissiveness of naval officers, as opposed to their counterparts in the army, to tolerate internal debate on key policy and operational issues, especially in a semi-official journal. Bieg's courage in submitting such an idealistic article was in accord with an old Naval Academy dictum: "when principle is involved be deaf to expediency." However, a better guideline for the young submarine force was "go along to get along." Submariners had to expand their branch of the service without presenting the submarine as an anomalous technological challenge to the battleship. It was easy to see that the argument for the use of U.S. submarines against an attacking battleship fleet could be easily turned around. If such issues were pursued in Congress, the navy's battleship building programs could come in for rough weather.

The early submarine force was commanded by officers who were not submariners, such as Commander Yates Stirling. Stirling, along with William Sims, was one of four officers in the first, eighteen-month "long" course at the Naval War College in 1911. In 1914 Captain Sims commanded the destroyers and submarines of the Atlantic Fleet. Sims asked Stirling, then executive officer in the battleship *Rhode Island*, to command the soon-to-be-independent submarine flotilla. Stirling recalled he went into submarines expecting "a lot of hard work and disappointment." The small U.S. submarine force was in poor material condition because they had not "won naval recognition of their usefulness." Stirling's flagship as commander of the Atlantic Submarine Flotilla was the coastal defense monitor *Ozark*, another reflection of the status and defensive nature of the submarine. Officers and sailors were assigned to submarines without regard to previous training. According to Stirling, "like the air service the submarine personnel was said to be temperamental, unreasonable, and at times almost mutinous in its attitude toward control by surface officers. The opinion of the Bureau [of Navigation] of the Navy Department seemed to be expressed in the saying: 'The best way to exterminate vipers is in their nests.'"[49]

The incorporation of submarines into the battleship technological paradigm began with a 1914 fleet exercise. Admiral Fletcher directed Stirling to form his submarines in a line through which the "enemy" battleship force would pass. The submarines were handicapped by Fletcher's order to attack submerged but with decks awash, making them visible to the battleships. Stirling reported only one submarine safe to submerge due to deficiencies in training and material. This "failure" of the Submarine Flotilla to "hide" itself led to a letter of reprimand for Stirling and an appearance before the House Naval Affairs Committee.[50] The apparent lack of readiness of the Submarine Flotilla became the focus of a Republican attack on the Wilson administration. Neither party seemed too clear about what submarines did but were aware that submarines were doing *something* in the war in Europe.

In November 1915 Admiral Benson, recently installed in the new position of chief of naval operations, privately expressed his belief to Sims that a reduction in battleship size was warranted because of the threat posed by the torpedo: "I am opposed to the tremendously large ship, particularly due to the great development in the torpedo, and will say to you that it is my intention to advocate a policy of going backward in the size of ships rather than forward; that is to say, ships at the very most of not more than 30,000 tons, and preferably, 20,000 to 25,000 tons displacement."[51] Just why Benson believed that a reduction in tonnage would make a battleship impervious to a torpedo is unclear. He did not appreciate that a battleship's resistance to torpedo hits could be improved only by increases in displacement caused by the addition of reserve buoyancy (larger hulls) or through the adoption of antitorpedo blisters and wing tanks that also required larger hulls.

Benson's inclination toward regression was challenged by Sims:

As to the question of large ships, I think we must always be governed in that matter very largely by the military qualities that our probable enemies put in their ships. Undoubtedly one of the military qualities must necessarily be resistance to torpedo attack. This quality must involve larger compartmentation outboard of the real body of the ship. The last batch of notes from O.N.I. [Office of Naval Intelligence] gives a sketch of a particular form of hull proposed in England, and perhaps actually under construction. If this proves effective, it follows that a certain sum of money expended for ships of this kind would be more efficient than battleships of the ordinary type. It would therefore seem that we would have to build larger rather than smaller ships, because these results cannot be achieved without an increase of displacement.[52]

In January 1916 Lieutenant Commander Harry E. Yarnell argued that the rival of the gun was no longer armor, but the torpedo, since torpedoes could now strike at ranges of 12,000 yards (6 nautical miles). Yarnell believed that more efficient hull subdivision, to minimize the flooding after a torpedo hit, could easily be offset by the development of larger torpedo warheads. According to Yarnell, the only defense against the torpedo was speed, and when faced with a torpedo attack, the fleet would only achieve victory if led by "an admiral, who realizing the danger, has the moral courage to ignore it, accept the losses entailed, and push his fleet in to decisive gun range and victory."[53] Moral courage did not make battleships, or soldiers, torpedo- or bullet-proof. One might excuse Yarnell's bravado since he was removed from the European land war and writing before Verdun and the Somme, where courage, virility, and esprit d'corps failed to counter the effects of the machine gun and artillery.[54] While Yarnell (a future admiral) trusted courage, the Bureau of Construction & Repair placed its trust in steel subdivision bulkheads and delayed the construction of battleships 43 and 44 to make the hulls "torpedo-proof."[55]

In December 1916 Lieutenant (junior grade) F. A. Daubin first elucidated a submarine with a mission subordinate to the battleship. In an apt comparison, Daubin called for the construction of large submarines exclusively since small, coastal defense submarines were as unnecessary as coastal defense battleships.[56] Between August 1914 and February 1916, submarines had sunk 487 vessels — the vast majority of which were merchant ships. Daubin rejected the commerce raiding experience of the European War, surprising given the parallel between the island empires of Japan and the United Kingdom and American strategic concerns over Japan. Instead, he argued for submarine employment in guerre d'escadre targeting enemy warships. Daubin did offer guerre de course as a secondary mission, while characterizing the European War as strangely inverted: "Abroad, it would seem the secondary role has become the primary, or the submarine having failed in its more important mission is doing what can be done within its capacity against commerce."[57] Daubin argued for the placement of fleet submarines under the command of the fleet commander-in-chief, just like the torpedo flotilla, so that the submarines' "offensive power would be coordinated with that of the fleet."[58]

Daubin's article reflected the more realistic view that submariners could only flourish if they presented their technology, not as a counter to the battleship but as a supportive element within the hierarchy of the battleship technological paradigm. According to Daubin, fleet submarines would act as scouts searching out the enemy battleship fleet or populate

Construction of the battlecruiser Lexington *has ceased in this 1922 view. A circular, heavily armored barbette for one of the main gun turrets sits on the deck and transverse watertight bulkheads are visible in the foreground.* Lexington *and sister ship* Saratoga *were completed as aircraft carriers and their large size influenced U.S. naval aviation doctrine.* (Naval Historical Center, 19-N-11978)

"submarine zones" over which the enemy fleet could be lured. In addition, submarines could also serve as "mobile mine fields."[59]

Even after the U.S. naval hierarchy learned how close submarines had come to forcing Britain to capitulate early in 1917, its members minimized the effectiveness of the submarine threat. Unlike the United States, Britain was an island nation whose past policies forced her to import most of her food by World War I. The German submarine campaign had been directed against merchant shipping, not against battleships. As such, it was an improper effort to gain command of the seas in a Mahanian sense and easily dismissed. Although held hostage by the submarine, no dreadnought had been sunk by one and the battleship remained supreme. In belittling the submarine, naval officers were voicing their traditional distrust of a counterweapon made more aversive by its nature. Although written about naval officers fifty years earlier, Alex Roland's observation remained valid. Naval officers were expressing

> a combination of medieval aversion to clandestine weapons and a lingering association of gunpowder weapons with the devil. Underwater warfare always offended both of these prejudices, but as time went on the officers were less and less able to remember why. The association with the devil was forgotten (but not lost) in the epithet "infernal machine," and though this term came to be applied to fewer and fewer weapons, underwater warfare was always one of them. The prejudice against clandestine weapons, a persuasion derived from the medieval view of war as a judicial duel, could only be expressed in terms like "want of manly courage."[60]

Many naval officers dismissed the acoustic submarine detection devices developed late in the war. They were peripheral to the traditional naval tactic of the convoy that had "already" defeated the submarine. Only five or six enemy submarines were sunk by ships employing the acoustic detectors developed in a naval-academic-industrial alliance under the auspices of the Bureau of Steam Engineering. As a result, Admiral Sims gave credit to the convoy system: "The world might still clamor for a specific 'invention' that would destroy all the submarines over night . . . but the naval chiefs of the Allies discovered . . . that they could defeat the German [submarine] campaign without these rather uncertain aids."[61] While dismissed by Sims, the new acoustic detection devices pointed to a vulnerability in the submarine and supported the battleship's continued supremacy.[62]

Sims, and most of his colleagues, perceived the submarine threat to the battleship as minimal and easily countered by the battleship's higher speed and the new antitorpedo modifications incorporated into battleship hulls. Under Sims's presidency after the war, the Naval War College paid no attention to the submarine, except as it could support the strategy of the battleship.[63] The navy built larger fleet submarines and the submarine service survived because it was assigned to work with the battle fleet in the supportive role proposed by Daubin.

THE RISE OF NAVAL AVIATION

The first flight of an aircraft from a U.S. Navy ship took place in 1908, but aviation developed slowly. It was not until 1914 that a mission for naval aviation appeared within the professional literature. Lieutenant R. C. Saufley presented aviation as an airborne scout cruiser that would increase the scouting range of the battle fleet.[64] Although he did not make the point himself, it is interesting that one important advantage Saufley claimed for the airplane was the same as that claimed for the battlecruiser. Scouting would allow the fleet commander to obtain a superior fighting position.

Saufley linked an adequate national defense to a strong fleet, but to him a fleet meant "more than dreadnoughts, and the commissioned personnel of the navy would be remiss in its duty should it allow our development to proceed upon limited lines, until, with a blind strategic eye, we are forced into action with a modern aggressive opponent."[65] The United States, according to Saufley, was five years behind Europe in aviation. One year later (1915), Commander (later chief of the Bureau of Ordnance) Ralph Earle advocated a mission for naval aviation within the battleship technological paradigm: installation of seaplanes on scout cruisers.[66]

In 1915 the first honorable mention in the Naval Institute's annual essay contest was awarded to Commander Thomas Parker for his article, "An Air Fleet: Our Most Pressing Naval Want."[67] Parker reiterated Saufley's contention that the United States was lagging farther and farther behind the European combatants with respect to aviation. Presaging air power strategists, Parker warned that should an enemy be able to reach America through the air, "the navy fails in its purpose and the Monroe Doctrine becomes archaic."[68]

Writing prior to the massive 1916 naval appropriations, when "we so severely limit our military budgets," Parker characterized the airplane as an efficient adjunct to the battleship. According to Parker, the United States "must seize, more eagerly than other nations, inexpensive weapons of great and known efficacy, like the submarine, the destroyer, and the airship."[69] The airplane was, according to Parker, an ideal counter to the submarine. Parker advocated purchasing one hundred airplanes ($1.1 million) and four dirigibles ($1 million) per year for four years, for a total expenditure approximately equal to the cost of one battleship.[70]

Parker quaintly described the role of naval aviation within the battleship technological paradigm. Aircraft and airships would guide friendly submarines and destroyers to their "prey." By performing aerial scouting, the airplanes would allow the battleship crews, "racked by watching, to rest." Airplanes could protect the battleship by locating and dropping "mines" on enemy submarines, unless, of course, the submarine "porpoises and unmasks an aerial gun."[71] The airplane could also serve to carry messages when "the enemy are 'jamming' the radio"; or, remarkably, even as the post of the fleet commander-in-chief during a battle.

Just before the publication of Parker's article, the Navy Department had approved the report of the special Aeronautical Board that had studied the question of naval aviation.[72] Appreciating the growing aviation gap, the Board called for a vigorous aviation program "superior in all articles of aerial equipment—in dirigibles as well as in aeroplanes." It was no surprise that the Board members considered aircraft a technology augmenting the battleship and recommended aircraft carried by "both fighting ships and auxiliaries. *Every battleship should carry them* [original emphasis]."[73]

The problem with developing American naval aviation centered around materiel. In July 1913 Rear Admiral Bradley Fiske, serving as the secretary of the navy's aide for operations, convinced Secretary of the Navy Josephus Daniels to ask for a General Board study on aviation.[74] The General Board farmed out the study to Captain Washington Chambers, who, with a committee of several officers, came up with a complex, multibureau organization for naval aviation. Fiske objected to the Chambers Plan, which would reinforce the power of the technical bureaus while failing to provide for centralized line officer control. Fiske pressured the General Board to shelve the Chambers Report and to convince Daniels to establish an Office of Aeronautics under Fiske. This office was established on a pro tempore basis on 30 August 1913. Commander Mark Bristol became Fiske's naval aviation assistant and director of the Office of Aeronautics when it was established formally on 1 July 1914.[75]

Bristol estimated that the battle fleet would require at least one hundred airplanes to support it during hostilities.[76] To defend American naval bases, even more planes would be required. Bristol reported difficulty in convincing the naval hierarchy to allocate funds for such a new technology, a perfect example of senior officers' hesitancy regarding a potentially destabilizing weapon. He complained that the thought of buying "1000 launches [small boats] would not stagger us; but we talk in whispers of 40 aeroplanes!"[77]

By the end of 1914, the naval hierarchy's desire to expand naval aviation far exceeded its ability. Testifying before the House Naval Affairs Committee in November 1914, Rear Admiral Victor Blue admitted that the navy owned only "five or six" flying machines.[78] The navy had established an aviation training station at Pensacola, Florida, and increased the number of aviation students. However, as Secretary Daniels pointed out, the war in Europe made airplanes hard to come by: "we have been trying to buy aircraft; we have been trying very hard, but we could not get them. As I say, we ordered one from France and one from Germany. If we had $5,000,000 now I could not spend it wisely."[79] Daniels had $1 million to spend on naval aviation in 1913, but the money was turned back to the Treasury because there were no suitable aircraft to purchase. In spite of the dearth of aircraft, both the General Board and Fiske were enthusiastic and the Navy Department requested $10 million for aviation as a part of the 1915 Appropriations Bill.[80]

In keeping with past reliance on the private sector for technology development, Fiske and the General Board sought to use a large congressional appropriation as bait for the development of aviation technology by the domestic aviation industry. Fiske told the House Naval Affairs Committee that the $5 million for aviation requested by the General Board would "get people in the United States to put up plants to construct aircraft, so that we will not have to depend upon the manufacturers abroad."[81]

In addition to establishing a domestic source for aircraft, the naval hierarchy wanted to ensure that naval aviators remained loyal members of the battleship thought collective. The unique practice of aviation, like service in submarines, increased aviators' loyalty toward their specialty. A separatist movement began to form to the extent that some aviators even prevailed upon friendly congressmen to support a reorganization of the naval officer corps. A provision was included in the Naval Appropriations Bill for 1917 calling for the creation of a separate "naval flying corps." Such a potential schism raised the ire of the General Board. The Board was quick to call for an organization that would temper the independence of naval aviation:

> The General Board believes that the establishment of a special corps so very distinct from the rest of the Navy would not tend to promote efficiency . . . [and]

that the aeronautic service of the Navy will best be dealt with under the existing law and conditions; by detailing young officers and men to the aeronautic service for instruction in flying and observation at the aeronautic station and afterwards for aviation duty for a few years while maintaining their place in the Navy List and their connection with the general duties of naval officers. The aviation party on board ship should constitute a division of the ship's crew available for general service as necessary, like a torpedo division or a turret division.[82]

To the members of the General Board, shipborne aircraft were merely a component of the battleship "system" and aviation personnel formed just another specialty group within the battleship crew. To ensure aviator conformity with the remainder of the officer corps, the naval hierarchy sought to restrict naval aviation, as they had naval engineering, to graduates of the Naval Academy. A shortage of academy graduates led Secretary Daniels, to the horror of the admirals, to recommend commissioning civilian aviators directly into the navy.[83]

At the end of 1916, Secretary Daniels reported that U.S. naval aviation policy specified four missions for aircraft, all of which supported the status quo. Naval aircraft would provide "Scouting from ships at sea . . . Off shore scouting from coastal stations . . . Spotting [for battleship gunfire] . . . [and] Offensive operations against enemy aircraft and possibly against ships or stations."[84]

American entry into the World War in April 1917 resulted in the expansion of naval aviation. More than $5.1 million was requested as a part of the naval appropriations bill for 1918. These funds were to purchase eighty-six airplanes and two dirigibles, to expand the ground facilities at Pensacola, and to establish "aeronautic" stations at Pearl Harbor, at the Canal Zone, and on the Pacific Coast.[85]

Some officers saw more to aviation than scouting and spotting in support of the battle line. Speaking one month before American involvement in the war, Rear Admiral Bradley Fiske, now retired, called for the development of the torpedo plane, a type of aircraft capable, for the first time, of carrying a weapon that could damage a battleship. Fiske believed that the torpedo plane, "under favorable conditions, would make the $20,000 airplane a worthy match for a $20,000,000 battlecruiser."[86] Fiske, incidentally, held a 1912 patent for a torpedo plane carrying a Whitehead torpedo.[87]

In August 1917 Fiske presented his argument to the naval profession in the Naval Institute *Proceedings*.[88] The torpedo plane, Fiske argued, supported Confederate General Nathan Bedford Forrest's definition of strategy: "to get the mostest men there the firstest."[89] Fiske, of course, was not referring to men, but to torpedoes, delivered by an attacking force comprising a large number of torpedo planes. In 1911 Fiske had defined military power as "mechanical power which can inflict injuries upon the bodies of men, and upon any defenses that may be placed around them."[90] Naval power, according to Fiske, was superior to military power since the "properties of water enabled larger machines to be moved" to an enemy. Since water was a better medium for moving military machines than land, Fiske argued that the power and mobility attending aviation would make it the most effective military force. He drew a parallel to his old nemesis, Mahan:

> Mahan proved that the influence of sea power upon history has been great. But this, strictly speaking, was not because it was sea power, but because it was mechanical power, so harnessed that it enabled the nation which possessed the power to bring force to bear on another nation, across the joining medium of the sea. Now the air joins countries together more intimately than does the sea; so that a book may some day be written, called "The Influence of Air Power on History."[91]

Fiske and his torpedo plane would have to wait.

During the summer of 1918 the General Board attempted to draw conclusions from the war in order to define the proper employment for aircraft within the navy, even questioning a Lieutenant Colonel Porte, of the Royal Flying Corps, on British experience with aircraft carriers.[92] The General Board determined that airplanes should be carried on scout ships (cruisers) and also on the new battlecruisers. As for battleships, the Board believed that it would be better to carry its aircraft on a separate aircraft carrier on which the planes could land.[93] The General Board characterized this aircraft carrier, which would support the battleship fleet of 1925, as one carrying twenty to thirty scouting seaplanes and displacing about 20,000 tons.[94] The one drawback with this ship was that its large size would prevent the navy from acquiring more than one or two.[95]

The General Board also considered naval aviation as an essential component in American naval power, especially in the vast Pacific:

This war is not conclusive with regard to aircraft. The North Sea with its few hundred miles yields little of importance for our larger operations. When you imagine that our merchant marine will cover the seven seas, airplanes, in some shape, are going to play a very important part in protecting this merchant fleet. Otherwise, we shall be reduced to the impotency of the German merchant and war fleets. We must look forward far enough to protect ourselves, and that means a very large fleet with all needed components.[96]

After the Armistice in November 1918, the General Board held hearings to formulate a policy to govern the future of naval aviation. The Board heard testimony from the bureau chiefs, senior commanders of the battleship navy, and naval aviators. Considerable attention was paid to the question of who would fill the ranks of naval aviation. The Board also laid the groundwork for the subordination of aviation to the battleship that would last during the interwar period.

Admiral Henry T. Mayo, commander-in-chief of the U.S. Fleet, told the General Board in 1919 that aviation is "a subject which must not be neglected in connection with the upbuilding and maintenance of an efficient Fleet."[97] However, Mayo was quick to squelch any idea of aviation autonomy or the development of naval aviation in any role except supporting the battleship. He characterized aviation as an "essential fighting arm of the Navy," but "no more of a specialty than submarine or destroyer duty." Mayo believed that "aviation will be so closely allied in the future with all naval activities" so it "should be considered an integral part of the naval profession." Mayo wanted a homogenous naval officer corps since "separate corps within one fighting organization usually leads to jealousies, lack of cooperation and inefficiency." The solution was "similar training as far as practicable for all officers of the fighting arms of the Navy" including aviators. This would be critical to inculcation of the battleship thought style and supremacy of the battleship strategic paradigm, or as Mayo phrased it, "essential to the establishment of a common point of view with regard to naval doctrine, traditions and policy."[98] Mayo also recommended a reduction in the hazardous duty pay for aviators to 10 percent of their base pay, believing that this would preclude jealousy among nonflying officers. Mayo further endeared himself to aviators by proposing to compensate for the reduction in hazardous duty pay with an increased government life insurance policy.[99]

The Bureaus of Construction & Repair and Steam Engineering had taken the lead in developing naval aviation. To prevent a recurrence of the early history of the battleship, and its control by the technical bureaus, Mayo warned that "the development of Naval Aviation must be governed by the development of Naval Tactics and Strategy — a branch of the Naval profession in which the decisions of line officers must be supreme."[100]

By 1920 the navy was faced with growing pressure for the establishment of an independent air force that would include all army and naval aviators. This movement came primarily from army aviators and their sympathizers in Congress and the public. To defend naval aviation, Rear Admiral David Taylor, chief of the Bureau of Construction & Repair, lobbied friendly congressmen to establish a navy bureau of aviation to preserve naval aviation as a distinct entity. A bill was introduced that provided for such a bureau and a naval air corps to be filled with commissioned officers drawn from all branches of the navy. Naval aviation was still under the control of a director of aviation in the Office of the Chief of Naval Operations. However, Taylor felt that "the present arrangement is not satisfactory, particularly if the Director of Aviation in Operations is not up to the job, and we have got to make some change or the Army (in the disguise of an Independent Air Service) will gobble up naval aviation."[101]

Fiske championed the creation of a bureau devoted to naval aviation to his old ally, Admiral William Sims:

> Meanwhile — for the sake of the U.S.N. and the U.S. [of] America — let's get a Bureau of Aeronautics *p.d.q.* — as *p.d.q.* as possible. Everybody is agreed to it: the House Naval Committee agreed to it unanimously and it only just escaped passing, last session. If we don't get that Bureau next session, Gen'l Mitchell and a whole horde of politicians will get an "Air Ministry" established, and the U.S. Navy will find itself lying in the street, and the procession marching over it. — But, with that Bureau nailed down, we can answer all the arguments against the navy having navy aeronautics, by *accomplishing* something practical. There is a lot of power of many kinds behind this Air Ministry project: — it is the most powerful enemy the navy has. If the Air Ministry is established, the navy will rank with mediaeval institutions in about ten years.[102]

Sims, recently returned from his wartime command in Europe, agreed on the necessity of establishing a bureau of aeronautics, promising Fiske that "When this matter comes up, as it eventually must, before Congress, I intend to take a hand."[103] But Sims was embroiled in a heated dispute before the Senate involving Secretary Daniels and the chief of naval operations, Admiral Benson, and whether Congress would want to hear from him yet again on the separate aeronautics issue may have been doubtful.[104] David Taylor drew a parallel between this latest Simsian disturbance and the 1908 Senate investigation and undoubtedly believed that Sims's intervention might be more of a hindrance than a help to naval aviation.[105]

By the end of 1920, the naval hierarchy was largely in agreement that aviation had an important future in the maintenance of naval power. The chief of naval operation's Planning Division even recommended that priority be given to aircraft carrier construction over that of battleships. The Division recommended the construction of three types of "plane carriers": scout carriers with scout planes, battle fleet carriers carrying planes "primarily for use in co-operation with surface ships in action," that is, for gunfire spotting, and base carriers for the repair and overhaul of aircraft.[106] The Planning Division advised allocating first priority within the navy's building program to the construction of eight carriers — four scout carriers and four battle fleet carriers — at an estimated cost of $10 million per ship. The colliers *Jupiter* and *Neptune* would be converted to base carriers.[107]

The Planning Division specified the construction of more smaller carriers rather than a few large ships. More ships could scout a greater area (akin to Mahan's concept of "flexibility" in his failed argument against larger battleships) and would also allow the "development of the maximum strength in the air" since "the rapidity with which a given number of planes can take to the air is dependent upon the number of flying off decks."[108] Smaller aircraft carriers also tended to underscore the auxiliary nature of naval aviation.

The air complement on the proposed carriers included torpedo planes and offered an entree for naval aviation as a counterweapon. Each scout carrier would carry eighteen fighter planes and eighteen scouting planes; the latter would double as torpedo planes to harass an enemy force until American battleships arrived. Each battle fleet carrier would carry eighteen planes to be used to spot friendly battleship gunfire. The remaining aircraft on the battle fleet carrier would be either fifty-four fighters to be used to achieve control of the air or a mixture of thirty-six fighters and eighteen torpedo planes.

The inclusion of torpedo planes by the Planning Division provided an offensive mission for naval aviation and extended the navy debate over the relative merits of the torpedo and the battleship from submarines into the air. In an attempt to make torpedo planes more paradigmatically "normal," they were even characterized as "100 knot torpedo boats."[109] The General Board was told that a fleet protected by an adequate number of aircraft carriers with fighter planes would have "no cause at all for worry against attack by bombing planes or torpedo planes."[110]

William Sims was one senior officer who did appreciate the aviation threat to the battleship. As president of the Naval War College after the World War, Sims ordered war game simulations at the College involving aircraft, something that met with Fiske's approval: "I think your idea of having two fleets fight each other, which are exactly alike in all respects, except that whereas one fleet has sixteen battleships the other fleet has sixteen aeroplane and torpedo-plane carriers, is bully."[111] Sims kept Fiske apprised of the results of these games and regularly asked his advice on aircraft tactics and capabilities.

In return, Fiske was not shy in giving Sims his own ideas concerning the restructuring aviation would force upon the battleship navy: "I also believe that the aeroplane carrier is a *more powerful* ship than the present battleship, — or that it will be just as soon as we have worked out a few details in the way of launching torpedoes, and landing on the decks of ships or in the water. The torpedo plane carrier *is* a capital ship."[112] When asked by Sims if he would exchange the "airplane carrier of the future for the battleship of the present," Fiske replied, "Yes: — I think a 35-knot torpedo plane carrier (able to shoot 20 torpedoplanes 200 miles would be a 'capital ship': — a *more* 'capital ship' than the [superdreadnought] *New Mexico*."[113]

Fiske's use of a gun metaphor to describe aviation would be echoed by other aviation enthusiasts over the next decade. Most notable among them was Rear Admiral William Moffett, the first chief of the Bureau of Aeronautics, a former big battleship advocate who had received a Congressional Medal of Honor as captain of a cruiser during the navy's bombardment of Vera Cruz in 1914.

Sims, former champion of the big battleship, was convinced of the correctness of Fiske's appraisal of naval aviation. Sims railed against the stubborn defense of the battleship by other officers: "I know, of course, something of the psychology of opinion, but this seems to go beyond the theories of the psychological experts. Can it be that the Navy is reluctant to give up the big ships to live in?"[114] Believing that the first impression regarding any new technology was driven by conservative feelings that were always wrong, Sims thought that the navy should have the courage to strike out on its own path, forsaking the traditional crowd mentality that governed naval selection of new technologies: "If I had my way, I would arrest the building of great battleships and put money into the development of these new devices [aircraft carriers] and not wait to see what other countries are doing."[115]

The navy was successful in having legislation passed authorizing the Bureau of Aeronautics in 1921. Yet even with its own air bureau, the navy was under attack to establish a separate naval flying corps or to transfer its aviators into an independent air force where they would join with former army flyers. In 1922 the General Board held hearings on the question.[116]

Moffett defended his bureau and was adamant that the navy maintain its own air forces and that naval aviators be trained by the navy. He described the air training facility at Pensacola as "a small organization similar to the Naval Academy." Naval aviators, according to Moffett, had to be drawn from the "officer class," be mature, and be trained to obey. To ensure comprehension and conformance with both naval paradigms (technological and strategic), Moffett sent all aviators to sea for three years before they were trained to fly. This prevented "men eighteen and nineteen years of age running into the air" who had no sense of discipline or appreciation for the strategic mission of the navy.[117] Rear Admiral Charles B. McVay, chief of the Bureau of Ordnance, echoed Moffett's opinion, wanting only "sailor men" as aviators.[118]

In light of General Billy Mitchell's sensational claims that the battleship was obsolete and the bombing trials conducted in 1921, the General Board called for an analysis of the effect that aviation weapons had on battleships. Captain Robert Stocker, of the Bureau of Construction & Repair, assured the Board that aerial weapons posed no new threats to the battleship. Everything done to the old battleships during the bombing tests in 1921 could be done by battleship gunfire.[119] Stocker did warn the Board that subdivision of battleship hulls had reached its limit. To defend the battleship from any new form of attack (the aircraft not being considered as anything other than a manned projectile) would require meeting the attack with a counterattack, that is, destroying attacking bombers with fighter planes.[120]

The naval hierarchy, as represented by the fleet commanders, members of the General Board, and the bureau chiefs, saw aviation as a technology necessary to naval power. Naval aviation had the potential to threaten the technological status quo, but, if properly controlled, could be used to enhance the battleship and to maintain the battleship-based strategic paradigm.[121]

The World War and Technological Diversity

Before 1914, the dreadnought battleship was the uncontested measure of naval power. The initial success of the submarine forced the British to readjust their naval strategy and tactics, vis-à-vis the North Sea. American neutrality allowed the navy to observe and attempt to garner lessons from the experience of the combatants. Initially, the war experience seemed to endorse the decision, driven by British and Japanese policies, to embrace the battlecruiser, a type of ship that violated traditional cultural and strategic imperatives in American capital ship design philosophy. While the U.S. Navy eventually rejected the battlecruiser, the episode emphasized the importance that foreign design decisions had upon American technological selections.

The submarine posed the greatest challenge to the supremacy of the battleship. Yet the relative newness and "infernal" nature of the submarine meant that there were no senior submarine officers to plead its case within the navy hierarchy. In fact, Yates Stirling, who had championed the battlecruiser, was detailed to command the navy's submarine force in 1914 although he had no submarine experience.[122] Younger submarine officers astutely perceived that the survival of their community, and its eventual growth, could only come through the adoption of a mission for the submarine force that conformed to the battleship strategy. Coastal submarines, designed to sink enemy battleships, were forsaken, in favor of larger "fleet" submarines that would travel with the battleships and scout for and harass an enemy battle fleet.

The rise of aviation had commonalities with both the battlecruiser and the submarine. Aviation scouts were seen as enhancing the capabilities of the battleship, but interest in aviation may have been driven more by the fear that uncontrolled naval competition would extend from the known world of battleships, into the unknown world of aviation. This is a thread that runs throughout the postwar General Board hearings on aviation.

As in the submarine force, there were no senior officers who were qualified as naval aviators. As a result, the mission of aviation was defined by the naval hierarchy in terms of the battleship: frail, contemporary aircraft were good for scouting for the battle line and for spotting the fall of shot during a long-range battleship duel, but the advocacy of torpedo planes to sink battleships was left to futurists like Bradley Fiske and to William Sims, who were at odds with the conservative clique of admirals, "the Daniels Cabinet," appointed to lead the navy after the war.[123]

The officer-operators of submarine and aviation technology found, as had the private sector with the electric drive, that their successful marketing of themselves depended upon their consonance with the prevailing strategy and battleship technological paradigm. The establishment of a Bureau of Aeronautics preserved naval aviation as an entity, but one that supported the status quo.

The most serious threat to the battleship, and to the naval status quo, came from the repudiation of superior naval power by the Republican administrations of the New Era. This, coupled with the agitation of aviation enthusiasts, the postwar sentiments for disarmament, and the call for more "efficient" military technologies, meant stormy seas for the battleship.

CHAPTER 6

Controlling Aviation
after the World War

The 1924 Special Board and the Technological

Ceiling for Aviation

A fter the World War, the battleship's primacy was challenged on a broad front. In Britain, the battleship was attacked vigorously for having failed to prevent, and quickly win, the war. The Royal Navy actively campaigned for a new postwar building program to maintain British capital ship superiority in the face of American construction. This was an unpopular, and expensive, proposition in the wake of the "war to end all wars." Even the creator of the dreadnought battleship, Lord Fisher, publicly lashed out at the deficit spending of the British postwar plans: "It is incredible — it is uncalled for — it is a ruinous waste that the cost of the Fleet is now 140 millions a year! (In 1904 it was 34 millions!) So the whole national expenditure before the war was only a third more than the present Navy Estimates. Then a huge anti-German Fleet had to be ready to strike! Now that German Fleet is at the bottom of the sea!"[1]

In the United States, there was much less popular and professional doubt in the efficacy of the battleship. In 1920 Rear Admiral David Taylor, chief of the Bureau of Construction & Repair and a key member of the National Advisory Committee for Aeronautics, denied that any new lessons had come out of the fifty-one months of the World War. For the battleship to become obsolescent, "either in the near or distant future, it will be as a result of engineering progress and the invention and perfection of new weapons and not as a result of [technologies developed during] the World War."[2] According to Taylor, technical improvements in the design of American battleships had solved the challenge posed by the torpedo "so far as the torpedo has been developed to date."[3] Ignoring aviation's potential to concentrate torpedo power, the one commonly feared counter to the battleship, Taylor pointed out that a "torpedo from a ship in the air is no more deadly than from a submarine under the surface."[4] Taylor believed that a battleship fleet, protected by an adequate number of aircraft carriers with fighter planes, would have "no cause at all for worry against attack by bombing or torpedo planes."[5] Taylor was a key designer of naval aircraft, and his views may have comforted the technically oriented naval officers who read his comments in the *Journal of the American Society of Naval Engineers*.

Warren G. Harding's inauguration in 1921 began the Republican New Era which continued through the Coolidge and Hoover administrations. The Republicans pursued international arms control (meaning primarily control of naval armaments) to eliminate the "wastefulness" of military expenditures. Between 1922 and 1930, construction was begun on only ten of the thirty-one warships authorized — a manifestation of the Republicans' 1920 campaign for a "return to normalcy" with significant reductions in federal spending.[6]

According to Roger Dingman, Harding seized upon the issue of naval arms reductions as a means to regain the political initiative from an increasingly unruly Congress and to block renewal of the Anglo-Japanese alliance. At the time Harding invited the naval powers to Washington, the United States was challenging British naval supremacy in the same way Germany had in 1914. With the German fleet scuttled at Scapa Flow, the German threat, as far as the Royal Navy was concerned, had been replaced by an American one. By 1923–24, the United States would be close to Britain in numbers of capital ships in service or building (Britain 43, United States 35).[7] The technical superiority of the newer American superdreadnoughts reduced the quantitative difference to relative insignificance.

Harding's efforts resulted in the Washington Naval Conference and its resulting Five-Power Treaty, ratified in February 1922. The treaty measured naval power in capital ship tonnage, giving the United States parity with Britain while granting Japan a secondary status according to a 5:5:3 tonnage ratio. Battleships were limited to 35,000 tons' displacement, and a ten-year

The navy was hard pressed to counter images such as this spectacular, but ineffective, explosion of a white phosphorus bomb over the old battleship Alabama *during the 1921 bombing trials.* (Naval Historical Center, NH 57483)

building holiday for battleships and battlecruisers went into effect. Britain was allowed to build two new battleships after the conference and the United States was allowed to retain three superdreadnoughts of the 1916 Program. Aircraft carriers were also limited by a ceiling on total tonnage, with each ship to displace no more than 27,000 tons. This was later waived to 33,000 tons as a result of U.S. pressure. Auxiliary warships — ships with gun bore not greater than 8 inches and displacement less than 10,000 tons (cruisers, destroyers, and submarines) — were not limited in any way.

U.S. naval officers generally believed that the United States had been outnegotiated by the British since Britain was allowed to maintain 580,450 tons of capital ships against 500,360 tons for the United States. Four 1916 Program battlecruisers were canceled and the remaining two were converted into the large aircraft carriers *Lexington* and *Saratoga*, which did not enter service for another five years.[8] Reduced Republican naval spending, compared to the massive outlay of Wilson's 1916 Program, promised to keep the navy well below the maximum size allowed under the treaty.

With the battleship's evolution limited by treaty, the navy came under vigorous attacks from aviation advocates, who claimed that a $10,000 airplane could sink a $10 million battleship.[9] The sensational aerial bombardments against captured German and obsolete American battleships conducted in 1921 provided aviation supporters with ammunition for their fight. The bombing trials yielded dramatic photographs of phosphorus and high-explosive bombs detonating on the huge battleships as fragile, biplane bombers flew overhead. However, the battleships that were bombed were anchored, unmanned, unarmed, and, by virtue of their age, markedly less resistant to aerial bombing than later designs.[10] The German battleship *Ostfriesland*, sunk in the tests off Cape Henry, Virginia, in July 1921, had been subjected to repeated attacks over a two-day period. During the first day, sixteen of sixty-nine bombs ranging in size from 230 to 2,000 pounds scored hits. On the second day, three out of eleven 1,000-pound bombs hit the ship, while three of the six 2,000-pound bombs exploded in the water close aboard. With most of her watertight fittings not fully repaired after her battle damage at Jutland and no damage control teams on board to fight the flooding, *Ostfriesland* sank.[11] The trials did not provide conclusive proof of the battleship's obsolescence, but a significant segment of the public believed it was vulnerable to air attack.[12]

Faced with this public relations fait accompli, the naval hierarchy worked hard to restore a common sense of the battleship's invulnerability. The navy argued that a fully manned battleship with antiaircraft guns, steaming at full speed, could thwart an attack from the air. Even if a bomb should hit, the navy claimed that the resulting damage would hardly prove fatal in light of the "superiority" of U.S. armor designs over those installed on British and German ships which had survived numerous large-caliber shell hits at Jutland.

In testimony before the House of Representatives' Joint Military and Naval Affairs Committee, Rear Admiral William A. Moffett, chief of the Bureau of Aeronautics, maintained that the nation's "first line of defense is the main fleet . . . the second line of defense would be auxiliary vessels of the fleet [cruisers, destroyers, and submarines] and the third line of defense is our coast fortifications, augmented by the Army Air Service."[13] In a pejorative aside, Moffett doubted the ability of army flyers to defend against a seaborne attack since they have "never been to sea except as passengers."[14]

THE SPECIAL POLICY BOARD HEARINGS OF 1924

Despite the creation of the Bureau of Aeronautics, critics persisted in their claims that the navy was suppressing aviation. Amid this continuing controversy, Secretary of the Navy Curtis D. Wilbur was embroiled in a dispute with the director of the budget over what level of naval appropriations the executive branch should request from Congress.[15] Wilbur directed the General Board of the navy to consult "experienced officers from both the Army and the Navy" and to make recommendations as to "development and upkeep of the Navy in its various branches, i.e., submarines, surface ships, and aircraft" in preparation for the naval appropriations bill to be submitted to Congress early in 1925.[16] Although a Naval Academy alumnus, Wilbur wanted professional advice on the most efficient expenditure of whatever funds were allocated to the navy.[17] As a part of its study, the General Board, keeper of the battleship technological paradigm, would rule on battleship obsolescence.

During the fall of 1924 the Special Policy Board of the navy's General Board heard the testimony of seventy-six witnesses that included the chiefs of the navy bureaus, fleet commanders, junior officers from the Bureau of Aeronautics, civilian engineering representatives of the National Advisory Committee for Aeronautics (NACA), aviation industrialists, and army officers, including General Billy Mitchell.[18] In its 1925 special report, the General Board reaffirmed the supremacy of the battleship-based strategy and the battleship technological paradigm. Since the conclusions drawn by the General Board supported the status quo, it is tempting to categorize the Special Board hearings as a show trial. But with the hearings held in executive session and the minutes unpublished, the testimony was confidential and quite frank.

The General Board defined the navy's technological needs in terms of its two strategic missions: first, to gain control of disputed sea areas and sea communications; and second, to maintain control of these areas and the lines of sea communication.[19] The first strategic mission involved "fighting the battle fleet of the enemy." This required a fleet composed of "all the various types of combatant ships." Listed in order of their importance within the U.S. Battle Fleet were "battleships, light cruisers, destroyers, destroyer leaders, submarines, and airplane carriers."[20] The second strategic mission would be carried out by "attacks upon and interference with enemy commerce, the defense of our own commerce, the blockade of enemy ports, the escort of military expeditions, the repulse of invasions and, in general, the control of sea communications."[21] This mission required "less fighting strength" and would be assigned to cruisers assisted by submarines, and whatever aircraft carriers — the battlecruiser-turned-carrier *Saratoga* would not be completed until 1927 — or destroyers were available.[22]

On paper, the U.S. Fleet was divided into four forces composed of various types of ships. In addition to the Battle Fleet described above were the Scouting Fleet, Control Force, and the Fleet Base Force. The Scouting Fleet was considered an auxiliary to the Battle Fleet. In principle it contained light cruisers, aircraft carriers, submarines, and destroyers. The Control Force was charged with accomplishing the navy's secondary strategic mission: maintaining control of the sea. It was to be composed of light cruisers, aircraft carriers, destroyers, submarines, minelayers, and patrol vessels. The Fleet Base Force included the Fleet Train, auxiliary vessels needed to supply the fleet, and whatever combatant vessels were assigned to protect the Fleet Train and advanced naval bases.[23]

The Board recognized the aviation threat to the battleship, but the testimony of civilian aeronautical experts indicated that the future evolution of aircraft, which were not subject to any treaty restrictions, was more limited than that of the battleship. The experts included Massachusetts Institute of Technology (MIT) President S. W. Stratton, Johns Hopkins physicist J. S. Ames, the chair of MIT's Aerodynamic Department, E. P. Warner (who later became the first assistant secretary of the navy for air in 1926), and several important NACA figures such as W. F. Durand of Stanford University, George W. Lewis, director of NACA research, and Rear Admiral David Taylor, secretary of NACA. These foremost representatives of the nation's aviation establishment all agreed that the "present maximum performance of heavier-than-air craft may be increased about thirty per cent by future developments extending over an indefinite period of years. All of them consider it most unwise to base a policy of national defense on expectation much beyond present performance.[24]

After the experts placed a technological ceiling on the future of aviation, the Board addressed the considerable agitation to create a cabinet-level Department of Aeronautics to supervise civil and military aviation. This department would include a united air service, or air force, that would absorb all army and naval aviation. The Board argued that the testimony of its witnesses supported the position that such a department would hinder the navy's ability to defend the nation. The agitation within the army and navy for a separate air service was attributed to "(1) Unsatisfactory promotion from their [aviators'] point of view. (2) A desire to be always under the command from the top down of practical flying officers. (3) A belief that there exists a lack of sympathy for them on the part of the senior officers of the Army and Navy."[25]

The Board dismissed the case for promotion as the misguided belief by pilots that an air force would be expanded so that existing officers would enter the new service with "greatly increased rank and pay." As to the command structure of naval aviation, the time was fast approaching when all the "immediate commanders of flying squadrons will be practical flying officers." However, the Board pointed out that the existing situation was no different from the fact that officers specializing in naval engineering and ordnance were not always commanded by officers of their own specialty. The lack of sympathy for aviators, if it did exist, was attributed to the fact that the "higher authorities in both Army and Navy understand the serious limitations in war operations in the air."[26] That is, they disagreed with the ardent advocates of aviation who espoused the unlimited potential of air power.

Since the push for a united air service and attacks on the battleship had grown out of the 1921 battleship bombing experiments, the Board discussed these trials at length, concluding that battleship antiaircraft guns would make the already inaccurate aerial bombing even more so. For example, when the battleship *Iowa* was attacked while being maneuvered by radio control, only two of eight bombs hit the ship from an altitude of 4,000 feet. In July 1924 the British had conducted a similar trial using the radio-controlled battleship *Agamemnon*. One hundred fourteen bombs were dropped without a single hit. Trials of antiaircraft guns had yielded 75 percent hits on small aerial targets at altitudes higher than 4,000 feet, indicating that a battleship could fend off aerial attack.[27]

The inaccuracy of the peacetime bombing trials cast doubt on the aviation argument. The navy argued that the battleship was the superior means to defeat an enemy battle fleet, since gunfire from a battleship was quicker and more accurate than aerial bombing: "The turrets of our latest [battle]ships fire a projectile weighing 2,100 pounds to a distance of twenty sea miles [40,000 yards]. At 19,000 yards no armor afloat will withstand a normal hit from these guns. While it takes 28 seconds for an airplane bomb to reach the target, 12,000 feet beneath it, the 16-inch projectile fired from a gun traverses an equal distance in less than five seconds."

The mystique of air attack was dismissed as well: "There seems to be in the minds of most of us, the idea that there is some especially deadly quality pertaining to missiles dropped from above. This idea is probably instinctive and is kept alive by accidents which occur from time to time. The idea is not true of course, but if it were, what of the 2,100 pound [battleship] shell which, when fired at the maximum range as mounted, drops from a height of 18,700 feet?"[28]

The battleship was judged superior to aviation by what Admiral William Sims used to call "rapidity of hitting" coupled with the greater percentage of hits from battleship guns.[29] The *West Virginia* delivered a volume of carefully aimed fire which amounted to one shell every five seconds.[30] In peacetime target practice, American battleships successfully hit a relatively small target about 10 percent of the time at ranges between 19,000 yards and 20,000 yards. In battle, the navy estimated the rate of hits would be reduced to around 3 percent, which, according to reconstructions, was believed to be the case during the Battle of Jutland in 1916.[31] The reduction of hits during battle was attributed to "haste, smoke, nerves, et cetera." The Board concluded that such factors also would affect adversely the accuracy of bombing and torpedo fire, reducing its effectiveness from the results obtained during the "drill and experiment of peace."[32]

Battle drills indicated that a large number of projectiles, whether shells or aerial bombs, would be required to get hits on enemy ships: "If a battleship fires five hundred 2,100-pound shells in an engagement, she may count on making fifteen hits."[33] The Five-Power Treaty had set a ceiling on aircraft carrier tonnage, and even the large aircraft carriers *Lexington* and *Saratoga*, being converted from the 1916 Program battlecruisers, would each only carry seventy-two aircraft and not all of these aircraft were bombers.[34] If the navy built the maximum number of aircraft carriers allowed under the Five-Power Treaty, the total number of aircraft available at sea would be approximately three hundred.[35] If each one of these aircraft was a bomber, which was not the case, the number of bombs that could reach an undefended, motionless battleship would still be less than the number of shells fired from a single battleship's guns and the aerial bombs (based upon trials) would achieve a smaller percentage of hits. When antiaircraft fire and ship maneuvering were added to the picture, the Board argued, the success of the airborne attack would be reduced even further.

Confidential torpedo, bomb, and gunfire tests conducted in November 1924 on the uncompleted hull of the 1916 Program battleship *Washington* indicated that the three newest American superdreadnoughts could withstand eight torpedo hits. During the torpedo trials, a 14-inch shell weighing 1,440 pounds was also dropped onto the *Washington* from an altitude of 4,000 feet and failed to penetrate the thick armored deck. The members of the test board believed, however, that if the bomb had been dropped from a higher height, which was within the capabilities of contemporary aircraft, the shell would have penetrated. The Board argued that such an occurrence could be avoided by adding more horizontal armor as part of the defensive tonnage increase allowed by the treaty. *Washington* was also subjected to long-range gunfire, and the tests indicated that present U.S. battleship designs "justified expectations."[36] Additional trials were conducted to evaluate the effect of underwater explosions caused by near misses of aerial bombs. Because of their intense pressure wave, the latter were considered more dangerous than direct hits.

In developing a strategic policy for the secretary of the navy, the Special Policy Board had to decide whether the present capabilities of naval aircraft — or rather the present capabilities increased by 30 percent — posed a credible challenge to the battleship. Both battleship advocates and aviators saw limitations in each other's artifacts, but none in their own. To ensure penetration of a reinforced armored deck, General Mitchell threatened to double the size of aerial bombs from the present maximum of 2,000 pounds to 4,000 pounds. Mitchell's plan sounded good to the public, but in reality doubling the weight of explosive (which comprised about 50 percent of the total weight in an armor-piercing bomb) only increased the pressure effect of the explosive charge by 50 percent rather than 100 percent.[37] While such a large bomb could penetrate the armored deck of a battleship, the navy dismissed it as too heavy to deliver based upon the testimony of the aeronautical experts. The maximum load of contemporary bombers was 4,000 pounds, which included fuel and bombs. To carry a bomb larger than 2,000 pounds would mean a significant reduction in the bomber's radius of action, even if aircraft performance improved by the predicted maximum of 30 percent. Using newly developed supercharged engines, a bomber could only reach a ceiling of 8,000 feet with a 2,000-pound bomb. A larger bomb would reduce the aircraft's ceiling, bringing the aircraft within range of ship-mounted antiaircraft guns which, as mentioned, were believed to be very effective based on live-fire exercises.[38]

The Board's focus on bombs ignored the weapon long advocated by Bradley Fiske — the torpedo plane. Although the three newest U.S. battleships of the 1916 Program had been judged capable of surviving eight torpedo hits, no mention was made of earlier designs which remained in the battle line. The definition of survival was also vague. Modern naval weaponeers speak of varying types of kills, such as mobility, mission, weapons, electronic, or total. In 1924 the Special Board's term "survival" was less sophisticated; the ship remained afloat. In May 1941 the German battleship *Bismarck* survived two hits from aerial torpedoes. The second jammed her rudder and inflicted a mobility kill, preventing *Bismarck* from reaching port.[39] Mitchell may have fared better with his antibattleship campaign if he had focused on delivering torpedoes rather than bombs. However, the torpedo was not a normal technology within the army aviation technological paradigm.

"Normal" Strategy

In their testimony before the Special Board, naval aviators reflected their philosophical identification with the normal practice of the battleship-based strategic paradigm. Rear Admiral Moffett, chief of the Bureau of Aeronautics, typified the battleship thought style, characterizing aircraft at sea as "auxiliary to the fleet — as an auxiliary arm. Its functions being, I would say, spotting [battleship] gun fire, reconnaissance, scouting, torpedo and bombing."[40] Perhaps recalling his own fondness for big-gunned battleships — he had advocated a battleship with "the largest gun" possible in 1916[41] — and his Medal of Honor-winning shelling of Mexicans at Vera Cruz, Moffett's ultimate expression was his statement to the Special Board that "Aviation is a gun; it is a form of a gun."[42] The bureau's Lieutenant Commander Marc Mitscher, who would later achieve fame as a commander of aircraft carrier task forces against the Japanese during World War II, also testified from within the battleship framework. Basing his conclusion upon bombing and gunnery exercises conducted since 1922,

Mitscher told the Board that gunfire was more effective than aerial bombing.[43]

Submarine and aviation advocates were competing for reduced postwar naval funding and each claimed their technology better served the battle fleet. The submarine's future was not as a counterweapon to the battleship, as Ensign Bieg thought in 1915, but as a more stealthy and more economical scout than the rival aircraft carrier or light cruiser. Rear Admiral M. M. Taylor, commander of the Control Force, emphasized the efficiency of the submarine when he argued that a Hawaii-based, continuous naval patrol off southern Japan would only require four 2,000-ton submarines at a total cost of $16 million. The same thirty-day patrol, using highly visible and vulnerable light cruisers, would require five ships at a cost of $20 million. An aircraft carrier with its forty planes, along with four destroyers for protection, could only patrol for ten days at a cost of $22.4 million. An added consideration was the fact that the large and costly aircraft carrier would be extremely vulnerable to attack.[44]

Conversely, Rear Admiral Magruder, commander of the Light Cruiser Division of the Scouting Fleet, defended aviation as *the* technology that was perfectly consonant with the geographical realities of the navy's Pacific strategy.[45] The weather in the Pacific, according to Magruder, lent itself to successful air operations, and he recommended allocating a larger percentage of the shrinking naval budget to naval aviation.[46] He also argued that aviation could serve as an economical buffer against the "treacherous" Japanese until the United States reached war footing by "bring[ing] our air fleet up to . . . its proper strength," which was a quicker method of "strengthening the Fleet" and "would be cheaper also."[47]

After reviewing the extensive testimony of the special hearings, the Special Policy Board recommended a force structure in which the battleship remained "the element of ultimate force in the fleet." In a perfect expression of the battleship technological paradigm, the Board defined all "other elements" as "contributory to the fulfillment of its [battleship's] function as the final arbiter of sea warfare." The Board mimicked Rear Admiral Frank Fletcher's testimony in 1914 that "offense always beats defense":

From time to time apparent threats to the superiority of the battleship have appeared. Each has resulted in some modifications of its design and in the methods of its employment in war, but its supremacy remains. With the invention and development of offensive weapons has always come the counter invention and development of defensive means and methods, so that in the end a fair balance is struck between them. The history of the gun and armor, and of the torpedo and interior subdivision, merely repeats the process by which offense always begets defense. . . . [Aviation's] influence on naval warfare undoubtedly will increase in the future, but the prediction that it will assume paramount importance in sea warfare will not be realized. The airplane (heavier-than-air) is inherently limited in performance by physical laws. Airplanes have demonstrated their great value to the fleet in scouting, observation, and bombing. The use of torpedo planes, [poison] gas,[48] and smoke screens still is in the process of development. Airplane carriers are necessary elements of a properly constituted fleet to carry airplanes to the scene of action.[49]

In reaffirming the battleship's dominance in guerre d'escadre, the Special Policy Board ignored the warning of retired Admiral William F. Fullam, whose career had spanned the battleship era and who admitted that his "whole heart was given to the biggest gun and the biggest ship that could be conceived."[50] Fullam's allegiance to the battleship ended on Armistice Day in 1918, when his flagship was anchored in San Diego Harbor. The commander at Rockwell Field sent up 212 airplanes to celebrate the end of the war, and they flew over the fleet for three hours "in flocks." It became perfectly clear to Fullam that if "each of those lads up there had a bomb as big as a grapefruit, that wouldn't have been the place for us, or for any collection of ships. If we started out at fifteen knots, they would have pursued us at a hundred knots."[51]

Although greatly overplaying the effect of grapefruit-sized bombs (and ignoring the potential of the torpedo plane), Fullam, like Saul on the road to Damascus, became a convert. He conceptualized a " 'Three Plane Navy' with forces on the surface, below the surface and over the surface."[52] Fullam pointed out that the most powerful battleship fleet in the world could not keep Britain from starvation and the brink of defeat in the spring of 1917. Citing German records analyzed at the Naval War College, Fullam characterized the German blockade as the epitome of efficiency, comprising only nine to fifteen submarines at any one time. He emphasized that the entire German submarine force of ten thousand men had managed to circumvent the one million men that manned the five greatest battleship navies in the world.[53] Using the submarine as an example, Fullam accused the navy of learning nothing from the World War since antisubmariners within the Navy Department had prevented the director of submarines from testifying before the Senate Naval Affairs Committee in February 1921 regarding the need for submarines.[54] In addition, Secretary of the Navy Josephus Daniels's appointee as chief of naval operations, Admiral Robert E. Coontz, had deleted all submarines from the naval appropriations bill.[55]

Although critical of the limited naval appropriations of the Harding and Coolidge administrations, Fullam idealistically argued that whatever funds were given to the navy should be used to develop the three-plane navy to its fullest efficiency, so that all technological factions could work together like a Nelsonian "band of brothers."[56] Rejecting the singularity of the battleship thought style, Fullam demanded the creation of a *three idea Navy*," as it would win out over a "one idea navy every time." According to Fullam, "The nation that first does this, that first solves that very complicated problem, will win the next war."[57]

The American naval profession had no room for three disparate and roughly equal technological paradigms. Fullam's argument was for naught, and the Special Policy Board justified the retention of the battleship-based strategy for two reasons. First, the testimony of civilian aviation experts set a technological ceiling on aviation, predicting that airplanes had little potential for growth and, therefore, would never attain performance levels that would pose a threat to the battleship. Second, those charged with the advancement of naval aviation, such as Rear Admiral Moffett, were firmly rooted in the battleship thought collective and envisioned aviation as supporting the battleship rather than challenging it.

There was strong support for the development of naval aviation by senior naval officers, but only in roles that reinforced the normal battleship-based strategic philosophy. A different employment of naval aviation was only a possibility if a rival naval power should pursue a radically different aviation mission, for example, offensive operations utilizing torpedo planes, which would threaten, or in the case of hostilities, pose a serious presumptive anomaly to the existing technological paradigm.[58]

AFTER THE "AIRPHOBIA" OF 1924–1925

The Special Policy Board report was an internal navy document, and the release of its conclusions failed to end the criticism from those calling for the retirement of the battleship and the establishment of a separate air service. As a result, President Coolidge appointed a civilian panel, the Morrow Board, to study the best way to develop aviation for use in national defense.[59] The Morrow Board found no merit in a separate air force, but did recommend the adoption of a multiyear aircraft procurement program for the army and navy.[60] Believing there was merit in the contention that aviators were being suppressed, the Morrow Board recommended increased promotion opportunities and the restriction of aircraft carrier command to aviation officers. The Board also advised the president to authorize the appointment of a new assistant secretary of the navy for aviation.

Coolidge signed the appropriations act of 24 June 1926 incorporating the provisions of the Morrow Board report, including funding for a thousand new navy aircraft over a five-year period.[61] In addition, all aircraft carriers were commanded by aviators. The exclusion of nonaviators from command of the almost completed large carriers *Saratoga* and *Lexington* irked Admiral Samuel S. Robison, commander-in-chief, U.S. Fleet. He expressed his dismay to the Bureau of Aeronautics:

> The particular aircraft carriers now in mind are the largest ships we have ever built, with the greatest horsepower that has ever been installed, and to try to connect up their command with ability to fly is piffle (see Morrow Board's recommendation); and that no officers should serve on them except aviators is equally piffle and defeats the desire of the Department and of the Congress that knowledge of aviation should be disseminated as widely as possible.[62]

In the wake of the Special Board and the Morrow Board, Captain Yates Stirling captured the essence of the battleship thought style regarding the role of aviation in the navy: "The notion that the airplane is to replace that panoply of war called fleets came as a surprise to the Navy. The naval man saw in it merely a weapon to be used by the Navy for offensive and defensive action. . . . In addition to its value in scouting, the Navy's interest first was directed to the airplane for the purpose of spotting gun fire. . . . Spotting the [battleship] gun salvos from airplanes was the natural solution [to visibility problems at long range]."[63] Like, Moffett, the navy's premier aviator, Stirling also compared the airplane to a battleship shell: "Airplane carriers are to all intents and purposes battle cruisers with airplanes replacing the big guns in armored turrets. . . . The fact is that the naval airplane is merely a new sort of projectile, carried by a surface ship."[64]

Stirling compared the aviation threat to that posed by torpedo-carrying destroyers prior to the World War. The rapid-fire secondary armament of battleships was judged unable to reach attacking destroyers with sufficient accuracy to prevent a torpedo attack. The answer had been to advance the antidestroyer guns to the point of attack, that is, to place them on destroyers and use these destroyers as a protective screen around the capital ship. Stirling advocated the same course of action with antiaircraft guns: advancing them to the point of attack by carrying them in airplanes. This would ensure control of the skies, neutralization of the airplane, and protection of American capital ships. Naval air assets would cancel each other out, and war at sea would remain dependent on the battleship.[65] The airplane would join the expanding defensive rings protecting the navy's castles of steel.

The technological ceiling and pace of subsequent aviation development prevented aviation theorists from presenting a credible presumptive anomaly to the battleship technological paradigm. That would have to wait for the development and perfection of carrier air groups equipped with modern aircraft able to damage or sink a moving warship. This would not occur prior to the introduction of metal, monoplane dive bombers, torpedo bombers, and fighters at the end of the 1930s.

Naval officers tended to see conflict as intra-artifact with little, if any, crossover. In war games at the Naval War College, for example, commanders of opposing fleets regularly launched air strikes against enemy aircraft carriers, rather than against battleships, as soon as the enemy was within range.[66] Fleet exercise experience underscored the vulnerability of aircraft carriers to air attack. The consensus was that there was no effective defense against a well-mounted dive bomber attack save the destruction of the enemy carrier prior to its launching such an attack. The side whose carriers were discovered and attacked first lost its aviation force.[67] Once these preliminary bouts were completed, the ring was clear for the main event — the battleship duel. As the *Tactical Orders and Doctrine for the U.S. Fleet, 1941*, stated: "The surest and quickest means of gaining control of the air is the destruction of enemy carriers."[68] Control of the air meant freedom from destruction from above.

In justifying their existence, both submariners and aviators presented themselves as the "most effective" supporting technologies to increase the survivability of the battle fleet. Stirling typified the view that aircraft would only fight other aircraft, just as Admiral M. M. Taylor argued that submarines would primarily engage submarines.[69] Citing the confusion at Jutland, Stirling predicted a general melee in which one air fleet would be completely victorious while the other would be cleared from the air. The victorious side would then be free to use its airplanes for scouting and gunfire spotting in support of the battleships. After the battleship action had been concluded, "Bombing, gassing, and torpedoing of the enemy battle line [by aircraft] will give the *coup de grace*."[70] It was inconceivable to Stirling and most of his colleagues that bombing, gassing, and torpedoing alone might do the trick.

The Special Board and the Strategic Paradigm

The battleship reductions and building holidays of the Five-Power Treaty provided aviation enthusiasts with a truly "stationary" target. Technical improvements to the surviving battleships were restricted severely, while aviation technology developed at a quick rate. The Bureau of Aeronautics funded structural research at the Naval Aircraft Factory in Philadelphia and at the U.S. Bureau of Standards and NACA.[71] Wind tunnel experiments were conducted at the navy's Experimental Model Basin at the Washington Navy Yard and at the Massachusetts Institute of Technology.[72] At sea, aviators worked to improve the efficiency of aircraft carrier aviation by "speeding up operations, including take-offs, landings, and faster handling of planes both on deck and below. Brakes and tail wheels [were] . . . now standard equipment for all carrier planes."[73] However, Admiral W. T. Mayo, commander-in-chief, U.S. Fleet, stated the reality in 1919: "The development of Naval Aviation must be governed by the development of Naval Tactics and Strategy" — the thought style which supported the battleship-based strategic paradigm.[74]

According to Rear Admiral Moffett, the biggest drawback to aviation was the failure to construct aircraft carriers, a byproduct of the limited funds for naval construction during the New Era.[75] Yet even if the carriers allowed by the Five-Power Treaty had been built, there is little to indicate that aviation's role would have been different early on. Its strategic and tactical functions were delineated by officers like Moffett, working in the intellectual mainstream of the normal strategic and technological paradigm. The lack of aircraft carriers did not strengthen the battleship's position. The battleship remained preeminent because the Special Board Hearings painted a future in which aviation technology remained weak vis-à-vis the battleship. The technological ceiling predicted by aviation experts relegated naval aviation to scouting, gunfire spotting, and fighting enemy aviation, assignments that endured in naval doctrine through the 1930s.[76]

Despite the dissolution of the 1924 aviation technological ceiling, naval aviation remained more presumption than anomaly and the battleship technological paradigm remained dominant. In February 1931 the chief of naval operations, Admiral William V. Pratt, reported to the House Naval Affairs Committee that the consensus of officers attending Fleet Exercise XII off western Central America was that air attacks were "of less value as a means of defense against approaching fleets" than previously believed.[77] The commander of the aviation units charged with defending the Panama Canal from attack and invasion by the battle fleet, Rear Admiral Joseph Reeves, reported that aviation alone "cannot stop the advance of battle-ships."[78]

The rapid improvement in aviation technology during the mid-1930s did not translate into an immediate, viable presumptive anomaly. Carrier air-craft development mirrored the often confusing co-evolution of the world's armored warships during the 1870–80s. A Bureau of Aeronautics' report that considered 47 percent of aircraft in service "obsolete, obsolescent, or about to become obsolescent" was typical of a period of rapid technologi-cal advancement.[79] The addition of the New Deal carriers *Yorktown* (com-missioned 1937) and *Enterprise* (commissioned 1938) required a 40 percent expansion of naval aviation.[80] A consensus existed among younger naval aviators that their day was fast approaching.

To pose an effective presumptive anomaly, naval aviation required not only more aircraft, but fighters, torpedo bombers, and dive/scout bombers that were state of the art. The acquisition of this technology began in 1936 when the navy ordered 114 all-metal torpedo bombers, the Douglas TBD-1. The TBD-1 was the first monoplane ordered for carrier use and marked a turning point in carrier aircraft.[81] The TBD entered squadron service in October 1937. Its time on the leading edge of carrier-based aviation tech-nology was relatively brief. Unfortunately for its crews, the TBD's 105-knot cruising speed made it an easy target during the battles in the Coral Sea (May 1942) and near Midway (June 1942).[82] The first in the series of Doug-las SBD dive bombers that would later play a key role in the early years of the Pacific war did not enter squadron service until December 1940.[83]

With senior aviators sprouting from the battleship thought collective and perceiving aviation as a gun, naval aviation was unable to escape the social, institutional, and technological momentum attending the battle-ship technological paradigm. The authorization of fast aircraft carriers in the late 1930s, to work with the post-treaty fast battleships, marked the beginning of the fast task force concept. However, American naval aviation had to wait until after Pearl Harbor for opportunities to demonstrate its abilities in war at sea.

CHAPTER 7

Disarmament, Depression, and Politics

Technological Momentum and the Unstable Dynamics of the Hoover-Roosevelt Years

Most U.S. naval officers during the New Era (1921–33) would have agreed with Henry Steele Commager and Richard Morris's description of the period as one of "crisis and failure" and of the character of the Republican administrations as "pervasively negative."[1] The technological basis of the naval profession — the battleship — was threatened by aviation, international treaty, and the attempts of three Republican presidents (especially Herbert Hoover) to negotiate or budget it into extinction. Survival of the battleship during the Republican New Era, and the Democrat New Deal that followed, was part of the larger, often unstable, dynamics of disparate presidential strategic visions, international disarmament, economic depression, and domestic politics.

In almost every way, Franklin Roosevelt was the antithesis of Herbert Hoover, and U.S. naval officers welcomed the election of the former assistant secretary of the navy in November 1932. Roosevelt actively directed the technological aspects of the navy's building program to a far greater extent than had the nation's previous archnavalist president, Theodore Roosevelt. Between 1937 and 1940, Franklin Roosevelt oversaw the authorization of seventeen new battleships to bolster the existing battleship-based strategic paradigm. But such support had a price. Battleship construction became ensnared in New Deal labor legislation and ship construction was the subject of heavy political lobbying. In addition, Roosevelt routinely meddled in naval ship design.

In its New Deal legislation, Congress also abdicated its historical role in the definition of naval technology to the president. This was of minimal concern to naval officers since Roosevelt thought as they did and defined naval power in terms of the battleship. But he did force the navy to construct the anachronistic *Alaska*-class battlecruisers, which the admirals decommissioned soon after his death.[2]

Hoover's and Roosevelt's differing attitudes toward the navy reflected disparate strategic visions. Hoover favored a hemispheric defense while Roosevelt was a globalist member of the navalist thought collective. According to his admirers, Roosevelt's naval policy fell within the framework of the actions of "forward-looking men who saved the country from economic chaos and later spared western civilization from the hands of fascist outlaws." This contrasted with Hoover's "denial of global responsibility." Revisionist and New Left historians, on the other hand, have emphasized Hoover's pursuit of naval disarmament, his noninterventionist foreign policy, and his "non-coercive military policy."[3]

Economic issues also buffeted the interwar navy. Hoover and Roosevelt differed on the role of military expenditures in economic recovery, specifically on the use of naval construction as employment relief. While Hoover accepted fleet maintenance as a valid expenditure to maintain shipyard employment, he refused to accede to new warship construction. Initially, Roosevelt announced a naval program more conservative than Hoover's, but within months had expanded this program dramatically.[4] Linking military expenditures and public welfare — a policy favored in late-nineteenth-century Britain, as noted by William McNeill, and followed in the United States by Grover Cleveland in 1895 — Roosevelt made warship construction part of the 1933 National Industrial Recovery Act (NIRA), an event essentially ignored by the leading historians of the New Deal, but one which stimulated naval expansion in Japan.[5] According to Frank Freidel, "In the spring of 1933 Roosevelt would have liked to embark on a more vigorous American leadership toward disarmament and a countering of the threat from Hitler," but was checked by "strong tempering influences in Congress" and "backtracked" to save his New Deal legislation. But Roosevelt's signing of an executive order, on the same day that NIRA became law, for thirty-two new warships paid for with $238 million in NIRA public works money would seem to indicate otherwise.[6]

The "Gun Club" chief executive. President Franklin Roosevelt reviewing the Fleet in 1934. Secretary of the Navy Claude Swanson is at left; former Secretary of the Navy Josephus Daniels, under whom Roosevelt served as assistant secretary during the Wilson administration, is at the right. (Franklin D. Roosevelt Presidential Library)

Naval officers were pleased by the upturn in their fortunes during the New Deal, and the largely Democratic populations of shipbuilding cities welcomed and even lobbied for the employment resulting from Roosevelt's actions. But construction of the "Treaty Navy"—the lesser warships that supported the battleships—stimulated international suspicions and underscored Roosevelt's repudiation of Hoover's hemispheric defense policy. When the major naval powers failed to agree on continued naval limitations at London in 1935, the battleship technological paradigm was as strong and pervasive as ever. Within its framework, President Roosevelt, Congress, and the naval hierarchy authorized seventeen new battleships between 1937 and 1940.

The Navy and the Post-Harding New Era

After Harding's death in 1923, President Calvin Coolidge rejected the global strategic responsibilities (and attendant force structure) peddled by navalists and set out in the General Board's "Basic Naval Policy of the U.S." Among other things, the "Basic Naval Policy" stated that "The Navy of the United States should be maintained in sufficient strength to support its policies and its commerce, and to guard its continental and overseas possessions." The officers of the General Board justified their strategic focus on Japan because of the threat Japan posed to American Far Eastern trade. Yet only 10 percent of U.S. trade was with Asia and the majority of that with Japan itself.[7]

Coolidge was sensitive to the technological devolution of the naval arms race from the battleship to the cruiser and spoke out against cruiser construction to "catch up" with the British and the Japanese. Coolidge considered American naval power sufficient to defend the United States and refused to give in to those who advocated "nibbling the limitation treaty to pieces." He continued the Republican policy of minimal defense expenditures and reminded the public that the Five-Power Treaty had been motivated by a desire to make "aggressive war" impossible. In 1926, Coolidge deleted funding for three of eight new 10,000-ton cruisers authorized by Congress in 1924, maintaining that construction of these ships would inflame naval competition.[8]

The failure of the 1927 Geneva Conference led to renewed navalist efforts to build the navy up to treaty limits. Legislation passed the House of Representatives for the construction of seventy-one new ships, including five aircraft carriers, twenty-five new cruisers, and thirty-two submarines, at a cost of $740 million over nine years. The proposal stalled in the Senate until after the 1928 presidential election. Coolidge, frustrated by British intransigence at Geneva, reversed his position and endorsed limited cruiser construction on Armistice Day 1928 and again in his final State of the Union address in January 1929. This led the Norfolk *Virginian-Pilot* to opine, "Happily for international peace, the law of presidential succession will spare the world his friendly thoughts on the Fourth of July."[9]

In 1931, the Five-Power Treaty's ten-year moratorium on battleship construction would end. A resumption of capital ship competition, along with a cruiser race, threatened to end the New Era's pursuit of strategic arms reductions. It fell to Herbert Hoover to find a solution.

HOOVER AND THE NAVY

Herbert Hoover believed that there were limits to American power and defined the U.S. military's strategic mission as defense of the Western Hemisphere, not the projection of American naval power throughout the Pacific basin.[10] During the 1931 Manchurian crisis, Hoover told his cabinet that Japanese actions "do not imperil the freedom of the American people, the economic or moral future of our people. I do not propose ever to sacrifice American life for anything short of this." As president-elect, Hoover toured Latin America to improve relations. He laid the basis for the Good Neighbor Policy, removed the marines from Nicaragua, and refrained from the practice of gunboat diplomacy.[11]

Hoover also continued to pursue arms limitations in the name of fiscal responsibility. As secretary of commerce under Harding and Coolidge, and especially as president, Hoover earned the enmity of the majority of the naval officer corps because of his support for capital ship reductions and naval limitations.[12] Like Coolidge, Hoover took issue with the General Board's "Basic Naval Policy." When he entered the White House in March 1929, Hoover was determined to extend the capital ship limitations of the Five-Power Treaty to "auxiliary" warships, such as cruisers and submarines, which comprised 70 percent of the navy and whose growth was unregulated by any international accord.[13]

Hoover's engineering background influenced his approach to naval arms limitation. He realized that arms reductions based upon tonnage alone did not take into account qualitative differences caused by technological variations in gun size, armor design, speed, and age — factors that contributed to the overall efficiency of warships. Elmer Sperry, representing the American Society of Mechanical Engineers, had conveyed much the same opinion to Hoover during the Washington Naval Conference in 1921.[14] In the month following his inauguration, Hoover directed Ambassador Hugh Gibson to propose a "technical yardstick" to the League of Nation's preparatory commission at Geneva to break the deadlock over the number of cruisers. Hoover's yardstick promised a way to measure naval power based upon technical quality, rather than by the simplistic, and misleading, standard of tonnage.

The officers of the General Board disliked Hoover's proposal and thought that deviation from a strict tonnage measurement, for any category of warship, was impossible: "It is highly improbable that accurate determination of the fighting or combatant value of any unit can be made. Any attempt to establish such a value necessarily must be based upon highly technical assumptions and complex computations upon which general agreement is most improbable if not impossible." This was an interesting position since the war games fought at the Naval War College since the 1890s were based on the "combatant value" of different types of ships.[15]

The General Board believed that the United States had been "cheated" by the British at Washington in 1921. The Board's consensus was that the adoption of any type of formula would result in Britain's "inevitably" possessing a greater tonnage of cruisers than the United States. Hoover dismissed the Board's objections and directed it to develop an appropriate arms limitation formula with "coefficients" that reflected American defense needs. Reluctantly, the Board submitted a proposal to the secretary of the navy at the beginning of August 1929.[16]

Hoover's novel approach to measuring naval power interested the British enough to agree to a naval conference in London in January 1930. The new Labour prime minister, Ramsay MacDonald, like Hoover, was extremely anxious to reduce naval expenditures and traveled to the United States in September 1929 for a meeting with Hoover at his Rapidan retreat. In addition to discussions on cruiser parity, MacDonald proposed a reduction in the maximum size for battleships from 35,000 to 25,000 tons, a reduction in gun size from 16-inch to 12-inch bore, and an extension of battleship service life from twenty to twenty-six years (before replacement was allowed). Hoover allegedly went even further than MacDonald and broached the subject of scrapping all battleships.[17] Both men also discussed the possibility of reducing the total aircraft carrier tonnage allowed under the Five-Power Treaty from 135,000 tons to 120,000 tons.

The London Naval Conference opened in January 1930 with three main agenda items. First, negotiators would consider battleship reductions by delaying or eliminating replacement of older ships. Second, Britain and the United States had decided in principle upon parity in "auxiliary" classes of warships to be achieved by 1936. Third, both nations also agreed to push for the abolition of the submarine. The Japanese delegation traveled to Britain via the United States in order to lobby Hoover and the State Department for a change in the naval ratios from 5:5:3 to 10:10:7 in cruiser tonnage, a ratio critical to Japanese defensive plans vis-à-vis the United States. The British Admiralty vigorously opposed this change as did the General Board.[18]

In choosing the members of the American delegation, Hoover made sure that the failure of the 1927 Geneva talks would not be repeated. Sensitive to naval officers' strong attachments to their vessels, Hoover excluded them from the American negotiating team.[19] The delegation was led by Secretary of State Henry L. Stimson and Secretary of the Navy Charles F. Adams. The naval officers sent to Geneva were limited to an advisory role and included Rear Admiral Hilary P. Jones, a hard-liner at the failed Geneva talks; Chief of Naval Operations Admiral William V. Pratt, an open-minded veteran of the Washington Conference; and four rear admirals, including William Moffett, the chief of the Bureau of Aeronautics.[20]

The British government proposed an indefinite extension of the capital ship building holiday that expired in 1931 so capital ships would fade away with age.[21] The U.S. delegation, at the urging of its naval advisers, demanded the right to build one new battleship to offset the supposed qualitative superiority of the new British battleships, *Rodney* and *Nelson*, designed in 1922 and completed in 1927. In the face of British opposition, the American proposal was withdrawn and, as a compromise, all the powers agreed to extend the battleship building holiday until 1936.

The limitation on battleship construction was joined by a limitation on submarines and an agreement on cruiser strength. The three major powers (Britain, the United States, and Japan) agreed to parity in submarine tonnage (52,000 each). This was a victory for the Japanese, who considered submarines critical in opposing an attacking U.S. battle fleet as it crossed the Pacific. The maximum size of each submarine was set at a displacement of 2,000 tons, the smallest judged by the Americans capable of operating with the battle fleet in the western Pacific against Japan.[22]

The administration pulled out the stops in selling the London Naval Treaty to the American public. Before the signing ceremony in London, Hoover claimed the treaty provided for a savings of $1 billion compared to the expenditures that would have been authorized under the naval agreement of the aborted 1927 Geneva conference. While the London Treaty had widespread support among peace groups and in the Midwest, South, and West, it also faced significant opposition. Opponents in the Senate managed to prolong the debate until the end of the session on 3 July 1930.[23]

Hoover called the Senate back into special session to consider the treaty, writing: "The only alternative to this treaty is the competitive building of navies with all its flow of suspicion, hate, ill will, and ultimate disaster." Limiting U.S. strategic defense requirements to support of the Monroe

Doctrine, Hoover was quite willing to accede to Japanese influence in Asia. He pointed out that "No critic has yet asserted that with the navies provided in this agreement, together with our army, our aerial defense, and our national resources, we cannot defend ourselves, and we certainly want no military establishment for the domination of other nations." The economy offered by the treaty was set at $500 million in construction alone and the savings would increase when the operating costs for the larger, pretreaty fleet were included.[24] On 21 July 1930, the Senate ratified the treaty, fifty-eight votes to nine.

As the Depression deepened, Hoover sought to reduce government expenditures, especially in naval armament. In 1931, he cut the naval construction budget and suspended naval construction for the 1933 fiscal year, reinforcing the General Board's fears that the navy would be budgeted out of existence.[25] In 1932, Hoover pushed for further naval reductions at the Geneva Arms Conference. When the conference broke down over the complex issues of military disarmament in Europe, Hoover sought to revitalize it by floating an offer of even deeper naval armament cuts: a reduction in aircraft carrier tonnage by one-fourth and capital ship and submarine tonnage by one-third. Hoover's plan was accepted by the majority of the attending nations but blocked by Britain. The talks continued, but the prospect of further naval reductions died with Hoover's loss to Roosevelt in the fall election.

ROOSEVELT AND THE NAVY

In contrast to his three Republican predecessors, Franklin Roosevelt was a navalist. He was influenced by the writings of Alfred Thayer Mahan and, while recovering from polio in 1922, had agreed to edit, but never made any progress on, a new edition of Mahan's works. Roosevelt had been a member of the U.S. Naval Institute since 1927 and, prior to joining, had acquired a set of the Institute's *Proceedings* from 1887 through 1917. Roosevelt's naval worldview was grounded in his tour as assistant secretary of the navy (1913–20), where he grew to know the navy, and its officers, during the height of the battleship era. As a battleship aficionado, Roosevelt subscribed to the navy's battleship-based strategic and technological paradigms and during the 1930s appointed members of the "Gun Club" — battleship officers who had served as chiefs of the Bureau of Ordnance — to serve in the navy's highest post, chief of naval operations (CNO).[26]

Franklin Roosevelt's occupancy of the White House meant more than just renewed executive branch support for a larger navy. Roosevelt always found time to manage the Navy Department, and according to Robert Albion, selected the old and ailing Senator Claude Swanson to serve as secretary of the navy precisely so that the president could have a free hand. Unlike his predecessors, Roosevelt regularly reviewed designs for new ships; he even ordered the relocation of airplane catapults and specified the number of "smoke-stacks" to suit his taste.[27] With regard to the navy, Roosevelt was like a small boy given the run of a candy store. Although not hostile to naval aviation, he saw its strategic and tactical roles as subordinate to the battleship in much the same way as did the navy's senior admirals.

While an active "tweaker" of surface ship technologies, Roosevelt deferred to the battleship hierarchy on broader issues within the navy's strategic paradigm. In 1937, for example, he canceled the navy's lighter-than-air program on the advice of the chief of naval operations, Admiral William Leahy.[28] But Roosevelt's was a strong voice on technological issues within the battleship technological paradigm, and he forced the navy to issue contracts for six battlecruisers in 1940 despite the professional consensus that this type of ship was no longer viable. Roosevelt's "meddling" resulted in a cold relationship with Admiral William H. Standley, who as CNO (1933–37) resented the president's "inflated opinion" of his own knowledge of naval matters. All in all, Franklin Roosevelt had more to do with the navy than any chief executive since Theodore Roosevelt.[29]

The naval officer corps welcomed Roosevelt's interest after the antinaval policies of the New Era. Roosevelt's campaign courted the naval vote during the 1932 presidential race, spreading rumors that President Hoover intended to reduce the navy further and abolish the Marine Corps if he was reelected. Hoover had canceled the salaried annual reserve training in 1932 in the interest of fiscal responsibility, and naval reserve officers in the 7th Naval Reserve Area (Great Lakes region) were warned that the Navy Department was planning on removing all training ships from the Great Lakes. After meeting with Roosevelt, a naval reserve officer, Commander M. R. Whortley, informed his colleagues that "Governor Roosevelt assured me with considerable force that if he is elected, and we can carry on until

March 4th [inauguration], that we need no longer worry about the security of the Naval Reserve and its training vessels on the Lakes."[30]

The election of 1932 hardly turned upon the votes of the relatively small number of active and reserve naval officers, but the naval officer corps strongly identified with Roosevelt and saw him as a chief executive who would support naval expansion and the construction of the maximum number of ships allowed under the naval treaties. The U.S. Naval Institute's prize essay for 1933, written after the November 1932 election, reflected the expectations engendered by Roosevelt's membership in the navalist thought collective, predicting that "the naval policy of the United States under the new Democratic administration is necessarily the realization of President Wilson's vision of a world dominant American navy establishing world peace." The editor of the U.S. Naval Institute Proceedings characterized the forthcoming years as yielding a "new Pax Romana based on the American Navy."[31]

Naval officers' exuberance over Roosevelt was typified by Harold Stark (later CNO when Pearl Harbor was attacked in 1941), a battleship captain and compatriot from Roosevelt's days as assistant secretary of the navy.[32] Stark asked permission to visit the White House to see Roosevelt "in those surroundings where we have so long wanted and needed you." In a second note, Stark told Roosevelt: "Take good care of Uncle Sam's Navy! We have long been in need of a friend; . . . it is good to have a one hundred per cent American in the White House."[33] If patriotism was measured by support for naval expansion, then Roosevelt would soon become Uncle Sam incarnate. During the seven and one-half years between his inauguration and the Japanese attack on Pearl Harbor, Franklin Roosevelt rebuilt the American navy. But the navy he rebuilt was based on authorizations for seventeen new battleships and reflected Roosevelt's commitment to the battleship technological paradigm and the strategy it embodied.

Naval Rearmament as Public Works

In November 1932 Roosevelt met with the Democratic chairman of the House Naval Affairs Committee, Carl Vinson of Georgia, at Warm Springs to develop a consensus regarding the naval program for the next four years. Vinson had fought unsuccessfully against the congressional leaders of his own party and Hoover to authorize naval construction up to the treaty limits. After meeting with Roosevelt, Vinson shocked naval officers when he announced a "Democratic naval program" that was more conservative (cutting the naval budget $100 million — 28 percent below fiscal year 1932) and provided less money for naval construction for the next five years ($30 million per year) than Hoover's.[34] The naval retrenchment proposed by Roosevelt was most likely a passing fancy attributable to early interregnum, conservative fiscal advisers who saw balanced federal budgets as the way to economic recovery. With the major private-sector and navy shipyards located in Democratic strongholds such as Philadelphia, New York City, Boston, Camden, and Charleston, such cutbacks were unlikely.

Once back among his congressional colleagues, Vinson repudiated this initial naval plan and led the fight for public works funding for naval ship construction. In December 1932 he requested a navy study on the economic effects of naval ship construction programs. The previous year, the navy had tried to promote itself by touting its relation to the "industrial, scientific, economic, and political development of the Nation." Outgoing navy secretary Charles Adams, frustrated and perhaps embittered by Hoover's antinavalism, informed Vinson that an annual appropriation of $400 million would "keep at least 1,290,000 of people of this country out of unemployed status . . . [and] In discharging its constitutional function of 'providing for the common defense' the federal government would thus manifestly be 'promoting the general welfare.' " Rear Admiral Emory S. Land, chief of the Bureau of Construction & Repair, began lobbying Senator Robert Wagner, Democrat from New York, for public works appropriations for ship construction as part of what became the NIRA.[35]

By March 1933, the fiscally conservative Roosevelt naval program of the previous November was dead. Within three weeks of his March inauguration, Roosevelt was encouraging naval rearmament as a part of public works since approximately 85 percent of shipbuilding costs went to labor. Secretary of the Navy Swanson boasted that "every State will benefit" since ship construction drew on materials produced throughout the United States and ship production required the skills of more than 125 trades.[36] On 21 March 1933, Swanson announced, not surprisingly, that the navy would support the inclusion of naval construction as a part of the public works included in any legislation enacted for unemployment relief. The following day, Swanson told the chief of naval operations and the chiefs of the navy's bureaus that "the Navy could rely heavily upon the friendly attitude of the President" in their efforts to include naval construction in unemployment relief.[37]

Senator Wagner's original public works relief package was expanded and incorporated into the compromise National Industrial Recovery Act hammered out by Wagner and Commerce Department undersecretary John Dickinson in meetings with budget director Lewis Douglas and labor lawyer Daniel Richberg. The $3.3 billion public works funding authorized by NIRA appealed to those who advocated government "pump priming," as well as to navalists who had fought, without success, to include naval construction as federal public works during the Hoover presidency.[38]

While apparently aiding the ailing domestic economy, the naval aspects of NIRA generated keen attention abroad — especially in Japan. On 16 May 1933, one month prior to signing NIRA, Roosevelt sent his "Appeal to the Nations" message to world leaders in which he called for the elimination of "all offensive weapons." In reporting his message to Congress, Roosevelt wrote: "permanent defenses are a non-recurring charge against governmental budgets while large armies, continually rearmed with improved offensive weapons, constitute a recurring charge. This . . . is responsible for governmental deficits and threatened bankruptcy." Roosevelt's focus was on military, not naval, expenditures, and the inference is that he considered expenditures on the U.S. Navy as defensive — a view not shared by the Japanese. He had sent his message, he told Congress, "because it has become increasingly evident that the assurance of world political and economic peace and stability is threatened by selfish and short-sighted policies."[39]

Was NIRA naval rearmament to stimulate the economy a selfish policy? To many, the plan for new naval construction seemed so. The president of the Women's International League for Peace and Freedom, Hanna Clothier Hull, castigated Roosevelt for tolerating any plans to build the navy up to the treaty limits. Hull was joined by a cross-section of academics, such as the president of Brown University and the dean of the School of Journalism at the University of Oregon, who characterized the Navy Department's policy as "insanely and dangerously reckless." Mary Woolley, president of Mt. Holyoke College, worried about the effect of naval construction on the Good Neighbor Policy.[40] But according to Raymond Moley — Roosevelt's confidant, de facto chair of Roosevelt's professorial "brain trust," and assistant secretary of state — Roosevelt did not believe that the

Depression could be ended by "international measures" such as reducing war debts or opening international trade. The core of the recovery program had to be domestic.[41]

Roosevelt's responses to his critics were sugary, and his support for the naval construction program, as a part of economic recovery, did not waiver. He did direct Henry L. Roosevelt, the assistant secretary of the navy, to muzzle the admirals who were calling too much attention to the NIRA naval construction program by their public attacks on the "professional pacifists."[42]

With battleship construction prohibited by treaty, the question for the admirals was how to use NIRA funds. Their answer was to shore up the battleship technological paradigm by replacing or adding supportive ships such as destroyers constructed during World War I and now reaching obsolescence en masse. To prevent block obsolescence from recurring, Chief of Naval Operations Admiral William V. Pratt proposed the construction of 132 ships spaced over the next 8 years. The ships included 76 destroyers, 9 destroyer leaders, 23 submarines, two 20,000-ton aircraft carriers, one 15,200-ton aircraft carrier, 8 cruisers, and 13 gunboats, for a total cost of over $944 million. First priority was assigned to destroyers, followed by aircraft carriers, light cruisers, submarines, and one heavy cruiser. To ensure broad political support, Pratt specified which private-sector and navy shipyards could receive particular contracts.[43]

In passing the NIRA, Congress relinquished its traditional role of defining the technological nature of American naval power as it had done with the "sea-going coastline" battleships of the 1890s. But the National Industrial Recovery Act, in accordance with Roosevelt's wishes, did not specify the level of funding or the numbers and types of ships to be built with public works money. In a complete break with precedent, naval construction now would be at the discretion of the president and begin by executive order. NIRA gave the president carte blanche to construct ships and procure aircraft as allowed under the terms of the naval treaties. The geographic diversity of NIRA public works blunted congressional opposition to placing so much power in the hands of the president.[44]

On 15 June 1933, the day Roosevelt signed NIRA into law, Secretary Swanson submitted a draft executive order for naval construction. In his annual report for fiscal year 1933, Swanson called for the creation, maintenance, and operation of a navy "second to none."[45] Toward this end Swanson's draft requested $253 million — $238 million for ship construction and

$15 million for new aircraft. Roosevelt approved the $238 million for new ships on the same day he signed the NIRA and personally approved the contracts for twenty-one NIRA ships to be built by private shipyards.[46] The majority of the thirty-two ships contracted by Roosevelt under NIRA were a boost to traditional Democrat areas. The Bath Iron Works in Bath, Maine, received contracts for two destroyers that returned 1,500 skilled shipbuilders to work. This contract fulfilled the promise made in Bath by Jimmy Roosevelt during the presidential campaign. As a result of the contracts, the president of Bath Iron Works, W. S. Newall, abandoned the Republican Party and became a "Roosevelt Democrat."[47]

Roosevelt's battleship orientation, coupled with the economic insignificance of the aircraft industry in comparison to shipbuilding, led him to reject the argument for new naval aircraft, and he cut the aviation procurement authorization by over one-third. The chief of the Bureau of Aeronautics, Rear Admiral Ernest J. King, argued that the battleship admirals had forced the bureau to pay wrongfully for the observation and scouting aircraft carried on new ships out of the 1,000-plane-navy appropriations stemming from the 1926 Morrow Board.[48] According to King, aircraft for new ships should have come from separate, annual "increase of the navy" appropriations, and not cut into the 1926 appropriations designed to beef up navy and army air power. "Increase of the navy" funds were extremely limited during the Hoover administration, and their expenditure was controlled by the battleship hierarchy, who considered aviation, although important for scouting, as just one of many areas competing for limited funds. Roosevelt referred the bureau complaint to Secretary of the Interior Harold Ickes, chairman of the president's Special Board for Public Works.[49] Ickes had no sympathy for King's request, but after several months of wrangling, naval aviation was able to wrest $7.5 million from Ickes for 130 new aircraft, half of $15 million allocated for army and navy aviation.[50]

House Naval Affairs Committee chairman Carl Vinson sought to build upon the momentum of the NIRA naval program by working for the enactment of a multiyear naval construction plan in 1934. In addition to the thirty-two ships authorized under NIRA, the United States was still allowed ninety-two ships (displacing 177,000 tons) under the terms of the naval treaties. In December 1933 Roosevelt expressed interest in a $100 million authorization bill to increase NIRA ship construction in 1935 from three ships to twenty-five. Opposition from budget director Lewis Douglas as well as a letter warning against expanded naval construction from Norman

Davis, the U.S. delegate at the Geneva arms negotiations, dampened Roosevelt's enthusiasm and he tabled his proposal, but Vinson and the navy pushed ahead. Encouraged by NIRA funding, the navy proposed its own ambitious 1935 program introduced by Vinson as H.R.6604. Vinson pried a grudging endorsement from the Bureau of the Budget and began hearings on his bill in January 1934. The naval construction proposed by Vinson became the target of pacifist and antinavy sentiment already inflamed by the naval construction included under NIRA. In February 1934, Secretary of State Cordell Hull informed the president that the government had been receiving over two hundred letters and telegrams per day, 99 percent of which opposed the Vinson Bill.[51]

The political infighting over the enactment of the Vinson–Trammell Act of 1934 is incidental to this study, but one point is worth noting. The Act confirmed the abdication of congressional authority over the technological composition of the navy, placing its determination in the hands of the president. After signing the bill into law on 27 March 1934, Roosevelt issued a statement attempting to pass the Vinson–Trammell Act off as a mere policy statement concerning the naval construction philosophy of Congress. But during the signing ceremony, Roosevelt had promised Vinson and Acting Secretary of the Navy Henry L. Roosevelt that the funds for the twenty ships of the first year's construction program would be taken directly from Works Progress Administration (WPA) funds, obviating the navy's potentially controversial $38 million supplemental budget request pending before the Bureau of the Budget.[52] Roosevelt and the navy were spared the publicity, and delay, of congressional hearings.

While some domestic attention was deflected, international interest was keen. The hierarchy of the Japanese navy, according to Arthur Marder, attached tremendous importance to the Vinson–Trammell Act, believing it provided the U.S. Navy the means to engage in offensive operations in Asia.[53]

THE END OF TREATY RESTRICTIONS

Japanese naval officers responded to New Deal naval programs with pressure on their government to increase overall naval construction and to withdraw from the London Naval Treaty to concentrate on the construction of superbattleships to ensure qualitative, technological superiority over the United States. The majority of Japanese naval officers had been offended by the termination of the Anglo-Japanese alliance after World War I and the "humiliating" Five-Power and London naval treaties. They equated "disarmament" and "naval limitation" with U.S. "oppression" of Japan.[54]

While the Republican New Era administrations had limited warship construction, the Japanese had built their navy up to their treaty limits and were close to achieving 80 percent of tonnage parity with the United States in all categories except battleships. The Japanese expected to maintain this ratio, and its favorable strategic implications, through 1936, but were concerned by renewed U.S. naval construction with NIRA funding. The navy minister, Ōsumi Mineo, lobbied for a larger naval budget in response to what appeared to be an open-ended American naval construction program. The army minister, sensitive to any increase in the naval budget, and the foreign and finance ministers argued for restraint. In the fall of 1933, Prime Minister Saitō Makoto ignored the advice of the finance minister and increased the military budget from 36 percent to 45 percent of total expenditures to fund the 1933 Emergency Program. The Second Replenishment Program (1934–37) authorized 90 ships totaling 221,096 tons in response to the 1934 Vinson–Trammell Act.[55]

Emboldened, the naval officer corps set their sights on ending Japanese naval inferiority rooted in the naval treaties' ratio system. Admiral Katō Kanji warned that if the government failed to end participation in the ratio system, it would be difficult to control the naval officer corps. With the navy's position "absolute," Prime Minister Okada Keisuke opted on 7 October 1934 to save his government and gave the two-year notice to abrogate the naval treaties.[56]

Prior to their pursuit of superbattleship technology, the Japanese relied on *yōgeki sakusen*, the "strategy of interceptive operations" put into place in 1933. This doctrine mandated repeated submarine and torpedo bomber attacks to reduce the U.S. battle fleet by 30 percent as it advanced toward Japan for a battleship duel.[57] Even before the advent of "interceptive" operations, Admiral Katō relied on Japanese "will power," based in the Yamato spirit and the warrior code of *bushidō*, to overcome the U.S. Navy's material superiority.[58] This was much like the Edwardian officer corps' planned reliance on "manly courage" and "virility" to prevail against the machine gun and modern artillery at the beginning of the twentieth century.[59] Admiral Yamamoto Isoroku woefully compared the Yamato spirit with the

Yankee spirit, and found the former "often verged on blind daredevilry" while the latter was "soundly grounded on science and technology — a case in point being Lindbergh's transatlantic solo flight."[60]

In justifying Japan's withdrawal from the treaty system, the navy general staff told Emperor Hirohito that "unrestrained naval construction would actually be *more economical*, for Japan would be free to concentrate on ship categories 'best suited to its national requirements' " — the superbattleships of the *Yamato* class, which would provide Japan with unassailable, qualitative, technological superiority over the Americans.[61] The argument resonates with one later advanced to support the development of the hydrogen bomb. The Japanese believed that even if the United States produced a superbattleship comparable to *Yamato* and her 18.1-inch guns, Japan would already be on its way to developing even larger superbattleships mounting 20-inch guns. Japanese planners estimated that if their battle plan was just 50 percent successful, the United States would require two years to rebuild and mount another attack which would fare little better without superbattleships. Presumably, rather than repeat the same mistake, or spend five to eight years to design, build, and deploy superbattleships of their own, the United States would recognize Japan's primacy in East Asia.[62]

Responding to domestic concerns over the Japanese naval appropriations of 1933–34, Roosevelt justified American naval construction in an October 1935 Navy Day letter by citing the "unsettled conditions existing throughout the world." He praised the Vinson–Trammell Act and its authorization to build up the navy to a "degree commensurate with America's needs, interests, and abilities." The chief of naval operations, Admiral Standley, echoed the president's support for naval construction in a speech before the New York State Chamber of Commerce. Standley argued that American national security could only be maintained through the "insurance of national military vigor" through continued appropriations to fund the ships authorized by the Vinson–Trammell Act.[63]

The 1935 London Naval Conference was the only hope of averting a new naval race. Roosevelt's appointment of the pacifist-baiting Admiral Standley (whom Roosevelt had previously had to muzzle) as a disarmament delegate drew widespread criticism. For example, Wyoming state senator Charles Bream, a Roosevelt supporter in 1932, wrote Roosevelt that he would work for his defeat in 1936 because Standley's appointment destroyed "all the confidence sincere Americans had in you." Calling Stand-

ley a "coyote among lambs" and a "bartender among drunkards," Bream
saw Standley's appointment as "an arrogant slap in the face to people who
have the good sense to see that this mad race in armament building *is* a
mad race, with the only possible out-come being more poverty, more war,
and the eventual destruction of civilized peoples by their own weapons."[64]

Cultural acceptance of the battleship as the measure of naval power was
reflected in the discussion during a presidential press conference in November 1935, one month prior to the London Naval Conference. When
asked if the United States would build new battleships to replace those
reaching obsolescence, Roosevelt admitted that new U.S. battleship construction had been considered, but only "for about a minute and a half,"
during a meeting with Assistant Secretary of the Navy Henry L. Roosevelt
and Admiral Standley the previous day. According to Roosevelt, no decision regarding a resumption of battleship construction had been made, but
he was being disingenuous about the extent of U.S. battleship plans and his
involvement with them. Two months earlier, Roosevelt had reviewed designs for new battleships with Admiral Standley and the chief of the Bureau of Construction & Repair, Rear Admiral Land. These designs originated with Roosevelt's December 1934 directive to Secretary Swanson to
"discuss with the proper officers . . . the development of new types of ships
on the theory that the Washington and London Treaty restrictions may be
entirely removed within the next two years." Roosevelt promised additional
appropriations for the design bureaus if special studies were required.[65]

Happily rid of Hoover, the navy responded quickly to presidential interest in new battleships. The navy had prepared design sketches for a
35,000-ton battleship in 1929 in anticipation of the resumption of superdreadnought construction in 1932. It is likely that these plans were updated
regularly by the Bureau of Construction & Repair since Roosevelt's naval
aide informed him that sketches and comments on existing designs would
be forwarded "at once" to the commander-in-chief, U.S. Fleet, for his comments, prior to being sent to the General Board and the Ship Development
Board.[66]

The Japanese were mistaken if they thought their interest in superbattleships unique. In January 1935, the Bureau of Construction & Repair
was directed to design the largest possible battleship with 20-inch guns limited only by the maximum beam (108 feet) that would fit through the
Panama Canal. The resulting designs were true leviathans: a 72,000-ton,
30-knot ship and a 60,000-ton, 25-knot design. But such monstrous ships

were merely the latest example of wishful, "bigger is better" thinking by battleship advocates dating back to 1903.[67]

When the London Naval Conference began in December 1935, the Japanese delegation rejected continuation of the quantitative limitations of the ratio system. Japan was willing to allow Britain to build above a "common upper [tonnage] limit" because of its worldwide commitments, but was insistent on exact parity with the United States.[68] With the talks at an impasse in January 1936, the Japanese delegation made a sweeping proposal for deep cuts in naval armaments. The head of the delegation, Admiral Nagano Osami, proposed a reduction in all "offensive" weapons, which he defined as battleships, aircraft carriers, and 8-inch-gun heavy cruisers. The Japanese further proposed retention of a common upper limit, that is, parity among the three major naval powers, with no restrictions on the type of warships allowed. This proposal was rejected in short order by the British and Americans, who believed that such a reduction in their naval superiority would cede Asia to Japan and threaten the security of Australia, New Zealand, and the Philippines. Admiral Nagano announced that Japan could no longer pursue negotiations based upon quantitative inferiority, and the Japanese delegation announced its withdrawal from the conference that afternoon.[69]

In spite of Japan's withdrawal, the Second London Naval Treaty was signed by the United States, Britain, and France on 25 March 1936 to run from 1937 through 1942. Separate agreements were signed with Germany and the Soviet Union. An escape clause allowed an escalation to 16-inch guns if either Japan or Italy failed to subscribe to the treaty by 1937.[70]

A new naval arms race was a certainty unless Japan signed the treaty and accepted qualitative parity along with quantitative inferiority. The Japanese declined two invitations to sign the treaty during 1937, preferring to be free to determine their own naval policy. The British Admiralty urged its government to threaten to use British financial and industrial superiority to outstrip the Japanese efforts through a more aggressive battleship building program. This was analogous to Carl Vinson's 1934 threat to build five warships for every three laid down by Japan.[71]

On 5 February 1938, France, the United States, and Britain sent Japan an ultimatum demanding details of Japanese battleship programs and assurances that they would not build ships exceeding the limits set by the Second London Treaty. Already committed to the construction of the *Yamato*-class superbattleships, the Japanese saw no reason to submit to the dictates of the other major powers. On 7 February 1938, the Japanese Foreign Ministry denied that Japan was building battleships of 43,000 tons as claimed by the British and Americans.[72] This was correct: the *Yamato*-class battleships displaced over 68,000 tons.

In March 1938, France, Britain, and the United States agreed to invoke the escalation clause in the Second London Treaty and, after discussion, signed a protocol in June 1938, raising the maximum battleship displacement to 45,000 tons and setting the maximum gun bore at 16 inches.[73] A new battleship race was on.

The Politics of Naval Construction

Japanese repudiation of the international naval treaties provided Roosevelt perfect justification for the general naval buildup authorized by the Vinson–Trammell Act. Japanese secrecy concerning their battleship designs also aided the political case for resumption of U.S. battleship construction. The first U.S. "treaty" superdreadnoughts, the two ships of the *North Carolina* class, were authorized for $50 million each in January 1937.[74]

The battleships *North Carolina* and *Washington* were the first capital ships to fall under the control of New Deal legislation. A provision of the Vinson–Trammell Act limited corporate profit from the construction of government ships to 10 percent. The Walsh–Healey Act of 1936 required the payment of prevailing wages and specified minimum standards for working conditions. According to Roosevelt, the government wanted to build one ship in a navy yard and one in the private sector. Given the economic climate, Roosevelt did not think the increased labor costs of the Walsh–Healey Act would deter private-sector bids. If it did, then Roosevelt said the government would build both ships in navy yards.[75]

Even though the New Deal had provided the vehicle for the navy's reconstruction, naval officers were learning that there was a price to be paid for their new affiliation with public works.[76] While the contract-starved private shipyards may have had no objection to the provisions of the Walsh–Healey Act, the steel industry was another matter. Steel company executives resented the Walsh–Healey provision that limited the maximum work week to forty hours and the prescription of a minimum wage as set by the defunct National Recovery Administration (NRA). Further compounding the problem was the dispute between the American Federation of Labor (AFL), representing skilled workers, and the Committee for Industrial Organization (CIO), which was attempting to organize the unskilled workers in the steel industry. The navy, which had been riding the crest of the New Deal, was caught in the middle and unable to obtain the 18 million pounds of steel needed for the construction of three submarines and six destroyers, let alone the massive quantities that would be required for the new battleships.[77] Labor problems also existed in the shipyards. In May 1937, the CIO halted construction of two destroyers by forcing a strike at the Federal Shipbuilding and Drydock Company in Kearny, New Jersey. The strike ended when management recognized the CIO as the collective bargaining agent for the company's workers and agreed to a pay raise of 5 cents per hour.

Industry concerns over labor and steel costs escalated the price of the *North Carolina*-class battleships from $50 million to $60 million per ship by June. The navy's review of the construction bids was delayed repeatedly by design modifications. When the bids were finally reviewed, the navy thought those from the private sector to be inflated unjustly, and the New York and Philadelphia navy yards were declared the winners. But the award of $120 million in battleship contracts to government yards drew strong criticism from the citizens of Camden, New Jersey, whose livelihood depended upon the New York Shipbuilding Company. In January 1937, Mayor George E. Brunner of Camden warned President Roosevelt that New York Shipbuilding's work would soon run out and result in massive unemployment. Recalling Roosevelt's personal approval of the contract awards made under the initial NIRA navy program, the mayor asked the president, without success, to direct new ship construction to New York Shipbuilding. The president of the Camden Chamber of Commerce, J. W. Burnison, appealed directly to Roosevelt to shift the $60 million battleship *North Carolina* from the nearby Brooklyn navy yard to New York Shipbuilding. The president responded that the battleship awards had been made on the basis of sealed bids and that the Navy Yards' were far lower than any submitted from the private sector. Roosevelt advised Burnison to tell the private shipyards to sharpen their pencils when subsequent navy contracts went out for bid.[78]

While the bids submitted by the navy yards were cheaper, at least on paper,[79] there were other factors that influenced the award of both battleship contracts. Although the New York Navy Yard had some labor troubles, they were minimal compared to the problems that private shipyards had with labor-management disputes and intralabor friction between the skilled trades, represented by AFL craft unions, and the CIO, which was attempting to become the sole collective bargaining agent for shipyard workers. Due to the increased labor and production costs from the Walsh–Healey Act and the 10 percent limit on profits from navy contracts, private shipyard management tended to inflate their contract bids to maintain their previous profit margin. It is possible that the award of the two new battleships to navy yards was a message from the administration to the shipbuilding industry, and indirectly to the steel industry, to accommodate labor and to submit bids in the future that reflected realistic construction costs. There was still a feeling within the naval hierarchy that private shipyards, building the *Maryland*-class superdreadnoughts after World War I, had used the cost-plus-fixed-profit contract as a means to prolong construction during the postwar slump. Given the strategic threat posed by the secret Japanese battleship program, the admirals wanted no delay in the construction of their long-awaited superdreadnoughts.

When local politicians had no success shifting the battleship contracts, New Jersey congressmen entered the act. Representative D. Lane Powers asked Roosevelt to reconsider the award to the New York Navy Yard. Senator A. Harry Moore warned the president that New York Shipbuilding would be forced to close unless it received navy business. Roosevelt told Moore that nothing could be done regarding the battleships since wide discrepancies existed between the private-sector and navy yard bids and the material had already been ordered. He advised Moore that the navy would be soliciting bids for new auxiliary ships the following month.[80]

In response to this enticement, the Camden area launched a campaign to pressure the White House into awarding one of the new navy contracts to New York Shipbuilding. Thirty thousand Camden children competed in an essay contest on shipbuilding. One hundred and twenty-five prize winners were brought to Washington and they, along with members of the Camden political and business community, visited the White House to plead their case. Roosevelt's secretary, Marvin McIntyre, asked Assistant (de facto Acting) Secretary of the Navy Charles Edison if "something could be done where they [New York Shipbuilding] would sharpen their pencils" according to the president's advice. Speaking for the president, McIntyre asked Edison to consider the matter and see what he could do.[81]

The concerted lobbying effort by the Camden area and New York Shipbuilding paid off in December 1937 when Roosevelt approved the award of two auxiliary ships to New York Shipbuilding. A relieved Edison told McIntyre that this action should finally "put a period" on the Camden matter. The president of the Camden Chamber of Commerce fully appreciated the debt his area owed to the president: "While as low bidder, the NY Shipbuilding Corp. was entitled to one of the navy tenders, we fully appreciate that the authority to award both of them was vested entirely in the President of the United States . . . therefore Camden County owes you a lasting debt of gratitude for this great boon to our community."[82]

Tenders, though, were in a different league than battleships. The admirals, recalling the labor trouble at Federal Shipbuilding in May 1937, were more comfortable with capital ship construction in navy yards. But by the end of the 1930s, American capital ship construction had expanded to the point where the navy was forced to build three of the four "1939" battleships (the *South Dakota* class) in private shipyards. But when the first of the six *Iowa*-class battleships was laid down in 1940, the navy was able to build all ships of this class in navy yards. The navy also allocated all five of the monstrous 60,000-ton *Montana*-class battleships to navy yards as well. Whether constructed in navy yards or in private-sector shipyards, these expensive capital ships provided a strong economic boost for their local communities.

Unstable Dynamics and Proportional Power

In a speech on 28 December 1933, Franklin Roosevelt said, "The blame for the danger to world peace lies not in the world population but in the political leaders of that population."[83] In a conventional sense, Roosevelt's truism refers to statesmanship and diplomacy. But in a market-based society in which significant segments of the economy rely upon — or can be stimulated by — national security expenditures, the initial danger was not diplomatic failure. It lay in political leaders exacerbating international rivalries by buying votes with military contracts.[84] Franklin Roosevelt viewed the use of NIRA funds for naval construction as relatively innocuous and defensive and a proportional response to Japanese naval policy, but it was an additional catalyst to an already-hardening Japanese position vis-à-vis Asia and the role of the United States there. Unlike bridges, dams, and highways, the effects of warship construction extended abroad. This was clear to the opponents of naval construction as public works in 1933–34, but with a large segment of the industrial base capable of providing products to the military, the desire for employment won out over moral sensitivities.

A recurring proposal — drawing on a New Deal precedent — during the 1992 presidential election was federal rebuilding of the national infrastructure to stimulate the sluggish economy. But five decades into the national security state, public calls for widespread increases in defense acquisitions to maintain employment were noticeably absent.[85] In fact, there has been a clamor to shift away from heavy development and production of military technologies in the early, post–cold war world. But things have not changed since the establishment of the armory at Harpers Ferry in 1794 or warship construction at Camden, New Jersey, in 1937; local military contracts and expenditures are political prizes. If cost-cutting has to occur, politicians try to ensure that it occurs outside their districts.[86] Thanks to sophisticated geographic diffusion by modern defense contractors and subcontractors, broad political constituencies now work against efforts to reduce defense expenditures — just as they worked to win support for naval construction in the 1930s.[87]

It would be extreme to argue that the U.S. economy during the New Era or early New Deal lived by the sword, although in many areas like Camden, New Jersey, military spending was considered essential to the local economy. The pervasiveness of the national security state in the national economy (among other aspects of U.S. society) would come with World War II. But its advent was eased by Franklin Roosevelt and the step he first took in 1933 as an important member of the navalist thought collective. Herbert Hoover was more moralist than politician and would not sanction naval construction as a means to improve the depressed national economy.[88] Roosevelt, the consummate politico, tapped into more plebeian instincts while fulfilling his international and domestic agendas.

For the navy, the technological outcome was little different. Hoover never directly challenged the battleship-based strategic paradigm — he merely tried to negotiate battleships and other offensive naval platforms out of existence. Luckily for the navy's battleship hierarchy in the 1930s, it had a friend and champion of the battleship thought style in Franklin Roosevelt, who must have been pleased to receive a model of the new battleship *North Carolina* in December 1937, sent as a "small token of the high esteem and affectionate regards of the entire Naval Establishment."[89]

While not hostile to naval aviation or submarines, Roosevelt saw them in supportive strategic and tactical roles in exactly the same manner as his "Gun Club" admiral acquaintances. From 1933 on, Roosevelt set out to reinvigorate the battleship technological paradigm, first with the ancillary warships allowed by treaty and, ultimately, with the seventeen new battleships authorized between 1937 and 1940. Unlike Hoover, Franklin Roosevelt perceived the U.S. naval mission as a global one requiring a large navy. New Deal construction programs provided the venue for this and reversed the navy's decline during the New Era while expanding domestic employment and solidifying political allegiances to Roosevelt.

The Japanese viewed the American naval buildup as directed at them which, in fact, it was despite Roosevelt's protestations. New Deal naval construction reinforced the Japanese admirals' determination to demand naval parity at London in 1935 and resulted in their leaving the conference when Britain and the United States refused to accede to their demands. The Japanese naval hierarchy viewed repudiation of the quantitative ratio system as the means to achieve economically qualitative, technological superiority by building superbattleships. New Deal naval construction did not spawn World War II, but it did exacerbate civil-naval relations in Japan. It bolstered the hard-line naval position and contributed to a military-driven, confrontative Japanese foreign policy which resulted in the Pacific Ocean War.

In *Vom Kriege* (1832), Carl von Clausewitz discussed the uncertain dynamics of war in which there is a real danger — thanks to the "collision of two living forces" — of employing force beyond proper proportionality.[90] The politics of U.S. naval technology during the 1930s illustrate the unstable dynamics which can accompany the "proportional" pursuit of military power — if only as part of the social welfare of the state.

CHAPTER 8

War and a Shifting Technological Paradigm

Fast Task Forces and "Three-Plane" Warfare

During the 1930s, naval aviation developed beyond the technological ceiling predicted in 1924. Yet the technological basis for an effective presumptive anomaly to challenge the dominant battleship technological paradigm did not exist until the end of the decade. The advances in naval aviation during the 1920s resulted in early elucidations of an aviation presumptive anomaly bolstered by the empirical data of fleet exercises. At the Naval War College in 1933, Commander Hugh Douglas, former executive office of the aircraft carrier *Saratoga*, sketched out the essence of what would occur nine years later during the Battle of Midway. Based on his experience in recent fleet exercises, Douglas predicted that if "an enemy carrier is encountered with planes on deck, a successful dive bombing attack by even a small number of planes may greatly influence future operations."[1] While Douglas was prescient, it is important to note that he was speaking in terms of intra-artifact combat — attack on lightly armored, vulnerable aircraft carriers, not on heavily armored battleships.

Since the London Naval Treaty of 1930 extended the moratorium on battleship construction through 1936, the navy continued to build aircraft carriers up to its allowed treaty limits. The navy also constructed submarines, not in the image of the counterweapon of the World War, but improved "fleet" submarines to support the battle line. Both submarines and aircraft carriers became important supportive ships within the interwar battle fleet.

The value of airships to the battle fleet was their ability to serve as effective scouts. In good weather an airship and its aircraft could search 62,500 square nautical miles in five hours. Here, two Curtiss F9C-2's approach Macon's *recovery trapeze in July 1933.* (Official U.S. Navy photograph, courtesy of the National Museum of Naval Aviation)

The development of the "task force" concept by the end of the decade marked a shift away from the battle line, but not the battleship, which remained the arbiter of war at sea. The transfer of three battleships to the Atlantic to bolster U.S. forces in May 1941 reflected its continuing dominance.[2] The European naval war reinforced the battleship's preeminence, especially the British operations which tracked, harassed, and sank the German battleship *Bismarck* in May 1941.

After the Japanese attack on Pearl Harbor in December 1941, the Pacific naval war remained firmly under the control of nonaviators until late 1943. Admiral Chester Nimitz, a longtime submariner and cruiser captain, and the former rear admiral commander of Battleship Division One, became the Pacific Fleet commander on 31 December 1941.[3] Nimitz's support for nonaviator commanders, such as Admiral Raymond A. Spruance, became the focus of aviator complaints. These criticisms came to a head when the escort carrier *Liscome Bay* was sunk, with 644 dead, while assigned to a defensive position off Makin Atoll in November 1943.

Many senior naval aviators, such as Vice Admiral John H. Towers, were ardent supporters of carrier-based aviation as the new technological foundation for the navy's guerre d'escadre strategy. They criticized Spruance's use of carrier aviation and advocated more offensive operations. According to his biographer, Thomas Buell, Spruance "believed the Japanese would be defeated primarily through amphibious warfare." Aviators, however, saw themselves as the new offensive warriors of guerre d'escadre and resented the "cost and drudgery" of protecting amphibious forces. Prior to operations in the Gilbert Islands in 1943, aviator Vice Admiral John S. McCain advocated carrier operations into Japanese waters to force a fleet engagement. Spruance advised Nimitz that the logistics requirements were too great, Japanese air power too strong, and the chance of a fleet engagement unlikely unless the Japanese saw a clear chance to win.[4] The Japanese would not be complaisant enough to "come and be killed," to paraphrase the French minister of marine Admiral Jackie Fisher had been fond of quoting. The post–*Liscome Bay* furor eventually led to a dual command structure in the Pacific in which aviators and nonaviators were teamed in the top two positions in every major Pacific naval command.

The destruction of a significant portion of the battleship navy at Pearl Harbor offers a seductive, but false, demarcation between the battleship and aircraft carrier-technological paradigms.[5] Clark Reynolds, in his history of the rise of carrier aviation, wrote that, after Pearl Harbor, the battleships "played a very minor part in the Navy's construction plans." In terms of numbers of hulls this was certainly true. However, Reynolds's attribution of the retention of the battleship to "conservative forces" ignores the battleship's continued relevance through the end of the war.[6]

Many of the weaknesses attributed to naval aviation by the conservative Naval War College staff in the late 1930s did not disappear with the Japanese or American aviation successes at Pearl Harbor, Coral Sea, or Midway. Naval aviation remained hampered by poor weather and darkness. There is ample precedence for weather adversely affecting battleship performance — Jutland in 1916 being the best example. Aircraft carriers, designed only to survive gunfire from cruisers, were extremely vulnerable to surface ship attack, especially at night or in bad weather. During 1942, the Japanese sent their surface ships carrier hunting during three of the four major battles that year.[7] U.S. carrier vulnerability to battleships continued in the Pacific until the last of the two *Yamato*-class superbattleships was sunk during the Okinawa campaign in April 1945.

On 7 December 1941, the two "1937" battleships (*North Carolina* and *Washington*) were in service and fifteen fast battleships had been authorized. By 1944, the four "1939" battleships (*South Dakota* class) and two of six *Iowa*-class fast battleships were with the fleet. Five superbattleships of the *Montana* class authorized in July 1940 were canceled in July 1943. This was done to free engineering design assets for the *Midway*-class large carriers (CVB) and to shift material and manufacturing resources needed for the broad spectrum of ship construction to support the United States's two-ocean war.

The nineteen months between Pearl Harbor and cancellation of the *Montana* class can be attributed to more than the residual momentum of the battleship technological paradigm. When the Pacific War began, the navy had seven aircraft carriers in service. In demonstrating their effectiveness to concentrate what Bradley Fiske termed "mechanical power" to bear at great distances during the major engagements of 1942—Coral Sea (May), Midway (June), Eastern Solomons (August), and the Santa Cruz Islands (October)—only three were left afloat by that autumn.[8] Aviator complaints about the lack of offensive use of carriers during 1943–44 by "black shoe" (nonaviator) admirals under Nimitz tended to minimize these losses and ignored the desperate night battles fought in the Solomons, where the effect of carriers was limited.[9]

By the end of the war, the chief of naval operations, Fleet Admiral Ernest J. King, was attributing the "successful application of our sea power" to the "flexibility and balanced character of our naval forces."[10] Surface forces, which before the war were defined myopically in terms of the battle fleet arranged around the battleship, were characterized in 1945 as "fast task forces comprised of aircraft carriers, fast battleships, cruisers, and destroyers."[11] The American experience during the Pacific Ocean War came closest to Admiral William Fullam's 1924 call for a "three-plane navy" capable of dominating above, on, and under the sea.

Fleet Admiral Ernest J. King's flexible and balanced navy of 1945 involved the refinement of aviation during the 1930s and then during almost four years of war. An important step in the evolution of naval aviation, as the navy moved toward the balanced fleet of 1945, was the termination of the promising lighter-than-air ship program.

STREAMLINING AVIATION: THE END
OF LIGHTER-THAN-AIR SHIPS

Interwar naval aviation was divided between traditional aircraft (classified by the navy as "heavier-than-air-craft") and lighter-than-air ships. The 1,000-plane navy act that followed the Morrow Board in 1926 included two rigid airships and an airship base on the Pacific coast. Airships and regular aircraft would both serve as scouts in support of the battleship. Airships were to act as land-based, independent, long-range scouts. Traditional aircraft scouted at shorter range and operated from aircraft carriers that required cruisers and destroyers for protection. Rear Admiral William Moffett, chief of the Bureau of Aeronautics, characterized the airship as an efficient scout cruiser for the battle line: "The rigid airship is primarily a scouting ship, the purpose of which is to travel long distances at high speed; carry observers who can see what is going on and report back by radio."[12] Unlike carrier-based scout-bombers, which could carry torpedoes and bombs, airships posed no direct threat to the battleship.

In many ways, the airship was a technology that was familiar, if not comforting, to battleship officers. As Douglas Robinson and Charles Keller pointed out, the airship was even perceived, not as an airplane, but as a flying ship:

> the big rigid airship, when compared with the flimsy and dangerous airplanes
> of the day, appealed to senior officers of Moffett's generation . . . [who] could
> identify with an aerial craft longer than a dreadnought, ponderous and stately
> like the ships they had grown up with at sea. As in surface ships, the airship crews
> stood watches . . . and it was the officer of the deck [not pilot] who gave orders
> to the helmsmen in the control car, also called "the bridge."[13]

The navy's airships were commissioned ships, for example, USS (United States Ship) *Los Angeles*, and the later versions even served as floating aircraft carriers, carrying their own complement of fighter planes to defend themselves during scouting missions. Unfortunately for its advocates, the navy's lighter-than-air program ran afoul of funding disputes in the wake of post-1936 naval rearmament. Ironically, these dreadnoughts of the air were done in by the leading member of the battleship Gun Club, Admiral William D. Leahy, and Rear Admiral A. B. Cook, chief of the Bureau of Aeronautics.

The navy airship program did not have a smooth start. Construction on the two airships authorized in 1926 was delayed almost two years. Moffett's reputation as a politician and a showman — he drove a golden rivet into the *Akron* in front of thirty thousand spectators — hindered the airship's acceptance into the fleet. Moffett emphasized public relations flights, which led many officers to dismiss the airships as "show" boats of little value. *Akron* was as large as a battleship, almost 800 feet long with a maximum design speed of 72 knots, and carried five Curtiss fighter/scout planes in a vertical hangar within her giant hull. Moffett's early emphasis on publicity flights delayed installation of *Akron's* aircraft recovery equipment and radio direction finding gear by the time the ship was called upon to take part in a scouting exercise in January 1932.

Akron's introduction to the Scouting Force was less than impressive.[14] She eventually located part of the fleet, but *Akron's* observers missed two destroyers, which had not failed to identify the huge target.[15] *Akron's* luck was no better in subsequent exercises. The chief of naval operations, Admiral William V. Pratt, wanted *Akron* to take part in joint fleet exercises in the Pacific beginning in March 1932, but *Akron's* tail was damaged while she was being removed from her hanger in Lakehurst, New Jersey. After two months of repairs, *Akron* proceeded to the West Coast.[16]

In June 1932 *Akron* participated in a small portion of the fleet exercises. Operating near Guadalupe Island, *Akron* located "enemy" cruisers, which launched their own observation/scout planes to attack the giant airship. Without her own aircraft, the airship could only present a passive defense. Ignoring the fact that *Akron* was filled with nonflammable helium, the scout-plane pilots reported that the airship was a sitting duck and could have easily been destroyed by their machine guns.[17]

Even though *Akron* worked only two of the thirty-eight days she was deployed to the West Coast, the navy's assessment remained optimistic. Airships remained attractive as high-speed scouts, especially against the Japanese navy in the western Pacific. Senior officers recognized the vulnerability of airships when confronted with the numerous aircraft carried by an enemy battle line. But in this, the airship was no different than a scout cruiser; neither was designed to engage strong enemy forces.[18]

Subsequent trials demonstrated *Akron's* ability, using two of her embarked aircraft, to scout a 250-mile-wide path at a speed of advance of 50 knots. With each aircraft carrying fuel for five hours, a remarkably large search area 62,500 nautical miles square could be covered — theoretically.[19] Further refinement of scouting techniques in support of the battle line was cut short when *Akron* crashed at sea during a storm off New Jersey on 3 April 1933, killing seventy-three out of seventy-six crew members, including Moffett.[20]

With the energetic and politically savvy Moffett dead, the airship fared worse at the hands of its detractors. *Akron's* sister ship, *Macon*, was commissioned three weeks after the crash of the *Akron*, but Franklin Roosevelt's new chief of naval operations (CNO), Admiral William Standley, was a staunch member of the "Gun Club" and cool to aviation and airships.[21] Standley ordered *Macon* to Sunnyvale to operate with the fleet (now concentrated in the Pacific) in order to determine the military value of the airship.[22] Admiral David F. Sellers, commander of the U.S. Fleet, continually assigned *Macon* to narrowly defined tactical scouting missions rather than to the long-range missions to which she was suited. Through the remainder of 1933, the airship supposedly was shot down several times by aircraft, cruisers, and even by the old battleships *New York* and *Texas*.[23]

In July 1934 Admiral Sellers condemned *Macon* for failing "to demonstrate its usefulness as a unit of the Fleet." Sellers recommended against further expenditure of funds for airships and called for seaplanes with a range of 3,000 miles.[24] Moffett's successor, another battleship-sailor-turned-aviator, Ernest J. King, considered Sellers's recommendation foolish. No existing aircraft, save the airship, had the capability to carry out a search over 500 miles from its base.[25]

When the sole surviving officer of *Akron*, Lieutenant Commander Herbert Wiley, assumed command of *Macon* in July 1934, the airship's performance improved markedly. Wiley delighted Franklin Roosevelt when aircraft from *Macon* delivered letters and newspapers to the president as he traveled aboard the cruiser *Houston* in the Pacific, 1,500 miles from the nearest land. In fleet exercises, *Macon* proved its ability to locate opposing ships and, in a bit of intra-aviation competition, even dodged an attack by dive bombers from the aircraft carrier *Saratoga*.[26]

The excellent performance of *Macon* converted some opponents of the airship. The General Board was so impressed that it recommended the immediate construction of a new training airship to replace the aging war-reparation zeppelin, *Los Angeles*. By the fall of 1934, Rear Admiral King had convinced Admiral Standley to allow the *Macon* to perform scouting services in support of Fleet Problem XVI to be held west of Hawaii.

The prospect of a bright future for the airship ended with the crash of *Macon* on 12 February 1935 off California. An exceedingly strong wind gust, acting on a known structural weak point, ripped off the upper vertical fin. Some of the metal framing punctured three gas cells in the stern, causing the tail to drop suddenly. To stay in flight, watchstanders dumped too much fuel and ballast, causing the airship to motor above her pressure ceiling. Automatic valves released helium to prevent the gas cells from rupturing in the low pressure. So much gas was released that the ship could no longer stay up. *Macon* struck the sea twenty-four minutes after the fin was lost. The navy had learned the value of carrying lifesaving equipment from the *Akron* disaster: only two men of the eighty-three on board died.[27]

The loss of *Macon* left the airship's strategic and tactical capabilities promising but unproven. The excellent demonstrations of the airship under Commander Wiley had convinced some senior officers that the airship program should be continued. The matter was referred to the General Board.[28] Airship veterans such as Commander C. E. Rosendahl defended the airship, but Rear Admiral A. B. Cook, chief of the Bureau of Aeronautics, was leery of expending limited aviation funds in pursuit of a peripheral technological trajectory. Cook told the General Board that "the building and operation of airships for solely military purposes is not justified. It is too costly." Cook advocated government aid to build commercial airships with "certain design features" which would allow the ships' crews to be augmented by "several naval officers" in time of war.[29]

In 1937 the General Board recommended the construction of another airship, equipped with aircraft, to continue the development of these flying aircraft carriers. President Roosevelt, perhaps recalling *Macon*'s newspaper delivery at sea in 1934, approved of the appropriation of $500,000 to start construction of a *Los Angeles* replacement in the spring of 1938. But the House Appropriations Committee deleted the training airship funds on 3 June 1938 since the testimony of senior naval officers had not shown "any desire or need for it."[30]

Acting Secretary of the Navy Edison warned Roosevelt that "exclusion of this item sounds the death knell of all lighter-than-air activity and development in this country."[31] Roosevelt referred the question of new airship construction to Standley's replacement as CNO, Admiral William D.

Leahy, writing, "we come back to the original question as to whether the expenditure of three million dollars is justified at this time, or should we not spend the three million dollars for some more useful purpose. A year ago the consensus of opinion in the [Navy] Department was adverse to building a lighter-than-air ship." Ignoring the General Board's recommendation, Roosevelt identified the "principal pressure" for a new airship as coming "from a very powerful lobby conducted by the rubber company [Goodyear] which is seeking to salvage fairly heavy speculative investment."[32]

Implicit in Roosevelt's memo was his opinion that money spent on airships precluded the acquisition of more "useful" technologies. Leahy disagreed with the secretary and assistant secretary of the navy as well as his own chief of the Bureau of Aeronautics, who had advocated a training airship. Leahy advised the president against future naval involvement with airships of any kind, writing that he did "not believe that [the] possible value of lighter-than-air ships for naval purposes justify at the present time the expenditure of $3,000,000 from the naval appropriation." Leahy thought that "all available funds can be used to better advantage for other purposes."[33] This was largely in line with Leahy's 1937 testimony before the House Naval Affairs Committee in which he reported that the "consensus of opinion of the high command of the Navy has been that up to the present time rigid airships have not been particularly useful."[34]

The testimony of aeronautical expert William F. Durand, professor emeritus from Stanford, contributed to that assessment and helped to turn the navy away from rigid airships. Durand had led the team investigating the crash of Macon, and informed the General Board that if the navy were to build another airship it should be a "flying laboratory" rather than an "adjunct to the fleet." The construction of this expensive flying laboratory would improve American airship design to preclude a repetition of the Akron and Macon tragedies. Durand told the Board that the Macon crash "led to the conclusion that the attack of gusts on airships or airship structure, and particularly the control structures, [indicated that] the seriousness of the consequences of such an attack have not been altogether fully appreciated."[35] Durand's back-to-the-drawing-board advice dismayed those who believed the Macon crash could be overcome with a quick-fix.

In addition to the general unwillingness to fund an expensive airship experimental program, Admiral Leahy saw no pressing need for airships. He expressed a classic argument for intraspecific weaponry development and

The battle cruiser origins of the large carriers Lexington *and* Saratoga *are evident in this October 1941 photograph of* Lexington. *The long flight deck allowed larger air wings and development of the concept of "full-deck strikes." The four 8-inch gun turrets, mounted to keep enemy cruisers at bay, can be seen to the left of the superstructure and to the right of the funnel.* (Naval Historical Center, 80-G-416632)

rejection of any technology not being pursued by other navies: "the Navy Department up to the present time has not felt it correct to expend our limited appropriation for construction on rigids, particularly in view of the fact that no other nation is developing rigid airships for naval use." However, Leahy worried — should a future airship gap develop — that "trained personnel will be dissipated" without rigid airships to operate. He offered no objection to an amendment to use some of the $15 million reserved for experimental vessels for rigid airships. Leahy held the line that the need for battleships to counter the Japanese buildup had first claim on nonexperimental funds.[36]

Roosevelt, swayed by Leahy's opinion, ended the navy's airship program. Like the *North Carolina* battleship contracts, the decision generated political complaints from the area most affected — Akron, Ohio. Ohio con-

gressman William Thom appealed to Roosevelt to go ahead with the airship program, as it would mean "much in the way of relief for [the] unemployment situation in Akron."[37] To minimize any political fallout, Roosevelt directed the Navy Department to develop plans and specifications
for a rigid airship and informed the Ohio congressman that bids would be
invited in the "near future."[38] Although legislation was introduced for civilian airships for transoceanic travel in 1940, no airships were ever constructed. The navy continued to dabble in blimps, smaller nonrigid airships
primarily used for antisubmarine and coastal patrols. However, the navy
program in rigid airships, despite Roosevelt's inferences to the Ohio congressional delegation, had ended with the crash of *Macon*.

In retrospect, the airship should have succeeded within the navy. Its
scouting capability supported the navy's battleship strategy, and its potential to scan 62,500 square nautical miles in five hours versus approximately
6,000 square miles for a scout cruiser gave the airship a marked advantage.
Unlike the aircraft carrier, with its dive bombers and torpedo planes, the
airship also posed no threat to the battleship.

Secretary Edison was correct that the death knell for the airship had
sounded in 1938. But Roosevelt, who took such an interest in the day-to-
day operation of the navy and its technology, was a battleship sailor at heart.
His deference to the CNO, Admiral Leahy, and cancellation of any future
airships led to a reallocation of airship funds to more traditional expenditures on the battle fleet — expenditures considered critical to regain an ad-
vantageous warship ratio vis-à-vis Japan. Leahy's action lends some credence to Herbert Hoover's belief that if the navy's admirals had their way
they "would spend their entire income on battleships" or at least on traditional supporting technologies.[39] However, the end of the airship program
had some help from the chief of the Bureau of Aeronautics. Streamlined
naval aviation increasingly developed into a presumptive anomaly capable
of successfully challenging the battleship technological paradigm.

IMPROVEMENTS IN NAVAL AVIATION

During the 1920s, many naval aviators saw themselves fighting the battleship establishment for operational independence as a means to develop a
viable aviation presumptive anomaly.[40] Their revolutionary spirit continued through the 1930s, and their efforts became more sophisticated and

fruitful as aviation technology improved. During the interwar years, scouting aircraft evolved into scout bombers and torpedo planes. Dive bombers — judged more accurate and lethal — came to dominate aviation tactics.[41] Air power zealots predicted a future in which aviation would eclipse other weapons.[42] However, compared with many army officers, naval aviators tended to be less radical.

Aircraft carriers, like battleships, were limited by naval treaties. Of the 135,000 tons the United States was allowed under the Five-Power Treaty, the large, battlecruisers-turned-carriers, *Lexington* and *Saratoga*, took up 66,000 tons. The navy debate during the 1920s centered on how best to use the 69,000 tons remaining: build three carriers displacing around 23,000 tons or six smaller carriers displacing half that size. The issue was confused further by a lack of consensus on how carriers would be employed — for example, in close support of the battle line or as advance scouts.[43] Employment affected speed requirements, and higher speed drove displacement up significantly. If operated close to the battle line, high speed became very important since a carrier had to steam into the wind for approximately thirty minutes to launch her air wing and sixty minutes to land it later. Between these evolutions, the ship would then have to rejoin the battle fleet at high speed since she was vulnerable to attack by enemy capital ships and cruisers. High speed also became necessary for carrier operations independent of the battle line.[44] During the late 1930s, independent carrier operations came to dominate U.S. naval thinking since carriers, tied to the battle line, were believed to be easy victims of enemy air attack.

The size versus numbers trade-off was reminiscent of the Sims–Mahan debate over more small mixed-battery ships versus fewer large all-big-gun battleships. Unfortunately for the bureaus and the General Board during the 1920s, no experience similar to the Russo-Japanese War offered guidance. In mid-decade, the chief of the Bureau of Construction & Repair, Rear Admiral J. D. Beuret, observed that smaller carriers would provide "greater deck area in the aggregate" and "greater hangar space." Small carriers also had the advantage of distributing airplanes "more widely, if that is a tactical advantage." He also pointed out that, as with battleships or any other type of ship, "larger vessels are more seaworthy" (recall the British squadron lost under Craddock off the Coronelles in 1914).[45]

The lack of operational experience with carriers during the 1920s complicated the decisions about carrier size and the number of aircraft needed. According to Beuret, "We do not know . . . how many airplanes can be delivered to the air for offensive purposes from a deck of a given area; nor do we know what should be assumed as a rate of wastage [number of replacement planes to be carried]."[46]

The result of the navy's carrier debate during the 1920s was the first American carrier designed from the keel up — the small, 13,800-ton *Ranger*. Her design began in 1922 and was solidified in 1926 for the fiscal year 1929 Program, and she was commissioned in 1934. After *Ranger*, the treaty limits allowed the United States to build carriers displacing 55,200 additional tons. The General Board considered four more *Rangers*, three 18,400-ton carriers, or two carriers of 27,000 tons. Pairs of similar carriers were considered tactically superior and, in September 1931, the General Board recommended construction of two 20,000-ton carriers (*Yorktown* and *Enterprise*) in fiscal year 1933 and one 15,200-ton carrier (*Wasp*) in fiscal year 1934 to operate with the smaller *Ranger*. These ships would then bring the navy up to its treaty limit for aircraft carriers.[47]

The small *Ranger* and *Wasp* proved ineffective. *Wasp* was sunk in 1942, and *Ranger* was relegated primarily to noncombat missions during the war. As with battleships, increased offensive and defensive capabilities required large size. Operations with *Lexington* and *Saratoga* led aviators to favor "full-deck" strikes — all airplanes spotted on deck, warmed up, and launched in a quick sequence. By 1938, the standard prewar air group in *Saratoga* consisted of four squadrons: one dive bombing, one scout-bombing, one torpedo, and one fighter. By 1939, a fifth squadron, a fighter squadron, was added to improve fighter defense of the carrier and defense for the air group during strike operations.[48] These larger, more capable air groups required longer flight decks and bigger carriers.

Although increasingly useful, the aircraft carrier's continued subordination within the battleship technological paradigm was reflected in resource allocation. The *Yorktown* design was a compromise and outdated by 1938. Little could be done to improve *Hornet* since design of the *Iowa*-class battleships was given priority in July 1938. The carriers that carried the brunt of the war in the Pacific were the *Essex* class, slightly improved over the *Yorktown/Hornet*. Improvements in carrier design were precluded by the need for more carriers sooner rather than better carriers later.[49] Yet, unlike Homer Poundstone's battleships, aviators did not consider extremely large carriers as the ultimate goal. There was a great deal of resistance within naval aviation to large carriers of the *Midway* class.[50] The complexity of carriers also required a great deal of engineering. For example, the *Essex* class, which bore the brunt of the Pacific War, required 9,160 plans (blueprints) compared to 8,150 plans for the larger *Iowa*-class fast battleships.[51] Ten *Essex*-class carriers were appropriated in 1940 and two more in December 1941.

INTERWAR DOCTRINE

As its technology and proficiency improved during the 1930s, naval aviation also enjoyed increased capabilities. The questions facing the naval profession were how naval aviation fit within the existing naval strategic paradigm of guerre d'escadre and whether it posed a presumptive anomaly sufficient to challenge the battleship technological paradigm. Fleet exercises were the experimental media for evaluating strategic and tactical anomalies. Usually held annually, fleet exercises were developed by the commander-in-chief, U.S. Fleet, who set up the hypothesis. Some fleet commanders gave naval aviation more latitude, allowing its strengths and weaknesses to surface. However, experimental replication, so critical to scientific validation, was impossible for fleet exercises. Umpires, themselves naval officers, assessed the exercises, assigned damage, and determined victors. Their judgments often met with scorn from the participants and feelings that the exercises were unfair.[52]

The Naval War College in Newport, Rhode Island, was the repository for strategic orthodoxy and the most conservative institutional defender of the navy's technological paradigm.[53] As Thomas Buell observed, "The fleet and not the college developed the strategy and tactics for air warfare in the Pacific."[54] There was a time lapse between strategic and tactical developments within the fleet and their incorporation into the War College curriculum. Presumptive anomalies remained at the periphery, hoping at best for a "nihil obstat" but never an "imprimatur" from the War College. Future admiral Richmond K. Turner was one member of the 1930s War College staff who dynamically espoused future naval warfare based on amphibious assaults and aviation. However, according to Buell "few listened. The battleship was still supreme at Newport."[55]

A record of more dynamic exchanges of ideas could be found in the pages of the U.S. Naval Institute's *Proceedings*. There, strategic and tactical dogma could be debated, and anomalies advocated, without real fear of institutional sanction. With this proviso in mind, one can glimpse the professional dialectic of the 1930s American naval profession.

In 1932, Commander Ralph Parker, a surface ship officer, wrote a satirical article in which a caveman, Ug, eliminated his Neanderthal opponent, Wok, by "caroming a rock off his ear at five paces, before Wok could close to fair swinging distance with his flint-headed mashie-niblick."[56] Pointing out that success in naval warfare goes to whomever delivers the most explosive on target the quickest, Parker attacked the battleship thought collective, those "die-hards of the old school who visualize naval warfare as a sort of fistic entertainment for which they have bought tickets, and which, after a few preliminary rounds between planes, submarines, and such ham-and-eggers, is bound to end with the championship bout between the heavyweight craft, or else money refunded."[57]

Parker warned of the new potency of aviation, but the technological ceiling predicted by aeronautical experts during the 1924 General Board hearings had become embedded in the upper levels of the profession and influenced the strategic and tactical doctrines taught to up-and-coming officers at the Naval War College. In 1937, the year that Franklin Roosevelt canceled the navy's airship program and the navy's first metal monoplane entered service, the officer students at the War College were told that aircraft would "affect naval operations to at least the same extent that in the past they have affected land operations [i.e., twenty years earlier during the World War]," but the war at sea would still be decided by a battleship duel.[58] The War College predicted that the combined air forces of the army and navy would swell to at least fifty thousand aircraft during any future war. In order to utilize this large air fleet properly in support of the battleship, the faculty informed the student officers of the good and bad points of naval aviation. Aviation could act quickly at great distances with massed forces, but was hampered by unfavorable weather, required highly trained personnel, and was "unable alone to accomplish unlimited military results."[59] The inference that the battleship could, on the other hand, achieve "unlimited" results was understood.

Defending the fleet against aerial attack might prove difficult. A prescient 1937 article in the Naval Institute *Proceedings* by Lieutenant Commander Logan Ramsey, an aviator, postulated torpedo plane attacks on fleets at anchor.[60] The photograph accompanying the article, showing the battle line anchored in a row in open water off the California coast, underscored the vulnerability of battleships away from their home port. Ramsey's article foreshadowed the carrier attack on Pearl Harbor during Fleet Exercises XVIII and XIX in 1937 and 1938.[61] Anxiety over this undoubtedly contributed to U.S. Fleet commander Admiral James O. Richardson's October 1940 challenge to President Roosevelt's retention of the battle fleet in the poorly defended and ill-supported base at Pearl Harbor. Roosevelt fired Richardson for his complaints. His replacement, Admiral Husband E. Kimmel, initially trusted Pearl Harbor's shallow depth to counter aerial torpedoes. Defense against dive bombers would be problematic.[62]

Just as the castle required expanded outer defenses to protect it from cannon, the battleship required rings of cruisers and destroyers and close support from aircraft carriers. The unique threat posed by aviation, when compared to other means of "naval action," was its ability to create "overlapping zones of hostile activity," that is, its flexibility and depth of operations.[63] In a back-handed compliment to the increasing power of carrier aviation, the War College faculty identified aircraft carriers as the primary targets for a battleship fleet desiring freedom of action.[64] Aviation's primary strategic mission was the destruction of enemy naval air power and the infliction of crippling damage on an enemy fleet in order to bring it within gun range of U.S. battleships.[65]

The tactical missions assigned to naval aviation underscored its subordination to the battleship. The primary tactical mission of the aircraft carrier was "to strike the enemy carriers as soon as possible."[66] Aircraft carriers also were charged with maintaining an aviation zone of control in support of the battle line.[67] In other words, aircraft would counter the threats posed by all nonbattleship technologies. Carrier aircraft were important for the early discovery of enemy destroyers and submarines, so that the battleship force could maneuver to avoid their attack.[68] In reality, aviation was not spending most of its flying time in battleship defense. According to Rear Admiral A. B. Cook, chief of Bureau of Aeronautics in 1937, carrier aircraft were spending "at least 70 to 75 per cent" of their flying time in support of battleship "gunnery work." Even Cook seemed amazed, telling the General Board that "the amount of flying hours for gunnery is terrific."[69]

In general, the orthodox — paradigmatic "normal" — mission for naval aviation during the late 1930s existed only to get the battleships into action. The gun remained the "most effective weapon for sustained action" because of its "great protection against destruction, and because of its great

range, accuracy, rapidity of firing, hitting power, and ammunition supply."
The "tactical effort in a fleet engagement" was "centered around the main
gun action, its aim . . . to bring about such coordination of effort of all
weapons that, by their concentration, they will destroy the enemy fleet."
Other weapons such as aircraft and submarines "would be brought into the
engagement to aid the gun or to take advantages of situations created by
it."[70]

The primacy of the gun subordinated submarines as well. The War College acknowledged that the "perfection of new instruments of warfare has
often threatened the conduct of naval warfare."[71] Yet prejudice against un-
dersea weapons kept the submarine a second-line weapon. War College
doctrine stipulated that submarines themselves could not control the seas,
but acknowledged that they could make control of the seas more difficult.
The presence of submarines was estimated to reduce a fleet's radius of action by almost one-third, due to the greater fuel consumption required by
high-speed, zigzag steaming to thwart submarine attacks.[72] This reduction
in steaming radius seriously affected U.S. planning for the long-range war
against Japan.

Submariners saw themselves as adherents to the guerre d'escadre strategy and supportive members of the battleship technological paradigm. The
primary mission of the submarine was as a covert, advanced scout that
would detect, report, and harass the enemy battle fleet. Pre-atomic-
powered submarines such as these were basically surface ships able to submerge for a short period of time. Their surfaced speed was relatively slow
and their submerged speed very slow and limited by the electrical capacity
in their rechargeable propulsion batteries. Submarines were vulnerable to
discovery by aircraft and, once driven underwater, had a hard time keeping up with higher speed surface ships.

Like aviators, submariners considered themselves most effective when
not tied directly to the battleship force. Submarine officers also lobbied
hard for increased speed for the fleet submarines of the 1939 Program during testimony before the General Board in October 1937. Commander
Charles A. Lockwood, later the vice admiral commanding the submarine
war against Japan, reported that his experience during the previous two
fleet exercises and "eight or nine fleet tactical periods during the last two
years" led him to conclude that submarines required even higher surface
speed.[73] Lieutenant Commander R. H. Smith underscored Lockwood:
"We are frequently forced down [by aircraft] in daytime. The demand is to

regain our position during night. High speed is essential for that. I was attacked by aircraft six times, having been stationed ahead of the fleet, and was forced down and back of our own battleships about 500 yards from our the main body. It was impossible to get back into position." Smith also complained of being tied too closely to the battle line by the fleet commander. Smith advocated a submarine scouting line 200 miles to 300 miles ahead of the advancing battleship force rather than at 20 miles, as in this failed exercise.[74]

SHIFT TO THE "TASK FORCE"

The mission-oriented task force became the organizational basis for operations against the Japanese during the Pacific War. Task forces, composed of fast carriers and fast battleships, destroyed the Japanese navy and made amphibious operations possible. In 1922 the navy was reorganized into four major task-oriented forces: Battle Fleet, Scouting Fleet, Control Force, and the Fleet Base Force. In 1931, the four forces became the Battle Force, Scouting Force, Submarine Force, and Base Force. During the 1930s, debate focused on whether the navy should be organized by type of ship or by mission.[75] A revised task force concept failed to win endorsement by the secretary of the navy when advanced by the General Board in 1937.

The task force issue resurfaced in 1940 during congressional debate over the reorganization of the navy. Rear Admiral Ernest J. King—former chief of the Bureau of Aeronautics, carrier commander, commander of Aircraft, Battle Force, and future commander-in-chief, U.S. Fleet, and chief of naval operations during World War II—wrote the General Board plan. King's goal was to "relieve the Commander-in-Chief, U.S. Fleet, and the commanders of major forces of the Fleet from administrative details in order that they may have more time to devote to the study and preparation of war plans, and the study of strategy and tactics."[76]

The General Board report provides an interesting glimpse into senior officers' perceptions of the fleet mission and technological organization on the eve of American entry into World War II. The Board distinguished between "operational" and "type" commands. The latter were to function as administrative, training, and maintenance commands to facilitate the "operative command" of the fleet. The "Fleet" was comprised of the "seagoing armed forces" whose "primary functions" were "to obtain, to maintain, and to exercise control of vital sea areas and sea-ways." The fleet would accomplish its missions in "either or both of two ways: by defeat of the enemy fleet in battle or by cutting vital enemy sea communications while maintaining its own." The "operating commands" of the navy included the commander, Scouting Force; the commander, Battle Force; and the commander, Base Force. According to the General Board, this organization of the fleet into "so-called permanent task forces" — the Battle Force and the Scouting Force — corresponded "somewhat in capabilities but not in practice with the two combatant operative functions" listed above. The Board complained that the task force concept had been used in a "somewhat indefinite way" for administration and "more recently, [in] some aspects of tactical development."[77]

The General Board, through King's report, clarified the role of the Battle Force and Scouting Force and freed the aircraft carriers from the battleships. The Battle Force, made up of battleships, light cruisers, and destroyers, was the "core" of the fleet. Submarines, patrol planes, and aircraft carriers had to be placed in "appropriate relative positions to support and to cooperate with this core; the same is, in general, true of heavy cruisers." Submarines and patrol planes were complements in an "advance force" in the "service of information." The section freeing aircraft carriers from the battle line was based on King's proposal in early 1938 to combine the high-speed carriers with the fast cruisers and destroyers of the Scouting Force:[78]

> Carriers, even when engaged in battle operations, cannot be held to any positive position in relation to the Battle Force, and so usually require cruiser protection readily afforded by appropriate numbers of heavy cruisers. This association of heavy cruisers and carriers is even more appropriate for the employment of carriers in detached or advance operations. In fact, they constitute a natural combination under all likely circumstances, as a "support force" or "striking force," whether to cooperate with the "advance force" or to reinforce the Battle Force, or even to operate independently.[79]

The General Board report further refined naval aviation as a presumptive anomaly based on its increasing offensive power.

Although freed from its ties to the battleship, aircraft carriers would also work with newer, faster battleships. The two 1937 and four 1939 battleships were designed when the naval treaties were still in force. The displacement limit of 35,000 tons forced a trade-off between adequate armor protection or speed over 30 knots. In keeping with American past practice that favored large guns and tough armor over speed, these first six battleships had a top speed of 27 knots, significantly faster than any previous American battle-ship. The 45,000-ton *Iowa*-class battleships had room for propulsion machinery and adequate armor for a full speed of 33 knots.[80] These "fast" battleships, in conjunction with new "fast" carriers, were intended to form the basis of the U.S. Navy during the 1940s. Fast carriers were also intended to operate against Japanese lines of communication and surface raiders while being protected by strong cruiser and destroyer forces.[81]

"NORMAL" WAR, 1939–1941

During the twenty-six months of war prior to the Japanese attack on the United States at Pearl Harbor, naval combat had conformed largely with the prewar battleship technological paradigm. Aviation, especially, had not departed from the role envisioned in prewar studies. In fact, air attacks against warships had not proven as disastrous as predicted. German bombers had a hit rate of only 0.7 percent against British warships during the 1940 Norway campaign, far below the dismal hit rate of Dewey's squadron against the mostly stationary Spanish ships at Manila in 1898.[82] In June 1940 the German battlecruisers *Gneisenau* and *Scharnhorst* reaffirmed the supremacy of the gun when they sank the British aircraft carrier *Glorious*. In the May 1941 search for *Bismarck*, the Royal Navy used aviation to locate and harass the German battleship.[83] The pursuing British battleships caught *Bismarck* when a carrier-based plane torpedoed her, jamming her rudder so she could only steam in circles until she was sunk.[84] This operation typified the prewar role of aviation within the battleship technological paradigm — carrier aviation influencing but not decisive in a battleship main event.

One potent, anomalous demonstration of naval air power occurred in November 1940 when British torpedo planes attacked Taranto and sank three older Italian battleships. Battleship defenders attributed the success of the attack to the laxity of the Italians and the antiquated designs of the battleships that were sunk. Taranto also was explained in ethnic terms — presaging subsequent historical presentations of inherently superior Anglo-

Saxons over Samuel Eliot Morison's "dago navy," an enemy Franklin Roosevelt dismissed as a "bunch of opera singers."[85]

American naval officers' pre–Pearl Harbor view of their Japanese counterparts was mixed. Intelligence reports that down-played the sophistication and ability of Japanese aviation may have reinforced low opinions of aviation in general. Many American naval officers probably agreed with the low British opinion of Japanese naval aviation, that is, that it ranked "similar to that of the Italians."[86] Another common belief among Royal Navy officers was that the epicanthic fold of skin, surrounding Japanese eyes, prevented their pilots from shooting straight or seeing in the dark.[87] Along the same lines, the U.S. Naval War College presented a series of lectures by civilian anthropologists during 1937 on the racial characteristics of the Japanese, a feature believed significant in developing appropriate tactical doctrine.[88]

On a certain level, the U.S. naval officer corps shared the British racial prejudice toward the Japanese navy and its technology. In general, American naval officers viewed the Japanese navy as a formidable opponent. Nevertheless, the self-referencing nature of the American naval profession allowed the navy to succumb to the "not-invented-here" syndrome. The Bureau of Ordnance, for example, refused to acknowledge "impeccable" intelligence regarding the Japanese "Long-Lance" — a 24-inch diameter, oxygen-propelled torpedo, with a 1,200-pound warhead (versus 800 pounds for U.S. torpedoes) capable of high speed and long range. According to Rear Admiral Arthur McCollum, "the tendency was to judge technical developments on the basis of our own technology and on the assumption that our technology was superior to any other. So if something was reported that the Japanese did have and we didn't then, obviously it was wrong. . . . They [Bureau of Ordnance] only came around to our point of view when the Japanese started blowing the tails off our cruisers down in the Solomons."[89]

Bureaucratic dismissal of any technology gap was complemented by the necessity to refrain from minimizing any foreign threats when Congress was considering naval appropriations. The dichotomous presentations of enemy capabilities continued through the cold war. Maximizing the threat served to engender larger appropriations but had to be countered within the navy to maintain morale. It would be incorrect to attribute the destruction of the battleship fleet at Pearl Harbor to American naval officers' dismissal of the Japanese navy as a minimal threat. But it is probably safe to say that there was more respect for Japanese battleships, battlecruisers, and cruisers within the U.S. Navy than for Japanese naval aviation.

THE PACIFIC OCEAN WAR AND THE RISE OF AVIATION

The Japanese attack on Pearl Harbor on 7 December 1941 marked the first demonstration of the efficacy of the aviation presumptive anomaly. This was bolstered by the sinking of the new British battleship *Prince of Wales* and the battlecruiser *Repulse* by Japanese aircraft off Malaya on 10 December 1941. Ten of the seventeen superbattleships on order at the time of the Japanese assault were completed and entered war service. Eventually, naval aviation's capability restricted the battleships' dominance to combat at night or during weather that precluded flight operations.

In December 1941 the burden of the Pacific War shifted immediately to the aircraft carriers, the only major combatants capable of effective offensive action against the Japanese navy. The aircraft carrier was on its way to becoming the navy's new capital ship, and the prewar strategy that envisioned a decisive confrontation between opposing battleship fleets was replaced by aviation duels, such as the Battle of the Coral Sea in May 1942, fought by fleets that never came within sight of one another. In the Pacific Ocean War, carrier-based torpedo and dive bombers destroyed enemy ships by bringing their "mechanical power" to bear through the medium of the air, as predicted by Bradley Fiske. Overall, the battleship increasingly became a subordinate technology devoted to the protection of the dynamically offensive aircraft carrier. In an ironic reversal of roles, the new, fast battleships spent more time as bodyguards protecting the carriers against night attacks by enemy capital ships. But the new, hierarchical relationship between battleship and aircraft carrier was a near-run affair dependent on the rate of development of early warning radar and proximity fuzes for antiaircraft projectiles.[90]

The Japanese attack on Pearl Harbor demonstrated that the older, slow superdreadnoughts were obsolescent. Nevertheless, they were not irrelevant. The six surviving battleships on the West Coast were no help in the Pacific in early 1942. In response to Admiral King's prompts to attack, Nimitz replied that "offensive employment of battleships does not fit in with hit-and-run operations and their independent or supporting use [is] precluded by lack [of] air coverage and antisubmarine protection. Such employment considered inadvisable at present. . . . Unless this fleet is strengthened by strong additions, particularly in aircraft, light forces, carriers, and fast fleet tankers, its effectiveness for offensive action is limited."[91]

The demonstrated lethality of naval aviation had established the aviation presumptive anomaly but had not ended the battleship technological paradigm. When the remarkable battleship salvage operations began at Pearl Harbor, the two "1937" battleships, *Washington* and *South Dakota*, were ready to enter the fleet and would soon sink two Japanese capital ships in night battles during the 1942–43 Solomon Islands campaign.

Speed became the criterion for battleship employment in the Pacific. The Solomons operations involved regular surface actions, usually at night when air operations were limited. Deployment of the limited number of battleships depended on the type of battleship available and the expected action.[92] The Central and Southwest Pacific offensive strikes involved carriers in consort with fast battleships or cruisers. As Wayne Hughes pointed out, the dilemma facing U.S. commanders before fast battleships arrived in the Central Pacific in 1943 was whether to pursue a Japanese force after dark with aircraft carriers and risk running into Japanese battleships or cruisers. By 1944, fast battleships were operating with the carriers and the question was then whether to keep the battleships with the fast carriers for mutual, day and night defense, or to let the battleships operate on the offensive independently.[93]

The small number of aviation admirals meant that nonaviators commanded the carrier task forces early in the war. This became a source of complaint for senior aviators, who chafed at the continued control of naval operations by adherents of what they considered a passé technological paradigm.[94] Vice Admiral John H. Towers, former chief of the Bureau of Aeronautics and commander, Air Force, Pacific Fleet, led the assault at Nimitz's headquarters in Pearl Harbor. The Battle of the Coral Sea in May 1942, fought with carrier aircraft, was the first naval engagement in which the ships of the opposing fleets never sighted one another. Towers thought it significant that the only carrier sunk (*Lexington*) was under the tactical command of a nonaviator.

The following month, three American carriers, operating under the command of two battleship admirals, Frank Fletcher and Raymond Spruance, inflicted a stinging strategic and tactical defeat on the Imperial Japanese Navy near Midway. Despite this, senior naval aviators in the Pacific, such as Admiral Towers, criticized Nimitz for not turning the carrier war over to naval aviators. Towers complained of a lack of aggressiveness and proper use of aircraft carriers, and even argued unsuccessfully that nonaviator command over aircraft carrier task forces violated the post–Morrow

Board legislation, which restricted command of aircraft carriers to avia-tors.[95]

By 1943, Towers was disparaging nonaviators with terms such as "battle-ship gang" and "black shoe sons of bitches" to his staff and subordinate avi-ators. Towers even referred to Nimitz's nonaviator planning officer, Cap-tain Bob Steele, as "Boob Steele." The aviators in Towers's quarters referred to the black shoe admirals as "buffalo hunters" reflecting their "outdated" war-fighting mentality. Nimitz was referred to as "Uncle Sink," a play on the acronym for his position as commander-in-chief, Pacific (CinCPac).[96]

Aviator complaints became pointed during the planning for Operation Galvanic, the seizure of the Gilbert Islands in the summer and fall of 1943. Towers spearheaded the aviation admirals' criticism of Spruance's planned use of carriers as part of the defensive screen to protect amphibious opera-tions. The catastrophic loss of the escort carrier *Liscome Bay* off Makin to a Japanese submarine-launched torpedo killed 644 of her crew and un-derscored the vulnerability of carriers, especially the smaller ones. Towers's views were representative of other senior aviation officers in the Yarnell Report.

On 6 November 1943, on the heels of the Gilbert Islands operations, re-tired Admiral Harry E. Yarnell submitted his "Report on Aviation" to the secretary of the navy. An aviator, Yarnell had been recalled to active duty in 1943, and the vice chief of naval operations, Vice Admiral Frederick Horne, authorized him to conduct a poll of naval aviators' feelings about their role in the war. The general tenor of the responses expressed anger over their perceptions of the continued suppression of naval aviation by nonaviators. Rear Admiral Fred Sherman had argued for a complete para-digmatic shift, lecturing Nimitz that "surface ships must be integrated and coordinated with naval air forces, rather than the reverse. . . . Naval avia-tors, as officers trained and experienced in naval aviation, should have the dominant voice in determining all naval policy and not just naval aviation policy." Sherman called for an aviator monopoly of the positions of com-mander-in-chief, U.S. Fleet, chief of naval operations, and commanders of all major fleets among other key positions.[97]

Yarnell's report to the secretary of the navy on the "means by which avi-ation may be made a more efficient arm of the fleet" was based on 127 let-ters responding to his survey.[98] Yarnell reported that naval aviators "have been steadfast in their loyalty to the Navy, and in their belief that aviation is an integral part of the Navy and an arm of the Fleet." In a statement largely true of regular officers in 1943, but one that would ring hollow in the postwar navy, Yarnell opined that "naval aviator's [sic] are seamen as well as airmen and have always prided themselves on that fact." Aircraft had "extraordinary powers of offense" and complex limitations imposed by "radius of action, wind, weather, and availability of landing areas." Because of this, Yarnell maintained that for operations "in which air predominates, the commander should be an aviator, or should have an aviator as Chief of Staff."[99]

Yarnell also argued for a restructuring of the naval profession by ending the distinction between regular and reserve commissions. At the time of his report, there were 3,275 regular and 19,686 reserve aviation officers in the navy. Some reserve officers would choose to remain in the service beyond the war, and Yarnell warned of the need to set up a method to retain only the best. In a statement reminiscent of the line's fear of engineering contamination in 1882, Yarnell worried about the effect of too many reservists in the postwar naval officer corps. The number of regular aviation officers in the ranks of ensign and lieutenant (junior grade) was small — 892 regulars versus 17,459 reserve officers. Yarnell wanted to maintain the purity of aviation officer corps and wanted Naval Academy graduates sent immediately to Pensacola for flight training so they would be competitive for promotion, having had "aviation battle experience before the war is over."[100]

Yarnell's recommendations proposed a recasting of the naval profession. If it proved impossible to have the commander-in-chief, Pacific, be an aviator, Yarnell proposed that he and all fleet and force commanders have "aviators of adequate rank" on their staffs. Yarnell also suggested establishment of an "effective air indoctrination school" for nonaviator regular officers. Undersecretary of the Navy James Forrestal, a former reserve naval aviator and supporter of aviation in general and Towers in particular, took an interest in Yarnell's report. To forestall Forrestal's intrusion into the high command structure, King and Nimitz agreed to the aviator/nonaviator command mix and moved Towers over to the new position of deputy commander-in-chief, Pacific. Towers's appointment placed him in line to replace Nimitz after the defeat of Japan. As Clark Reynolds observed, Towers's appointment meant that "naval aviation would at last gain control of the navy."[101]

One of Yarnell's most interesting proposals undermined the "every officer a seaman" concept. Instead of command of a large ship as a prerequi-

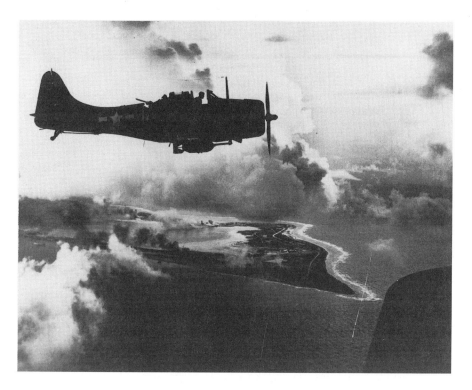

A viable naval aviation presumptive anomaly required an effective dive bomber. The Douglas SBD Dauntless, seen attacking Wake Island in December 1943, filled the role and entered squadron service in December 1940. (U.S. Navy Photo by Charles Kerlee, courtesy of the National Museum of Naval Aviation)

site for promotion to rear admiral, Yarnell offered command of a fleet air wing or service as chief of staff to an "area commander."[102] This new measure of professional competence was an important attempt to redefine the profession and, eventually, the U.S. Navy's technological paradigm. If enacted, Yarnell's recommendation might have saved the navy from subsequent collisions and groundings involving aircraft carriers commanded by nonseamen aviators.[103]

The aviator chief of naval operations and commander-in-chief, U.S. Fleet, Admiral Ernest J. King, dismissed most of Yarnell's recommendations. King thought aviators' special claims to senior commands improper: "All line officers, whether qualified as aviators or not, must be fitted for high command by being given instruction and experience in all arms." King agreed that the principle stated by Yarnell that "naval operations must be carried out *under one command* by officers trained in the *use of all arms*" was the "basis of naval organization . . . it has been proven abundantly in the current war that in seagoing operations, all craft — including aircraft — must be integrated into a team. This principle holds true from the smallest unit to the largest fleet [original emphasis]."[104] The war brought the question of who should be running the team to the fore.

King stood firmly against any move to replace Nimitz with an aviator: "The question of air representation on the Pacific Fleet staff has been left to Admiral Nimitz's discretion. An officer who bears the responsibilities which he does should not be circumscribed in selecting the members of his staff."[105] King also pointed out that no "rigid rule" concerning "capital ship command at sea" existed as a requirement for flag rank. King, the "triple threat" pilot, submariner, and surface ship officer, observed that "officers of wide experience are those best qualified for high rank and, consequently, officers are given such opportunities as can be provided to diversify their duties, particularly in command."[106]

According to Thomas Buell, King received concurrent advice from Vice Admiral John S. McCain, the new deputy chief of naval operations for air. Although King considered McCain "not very much in brains," he was prophetic, telling King: "Steam marked the passing of the profession of sail, and wiped out a profession. I do *not* share your apprehension that air marks the passing of the trained and experienced non-aviator, nor does it deny him adequate command outlets for his professional qualifications . . . [surface ships] are now, except the submarine, the auxiliaries of air. This is a patent fact, and like all facts, should be freely acknowledged."[107]

The dominance of naval aviation within the Pacific and its blunting of the German submarine threat in the Atlantic cleared the way for the rise of aviation admirals and the establishment of a naval aviation technological paradigm. One wartime expression of this shift was King's extreme break with naval tradition. He stipulated that the aviation admiral commanding a task force, even if junior to the commander of the surface forces operating with him, would be the officer in tactical command. Perhaps the most telling move away from a battleship-officer naval hierarchy was the promotion of aviators to flag rank. King expended a great deal of effort to limit the number of admirals in the navy, regardless of specialty. However, of the forty officers promoted to rear admiral between November 1942 and August 1943, twenty-six were aviators — almost a two-to-one advantage.[108]

The postwar command dominance of the naval aviation thought collective was clear when James Forrestal, now secretary of the navy, told the *New York Times* in December 1945 that the "Navy is becoming an air Navy." The war had drawn King beyond retirement age. Forrestal preferred an aviator as the next CNO, and fought hard to keep Nimitz from the job. Forrestal eventually accepted Nimitz, but held him to a two-year term. Spruance replaced Nimitz as Pacific Fleet commander, but soon went to the presidency of the Naval War College, at his own request, to serve until retirement. Towers replaced Spruance as fleet commander in the Pacific in February 1946. Aviator Admiral Marc Mitscher moved up to command the Atlantic Fleet. The reorganization of the office of chief of naval operations, tied to King's retirement, absorbed the position of commander-in-chief, U.S. Fleet. In the CNO's office, the vice CNO and two of the five admirals serving as his deputies were aviators. The special position of deputy chief of naval operations (DCNO) for air underscored aviation's supremacy. No comparable DCNOs for submarines or surface ships existed until 1971. The Naval Academy, so important to the profession, received its first aviator superintendent, Vice Admiral Aubrey Fitch in August 1945.[109]

The defeat of Japan left the United States with no maritime rival. The dominance of the aircraft carrier in Admiral William Fullam's "three-plane" warfare was something to which the battleship could only pretend. Survival of the navy in the immediate postwar atomic world would be difficult. However, survival would be easier for a navy based on the aircraft carrier with its offensive strike range of over 200 miles than on the battleship with its 20-mile reach.

In early 1946, CNO Admiral Nimitz reaffirmed the battleship's reduced status, making its mission, along with the rest of the surface combatants, to "support the amphibious forces and carrier forces and to furnish gunfire support for amphibious landings."[110] Yet a year later — in an echo of Sims's and his colleagues' dismissal of the submarine after World War I — Naval War College students were assured that no battleship "having a dual purpose secondary battery [for antiaircraft defense] was sunk by aerial attack."[111]

CONCLUSION

By the late 1930s, naval aviation was on the verge of becoming a viable presumptive anomaly to the battleship technological paradigm. Its capabilities had increased markedly since 1924, although assessments of naval aviation's ability were varied. Prior to the outbreak of World War II in 1939, the only measurements for the sufficiency of an anomaly were the fleet and tactical exercises. The officers who commanded the fleet set the boundaries for the exercises, posing the hypotheses and setting the experimental limits. Senior umpires were typically members of the battleship thought collective, and their rulings, and periodic prejudices, often worked against any weakening of the battleship technological paradigm.

War removed any prejudicial interpretations, being, in the broadest sense, impartial. As with the mercenary *condottieri* in Machiavelli's Italy, technological superiority was a significant factor in military success. The war during 1939–41 largely reaffirmed the existing battleship technological paradigm. However, the superiority of the aircraft carrier became more clear as the war progressed. The April 1945 sinking of *Yamato*, the largest battleship in the world, by carrier aircraft obviated the need for battleships. Their future became tied increasingly to missions such as amphibious warfare that were being characterized as superfluous by the air power advocates of the atomic age. By 1958, the last of the U.S. fast battleships had been taken out of commission and placed in mothballs.[112]

World War II provided an unbiased crucible to evaluate the competing naval technological paradigms. The battleship's dominance declined, and the effective German submarine campaign could not stand up to escort carriers and land-based aircraft, signals intelligence, and Allied technological advances. By mid-1945, the aircraft carrier was the dominant technology at sea. The operators of this superior technology in turn controlled the American naval profession, although Nimitz's successor as CNO was the nonaviator, Louis Denfeld. However, even Denfeld had become "airminded" and made the ultimate professional sacrifice. He was fired in 1949 when he fought the political establishment in defense of the canceled atomic-capable supercarrier *United States*, the exemplar of the U.S. Navy's new aviation technological paradigm.[113]

CHAPTER 9

Castles of Steel

Technological Change and the Modern Navy

In 1494 Charles VIII of France invaded the Italian peninsula to pursue his claim to Naples. The mobile cannon he brought quickly reduced a castle in eight hours that had previously withstood a siege of seven years.[1] Charles's actions threatened the foundation of contemporary warrior society and marked a nascent shift in the European way of war. European leaders were not about to abandon castles constructed at great cost to defend cities, towns, and strategic areas. The solution lay in modifying fortification design to preserve the castle from the cannon. Military architects added earthen redoubts outside masonry walls to absorb shot, developed the trace italienne, and generally extended exterior defensive rings to prevent investing artillery from coming within range.

Four hundred years after Charles, the steel battleship was the measure of naval power.[2] When the self-propelled Whitehead torpedo — carried on small, fast torpedo boats — threatened the established order, naval hierarchies pursued ways to preserve the battleship's preeminence. Naval architects modified battleship hulls to absorb torpedo hits. Naval tacticians countered the torpedo by expanding the defensive ring around battleships using torpedo boat destroyers. The same ideas of defense in depth, combined with technical improvements, governed later attempts to protect the battleship from submarines and aircraft.

After World War II, the navy created a technological hierarchy to protect the aircraft carrier just as it had with the battleship. Components of the aviation technological paradigm — even small ones like the ULQ-6 I encountered as a junior naval officer — contributed to preserving the new status quo and attempted to relegate counterweapons, and presumptive technological anomalies, to the periphery. As we begin a new century, there are strong indications that new presumptive anomalies threaten the aviation technological paradigm.

Despite its subsequent vilification by pundits, the battleship was not obsolete in December 1941. When the Japanese attacked Pearl Harbor—ignoring the fleet's oil storage and repair facilities and missing the aircraft carriers (all were at sea)—they only managed to destroy older battleships already slated for replacement. The navy's perceived future lay in the seventeen battleships on order and beginning to enter service, especially the six *Iowas* and five *Montanas*. Aviation-oriented historical assessments of the fast battleships minimize their relevance to the Pacific War and confuse their strategic mission with the obvious tactical use to which they were most often put: bolstering the antiaircraft defenses of *joint* battleship-aircraft carrier task forces.

During the Pacific War, the fast battleship and aircraft carrier became the exemplars of two, competing technological paradigms. The eventual ascendancy of carrier-based aviation was a near-run thing. Slightly earlier developments in sensor and gun technologies might have relegated aircraft to their 1924 status when towed aerial targets were being hit 75 percent of the time by ship-mounted guns during live-fire exercises. The earlier addition of radar-augmented defensive air patrols from fleet carriers would have blunted the effectiveness of attacking aircraft—in essence what happened by war's end when search radars, radar-directed guns, and the proximity fuze combined to enhance fleet defense. Even so, the navy suffered tremendous losses at the hands of Japanese kamikazes—essentially antiship missiles—off Okinawa in 1945.

By war's end, the U.S. Navy had fulfilled Secretary of the Navy James Forrestal's wish and had become an air navy. The aircraft carrier had proven its ability to dominate Admiral Fullam's "three-plane" war at sea.[3] Naval aviation could concentrate incredible destructive power against sea and shore targets at ranges over 200 miles. After Japan's surrender, no enemy battleships existed to challenge American carriers at night or during bad weather, and the battleship became difficult to justify both inside and outside the navy. The battleship soon became irrelevant in a navy desperately struggling to establish its own importance in the confusing early atomic era.[4]

Supporters of the aviation technological paradigm had two, immediate,

With no enemy battleships left to fight in August 1945, the necessity for fast battleships such as Missouri *(left) and* Iowa *receded.* (Naval Historical Center, NH 96781)

postwar missions. First, they had to elucidate a new naval strategic paradigm that moved beyond guerre d'escadre. They needed to justify a navy in terms of a national maritime strategy that fit into the new atomic strategies accepted by senior civilian policymakers. For the long term, this involved elucidating a politically accepted strategy for sea control. In the near term, the navy had to gain some responsibility for a part of atomic warfare. The second mission involved refining the rationale of the aviation technological paradigm built around the aircraft carrier. To do that, the navy denigrated the possibility of large-scale nuclear warfare and advocated a strategy of "strike-at-source" using carrier aviation to destroy the Soviet navy before it reached the open oceans and interfered with allied maritime operations.[5]

THE VICTORIOUS AVIATION TECHNOLOGICAL PARADIGM

Proven in war and with the postwar naval profession devoted to sustaining it, the aircraft carrier became the exemplar of the U.S. Navy's post-1945 technological paradigm. Assisted by the aviation-focus of atomic warfare, supporters of the aviation technological paradigm worked to place the navy on an aviation-related technological trajectory.

While submariners and aviators could both claim they won the Pacific War, the submariners could not sell themselves as a cold war military tool as effectively as naval aviation could. The Soviet Union was not a maritime nation nor a naval power. Once again, the submariners had to dance to someone else's tune. As a result, in 1947 the naval hierarchy considered perfecting "special purpose submarines" such as an antimissile "picket," a 34,000-ton "carrier" equipped with two aircraft with atomic bombs, and a "bombardment" submarine.[6] These programs encroached on naval aviation and met opposition.

Aviators, if equipped with larger carriers, could strike the Soviet Union more easily, but they had to fight the air force for a role in atomic bomb delivery. The air force was a tough rival, and naval aviators wanted no internal competition for the anti-Soviet strike mission. The submarine only achieved an offensive strategic role during the late 1950s with the development of the solid-fueled, ballistic missile with an atomic warhead.

Restricted navy budgets forced cancellation of development of shore bombardment and antiship guided missiles for surface ships and submarines. The last survivor was the Regulus II missile slated for installation on cruisers, cut in December 1958 since it competed directly with the purchase of carrier aircraft.[7] Missile development funds were steered to antiaircraft missiles to prevent an atomic air attack on the fleet. In 1950 Rear Admiral Daniel V. Gallery, head of the guided missile development branch in the chief of naval operations office, reported that 21 percent of the Department of Defense research and development budget went to guided missiles, with the navy receiving the "Lion's share." According to Gallery, the navy, equipped with cruisers carrying antiaircraft missiles and carrier-based fighters, would "be able to move in on any coast in the world and launch an attack without worrying too much about what is going to happen to our ships."[8]

The pervasive naval-industrial-academic network and the wars in Korea and Vietnam supported the aviation technological paradigm, increased its

momentum, and solidified its institutional structure and institutional inertia.[9] During both wars, the navy projected power ashore from aircraft carriers in littoral waters. This offense-minded aviation strategy persisted after
the dissolution of the Soviet Union and relates closely to the "strike-at-
source" strategy pursued for most of the cold war. According to one surface
officer, "the aviators saw everything in terms of carrier strikes, with nukes
[nuclear weapons], even. You have a submarine problem? Not to worry,
we'll clean out the Kola Peninsula. This sort of thinking led to a denial of
the need for conventional ASW [antisubmarine warfare] forces."[10]

From the beginning of the cold war, the naval hierarchy had to elucidate a legitimate and important strategic rationale to defense policymakers. For much of the period covered by this study, the battleship technological paradigm served the guerre d'escadre strategy. By and large,
Congresses and presidents considered the battleship navy a useful force
with which to control the seas. However, the absence of any rival fleet
after World War II made a traditional guerre d'escadre strategy ostensibly
irrelevant. Even if an enemy fleet existed to challenge the U.S. Navy,
air power advocates claimed that atomic bombs, delivered by air force
bombers, could destroy it.

For the navy, the postscripts of both World Wars were distressingly similar, with aviation advocates predicting the navy's demise. In the early 1920s,
General Billy Mitchell had called for an independent air service which
would obviate the need for a large army and navy.[11] His successors
achieved an independent air force through the National Security Act of
1947. The air force hierarchy argued for primacy within the reordered defense establishment because future wars would be deterred or ended by
atomic weapons under air force control. The army and navy would be
largely irrelevant. Fortunately for the navy, the Korean War demonstrated
the persistence of conventional war in the atomic era and provided a new
venue for naval aviation. The Pusan perimeter was stabilized in 1950 and
the U.S. 8th Army and South Korean forces survived thanks, in large measure, to tactical air power based on aircraft carriers.

The leaders of the American naval profession had to come to terms with
the early, transitional national strategies of the atomic age. Many politicians
believed atomic strategy was too important to be left to admirals or generals, a point underscored by the establishment of the National Security
Council in 1947 and the relief of General MacArthur in Korea in 1951.[12]
The air force argument seemed persuasive to many political leaders. If the

air force hierarchy had its way, the rise of the naval aviation technological paradigm would remain a brief footnote in the history of a navy that soon would "rank with mediaeval institutions," as Rear Admiral Bradley Fiske had warned during the navy's first skirmish with air power advocates in 1921.[13]

The naval hierarchy worked hard to discredit the "air force" future. The navy concept for World War III had it surviving a nuclear onslaught on the United States and providing the decisive advantage by which the United States would prevail. As Richard Hegmann observed, the navy thought the war against the Soviet Union would be long and argued that massive retaliation was a flawed strategy, that is, that nuclear weapons would not be employed in large numbers. In 1953 Chief of Naval Operations Admiral William Fechteler told the Naval War College that there would be a battle for the Atlantic, as in the previous two wars, since nuclear weapons were "evolutionary, not revolutionary." His successor, Admiral Robert B. Carney, considered the navy's mission to " 'dust off the atomic residue' and complete sea control and power projection missions 'identical in most respects to [the Navy's] job in World War II.' "[14]

The navy argued — with some justification based on the Bikini Atoll tests of 1946 — that a dispersed fleet formation could survive atomic weapons. However, the increased intervals between ships in larger fleet formations made the centrally located attack carriers vulnerable to the submarines that made up the majority of the Soviet navy. Part of the solution was the development of sonars with increased range and sensitivity and longer-range antisubmarine weapons.[15]

Dispersed fleet formations reduced the relevance of the naval gun. At the Naval War College in February 1949, Rear Admiral C. R. Brown characterized the aircraft carrier as embodying "all the principles of war" and pointed out that

> an average fast carrier task force of the last war had a concentration of over 1600 guns to use in its defense. When translated into fire-power, gentlemen, that means over 6,000 bullets per second or just under 200 tons of steel every minute. This, in its ability to deliver hot metal, surpasses any conceivable concentration of artillery ashore. It was positively brutal and it is small wonder that even those Japs who were not suicidally inclined grew to consider an anti-carrier mission as almost automatic enrollment in the Kamikaze Corps.[16]

However, in 1949, longer-range air defense required antiaircraft missiles, not wistful recollections of "hot metal." The costly development of anti-aircraft missiles forced an end to development of surface-to-surface missiles. Surface-to-surface (cruise) missile development would have provided submarines and surface ships with offensive capability and strategic and tactical enfranchisement. With surface-to-surface missiles, surface and submarine officers could have competed directly with naval aviation. However, without a surface-to-surface missile, submarines and surface ships remained supporting technologies. Development of surface-to-surface missiles eventually fell victim to the tremendous budget outlays during the 1950s for the Polaris intermediate-range ballistic missile for the nation's new, submarine-based, nuclear deterrent force.[17]

As exemplar of the aviation technological paradigm, the aircraft carrier revisited the path of the battleship. Bigger carriers were more capable and initially were necessary to carry the large aircraft needed to deliver atomic weapons. The large *Midway*-class carriers (45,000 tons), judged too big by aviators in 1943, gave way to the 59,900-ton, large attack carriers of the *Forrestal* class (1955) and eventually to the 88,900-ton, nuclear-powered *Nimitz*-class supercarriers (1975). The superiority of the aircraft carrier was as clear to postwar American naval officers as the big battleship had been to Sims, Fiske, Dewey, Poundstone, Fullam, and Stirling.

Like battleship officers during the 1930s, naval aviators considered themselves the final arbiters of naval warfare, still valid in the nuclear age and increasingly focused on war against the shore. The aircraft carrier proved its utility during the Korean and Vietnam Wars. However, neither enemy posed any appreciable naval threat. Aviators contended, with some justification, that the means for projecting power ashore — the carrier air strike — was also well suited for open-ocean sea control against surface ships. Submarines were a different matter.

Even before the advent of nuclear-powered submarines in the late 1950s, the navy considered Soviet submarines a very serious threat. Beginning in 1953, the navy converted World War II *Essex*-class carriers into antisubmarine support aircraft carriers.[18] These ships worked with new antisubmarine destroyer escorts in a continuation of the World War II Hunter-Killer (HUK) groups that had proven so successful against the Germans in the Atlantic.

Like the battleship officers before them, most aviators viewed the open-ocean submarine problem as someone else's. It belonged to the black shoes of the surface navy and to the operators of attack submarines. Unfortunately, stealthier conventional and nuclear-powered submarines were difficult to detect. In 1956, the commander of antisubmarine forces in the Atlantic reported that his forces could not "effectively oppose a nuclear submarine." By the end of the 1950s, submarines were so successful in exercises that the antisubmarine carriers usually were sunk.[19]

The answer to the difficult open-ocean submarine problem lay within the aviation technological paradigm. The solution was the "strike-at-source" strategy, in which Soviet submarines would be neutralized by "offensive action of fast carrier groups" attacking them in their bases within the Soviet Union. In 1962, the chief of naval operations, Admiral George W. Anderson, described the "best way to kill Russia's five hundred submarines" as "a precision attack by carrier planes against the subs before they leave their home ports."[20] Use of nuclear weapons was implicit.

The naval aviation hierarchy seemed as myopic as battleship officers, with their focus on battleship gun duels. Aviators increasingly focused on air warfare and power projection with carrier aircraft. The aircraft carrier and aviation technological paradigm was *the* answer. A classic representation of the increasingly narrow aviation thought style was the observation by one aviator chief of naval operations (CNO) that "if it can't fly, I don't want to hear about it."[21] Almost a decade of combat operations in Vietnam reinforced this.

Naval aviation paralleled the nine-tiered biblical hierarchy of angels. Fighter pilots were the seraphim, attack (bomber) pilots ranked second as cherubim, while the pilots of shore-based antisubmarine aircraft ranked last as mere angels. For aviators, the path to flag rank in the postwar navy included command of a carrier air wing and of a carrier — jobs usually reserved for fighter and attack pilots.

The naval air war in Vietnam led to an increase in the number of light attack squadrons composed of single-seat aircraft. By the early 1980s, almost half of the commanders of the navy's air wings were light attack pilots.[22] Whether seraphim or cherubim commanded the carrier air wings, carriers, or carrier battle groups, their focus was not on the drudge work of antisubmarine warfare. Carrier-based antisubmarine warfare — so successful during World War II — was given short shrift. Secretary of the Navy John Lehman reported that in the early 1980s the S-3 Viking, built specifically as a carrier-based, submarine-hunting aircraft, had "averaged only 30 percent mission capability."[23] The Vikings languished in large part because their antisubmarine mission was external to the thought style of the admiral and captain fighter and attack pilots who could have provided the spare parts, maintenance, and training to improve its performance. Not surprisingly, the navy-supported 1986 movie *Top Gun* did not feature the pilot of an antisubmarine aircraft tediously monitoring sonobuoys while trolling for Soviet submarines.

Defenders of the large supercarrier, such as Lehman, echoed advocates of the superdreadnought, equating size with strength and ability to survive attack. The modern supercarrier, according to Lehman, could "dominate the airspace within five hundred to eight hundred miles of the battle group." But domination of airspace — a problematic claim in some areas of the world — was not the same as domination of the underwater battlespace populated by capable nuclear submarines. Titanic size was no defense against torpedoes or antiship missiles launched from submarines and equipped with nuclear warheads. Naval officers, peering at the world through the service's "mask of war" — to borrow Carl Builder's term — conveniently preferred to think that war at sea would not involve nuclear weapons.[24] Yet even conventional weapons, such as wake-homing or acoustic-homing torpedoes, had a good chance of achieving a "mobility kill," thereby removing the carrier from combat operations.[25]

AVIATION MOMENTUM AND TRAJECTORY

Institutional reorganization within the navy after 1945 supported the aviation technological paradigm. The position of deputy chief of naval operations for air, with supporting bureaucracy, provided top-level stewardship for naval aviation and, until 1971, was the only CNO deputy representing a technology-defined warfare specialty. Then, submarine and surface warfare officers received similar representation by their own DCNOs with the rank of vice admiral.

In the early postwar period, the Bureau of Ordnance, the institutional bastion of the prewar "Gun Club," fought hard to remain relevant in the early postwar period by developing guided missiles. If it failed, it would become the "Bureau of Obsolete Weapons."[26] The Bureau of Aeronautics, however, became the agency on the cutting-edge of naval technology. It drew the best and brightest officers into its organization, and the resulting technologies — new aircraft, air-to-air missiles, and surface-to-air antiair-craft missiles — reflected the navy's new technological path.

Maintaining a proper technological trajectory was critical to the long-term supremacy of the aviation technological paradigm. This included acquisition of aircraft carriers and construction of a supporting surface ship force designed for defense of the carrier. The *Leahy*-class guided missile cruisers designed during the 1950s, and in service starting in 1962, were eventually replaced by the more sophisticated *Aegis*-class air defense cruisers, the trace italiennes of the early 1980s.

In addition to acquiring appropriate technologies, technologies that might pose potential presumptive anomalies had to be managed or suppressed. The naval aviation community exerted tremendous pressure during the late 1960s and early 1970s to maintain their monopoly over offensive operations. Surface warfare (antiship) capability rested with carrier attack aircraft; early supporting cruisers like the *Leahy* class were primarily defensive in nature and had no real capability to attack other ships. Senior aviators ensured that the Harpoon antiship missile eventually carried by surface ships was limited in range so as not to infringe upon tactical aviation.[27]

Development of the submarine-launched ballistic missile and the nuclear-powered ballistic missile submarine during the 1950s was a mixed blessing for naval aviation. These programs limited funds available for ship acquisition, including supercarriers. Navy budget problems became more acute when President John Kennedy expanded the ballistic missile submarine program to forty-one ships soon after entering office. However, development of nuclear-powered submarines and Kennedy's expansion of the submarine nuclear-missile deterrent force contributed to the tremendous congressional support and power of Admiral Hyman G. Rickover — father of the nuclear navy. Rickover also was a forceful advocate of nuclear-powered supercarriers.

During the 1960s, Rickover envisioned a new navy of nuclear-powered supercarriers, cruisers, and submarines. Such an expensive, high-technology acquisition plan had to compete against the financial drain of naval operations in Vietnam. The majority of the navy's surface ships had been built during World War II and were approaching the end of their useful lives. Yet the navy's worldwide cold war commitments — including the war in Vietnam — required a large number of ships.

The long-term implication of Rickover's nuclear program was a smaller navy. Congress, dealing with acquisition costs, was chary of the higher up-front price of nuclear-powered ships. When fuel and operating costs were amortized over a twenty-year period, a nuclear-powered ship cost one and a half times more than one powered by fossil fuel. Rickover's advocacy typically resulted in Congress approving fewer nuclear-powered ships at the expense of more fossil-fuel-powered ships.[28]

In 1970 the new chief of naval operations, Admiral Elmo R. Zumwalt, tried to shift the navy's technological trajectory. Zumwalt was a surface line officer serving as the commander of U.S. naval forces in Vietnam when selected to run the navy. Zumwalt forced a debate over what type of navy the United States should have and challenged the aviation technological paradigm and Rickover's all-nuclear navy.

The navy could not meet its global responsibilities without a tremendous increase in the navy budget — extremely unlikely — or a shift in the types of ships the navy acquired. Zumwalt believed the navy was putting too many resources into new, expensive, "high" technology programs like the *Los Angeles*-class nuclear-powered attack submarines, the *Nimitz*-class nuclear-powered supercarriers, and large amphibious assault ships. Sounding the alarm over the navy's future, Zumwalt publicly doubted the navy's ability to defeat the Soviets at sea.

Zumwalt emphasized a broad-based naval strategy of sea control within the context of the cold war. The navy's mission involved support for the North Atlantic Treaty Organization (NATO) and open-ocean antisubmarine warfare against a new Soviet navy that had demonstrated a global presence during its major exercise, *Okean*, in 1970. Open-ocean antisubmarine warfare contradicted the aviation strategy of "strike-at-source" and did not appeal to carrier aviators. It also challenged the power projection focus of the aviation technological paradigm.

Zumwalt proposed a reduction in the number of *Los Angeles*-class submarines to be built, continued purchase of fossil-fueled ships, and more "low" ships for a "high-low" technology navy. In his Project 60, a fiat developed outside the CNO bureaucracy, Zumwalt proposed four low-end ships for sea control: the *Pegasus*-class hydrofoil missile ship; continuation of the Patrol Frigate design that evolved into the FFG 7 class; a small aircraft carrier called the Sea-Control Ship costing one-eighth the price of a supercarrier; and a large, cruiser-sized, 80-knot surface effect ship — a kind of huge hovercraft.[29]

Zumwalt's low-end navy never materialized to the extent he wished. His forceful efforts on behalf of personnel and social changes in the navy alienated a large number of potential senior officer allies. Zumwalt was unable to overcome the momentum of the three, powerful, technologically defined groups within the navy: Rickover's nuclear power organization, submarine officers (a distinct subcollective of Rickover's empire), and naval aviators. The surface effect ship went the way of other radical technologies: numerous articles in the Naval Institute *Proceedings* and two small prototypes built and rejected for real technical shortcomings.[30] When the Germans and Italians backed out of the *Pegasus* hydrofoil missile ship, the U.S. Navy built six and groused about its components built to metric standards. The surface navy never liked them and strike aviators perceived their Harpoon antiship missiles as poaching on carrier aviation. The *Pegasus* class ended up in Key West fighting the drug war.

Naval aviators and Rickover's surface-ship nuclear officers combined to kill the small, fossil-fuel-powered, sea-control aircraft carrier. This ship harkened back to the antisubmarine escort carriers of World War II. It was intended to show the flag in littoral waters too risky for a supercarrier during crises. During hostilities, the sea-control carriers would trade places with the supercarriers and move to the open ocean to keep the sea lanes to Western Europe free of Soviet submarines.

The sea-control carrier was completely dissonant with the aviation technological paradigm. It carried no fighters or attack aircraft, had no catapult, and would be equipped with antisubmarine helicopters and a few subsonic Harrier vertical/short take-off and landing jet aircraft. Zumwalt was unable to muster sufficient organizational, institutional, and political leverage to counter the aviation technological momentum, and the sea-control ship failed to materialize.

The only low-end program that survived was the FFG 7-class frigate. With a Dutch fire-control system, a single Italian Oto-Melara 76mm gun, one short-range surface-to-air missile launcher, a Canadian sonar, and one propeller, the FFG 7 was relatively cheap, culturally diverse, and appealing to international customers. Its low cost meant it could be produced in quantity for use in convoy duty to reinforce Western Europe. Many surface officers, accustomed to aviation-related high technology — such as the $1 billion *Aegis* air defense system cruisers — pegged the FFG 7 as a lemon, a "square peg in a round hole."[31]

In July 1974 aviator James Holloway succeeded Zumwalt as chief of naval operations. Holloway's challenge was to find a strategic justification for the aircraft carrier in a post-Vietnam political environment less than en-amored with U.S. intervention in the Third World. The answer lay in the sea-control mission against the Soviet navy and a shift from wars in support of client states back to the "strike-at-source" offensive strategy to destroy the increasingly powerful Soviet navy in its home waters. In 1975 the navy dropped to less than 500 ships, comparable to its 1939 strength. That year, 220 Soviet warships conducted a worldwide exercise, termed *Okean 75*, in the West. A combination of factors within the Soviet Union made this the high point of Soviet open-ocean adventurism. However, in 1975 leaders of the U.S. Navy only saw a threatening and increasingly capable adversary.[32]

The navy continued to decline during the presidency of Jimmy Carter. To quote George Baer, Carter and Secretary of Defense Harold Brown di-verted resources from a "maritime strategy to a continental commitment; from the Pacific to Europe; and from the maintenance of a two-ocean Navy to the development of a one-ocean Navy. The Navy, Brown declared, was for antisubmarine warfare, convoy in the Atlantic, and 'localized con-tingencies outside Europe and peacetime presence.'" Given the post-Vietnam hostility to regional warfare, the Carter administration felt free to limit a navy with no central strategic role. In 1978 President Carter cut the navy's shipbuilding program in half and canceled construction of a *Nimitz*-class supercarrier, believing that their day had passed.

When Carter resurrected Zumwalt's smaller, fossil-fueled carrier, the next chief of naval operations, aviator Admiral Thomas B. Hayward, dismissed the president's plan as a "Third World strategy" reflecting "the convoy syndrome" to save Western Europe.[33] Many senior naval officers doubted that war in Europe would be short or confined to that continent. Yet naval leaders failed to elucidate their doubts and continued to argue for a navy based on Soviet capabilities without clarifying why those capabili-ties mattered. As a result, the navy remained on the periphery of national strategy and limited in its budget.

The navy regained its strategic focus with the Maritime Strategy of the early 1980s that rode the crest of the defense buildup during the Reagan presidency. The efforts spearheaded by Admiral Hayward, his successor, Admiral James D. Watkins, and Ronald Reagan's first secretary of the navy, naval reserve flight officer John Lehman, moved the aircraft carrier back onto center stage. The Maritime Strategy embraced the strike-at-source, carrier-based strategy for which the aviation technological paradigm was suited perfectly. The Maritime Strategy minimized the "NATO-centered, reactive sea control and limited Third-World interventionism," and moved "toward a worldwide offensive in . . . a long, conventional war of global scope."[34]

In refining the Maritime Strategy, the navy's leadership disagreed about the Soviet navy's wartime strategic mission. Would it attack the U.S. Navy and challenge Western naval supremacy, or act defensively to protect its own technological exemplar — the strategic ballistic missile submarine — located in ocean areas near the Soviet Union? In either case, offensive action would, theoretically, buy success for the United States and its allies. Attacking the Soviet navy in its home ports — assuming the Soviets would wait for war before sailing — would blunt its challenge to Western naval supremacy. Attacks would also threaten the Soviet Union's reserve force of nuclear missile submarines. When intelligence analysis indicated that the Soviets were concentrating the bulk of their naval forces to protect their missile submarines within their ocean bastions (in the Sea of Okhotsk and in parts of the Barents Sea and Arctic Ocean), the U.S. Navy considered the bulk of the high seas its to control by default. The geographically confined Soviet navy could be attacked and overwhelmed.[35]

The Maritime Strategy became the subject of controversy within the defense community during the 1980s, primarily over Lehman's advocacy of a naval strategy against a continental power.[36] In arguments reminiscent of those of battleship officers, supporters of the aviation paradigm rejected criticism of the carrier's vulnerability, especially in the waters contiguous to the Soviet Union. Critics argued that attacks on Soviet bastions containing ballistic missile submarines would force a "use or lose" decision on the Soviet leadership, resulting in a cataclysmic nuclear assault on the United States.[37]

In their formulation of what are known as "service strategies," Carl Builder accused all the services of ignoring the "national strategy" generated by the National Security Council, and the "defense strategy" promulgated by the Department of Defense and the Joint Chiefs of Staff. Builder maintained that each service develops a strategy that is based on its "expertise" and which perpetuates its current or desired force structure, based on its preferred technology rather than on national strategic goals. In other words, technological paradigms drive service strategies developed during peacetime. The navy's "strike-at-source" doctrine of the 1950–60s, resurrected in large part in the Maritime Strategy of the 1980s, reflected this.[38]

According to Richard Hegmann, the post-1945 navy, rather than squabble with the air force for limited atomic bomb missions, created a broader mission based on the Pacific Ocean War: conventional power projection and ocean dominance through naval air power. If it had to support NATO, the aviators preferred not to emulate the drudge work of the Atlantic War — convoying merchant ships and chasing U-boats. The way around that was established by the aviation paradigm: use carriers to strike at Soviet naval facilities and destroy Soviet naval forces before they put to sea.

Early in the Reagan administration, Secretary Lehman pushed for the buildup to a six hundred-ship navy to carry out the latest incarnation of this anti-Soviet sea-control strategy.[39] Lehman's navy was centered around fifteen aircraft carrier battle groups and four resurrected *Iowa*-class battleships. He became the spokesman for the Maritime Strategy and defender of the aircraft carrier as the decisive player and, perhaps, final arbiter in a general war with the Soviet Union.

One naval critic described the Maritime Strategy as more of war plan than a long-term naval strategy for the United States.[40] If war came, the navy would destroy all deployed Soviet naval forces, and aircraft carrier battle groups would position themselves on the doorstep of the Soviet Union and attack the remaining Soviet naval forces, threatening their nuclear missile submarines and preventing significant Soviet forces from reaching the high seas.

The Maritime Strategy depended on the blithe assumption that the Soviets would not use nuclear weapons. Another problem was that the Soviet Union was not Vietnam. At Okinawa in 1945, very capable fleet defensive capabilities were battered by Japanese aerial attacks.[41] The Maritime Strategy would subject American naval forces to concentrated three-dimensional (aerial, surface, and submarine) attacks that many thought they would be hard pressed to survive.[42] Secretary Lehman's response to intimations of the vulnerability of American carrier battle groups was to call for the acquisition of more aircraft carriers and aircraft and reiteration of the carrier's supremacy.

What Lehman and other staunch supporters of naval aviation would not admit was that modern submarines posed a significant presumptive anomaly and had moved far beyond the reach of Robert Mitchum in *The Enemy Below*, Richard Widmark in *The Bedford Incident*, and James Earl Jones in *The Hunt for Red October*. Open-ocean antisubmarine warfare remained difficult and the success of strike-at-source remained problematic. Those aviators who acknowledged the submarine problem dismissed it as temporary. The solution would be blue-green lasers, satellite wake detection, or some other new, nonacoustic detector—the technological Holy Grail of modern antisubmarine efforts. Such wishful thinking was captured by fighter pilot Captain Gerald O'Rourke in his 1988 article "The End of the Submarine's Era?" in the Naval Institute *Proceedings*.[43]

Like Bradley Fiske's "compromiseless" battleship of 1905, Lehman defended the *Nimitz*-class supercarriers as just the right size, especially for forward operations on NATO's northern flank. There, the carriers would threaten the Soviet rear areas and Soviet army flanks, destroy the Soviet surface fleet, and aggressively attack Soviet submarines in their home waters. According to Lehman, the supercarriers' 31-knot speed made them "less vulnerable" than immovable land bases. Lehman's argument was based in the carrier's operational and strategic success during World War II and was enhanced in Korea and Vietnam. However, even a fourteenth-century Mertonian kinematicist probably could deduce that, to a mach 2.5 missile, there was no appreciable difference between a stationary target or one moving at 31 knots. Immovable land-bases were easier to target, but modern sensors, including those in orbit, made carrier battle groups more difficult to hide.[44]

The aviation technological paradigm and the Maritime Strategy embodied certain notions regarding the warrior's relation with technology and the naval profession's concepts of warfare. During the battleship era, its crew was comparable to the workforce at Henry Ford's River Rouge Plant. The battleship's output was not automobiles but the delivery of high-explosive shells on target. This modern industrial mode of warfare was rooted in the traditional, cooperative nature of naval warfare. Naval aviators, however, fought generally as individuals—masters of technologically complex artifacts. The naval profession had garnered social status as operators of complex machines in late-nineteenth-century America. As masters of complex technologies, modern aviators, at some level, seemed to prevail over people-dominating technological systems of the post-Hiroshima era.

To repudiate the quantitative, industrial nature of total warfare that arose between 1914 and 1918, militaries and governments propagated the myth of individual worth and the triumph of individual skill—augmented by technology—that generated the culture of the "knights of the air." In the end, the characteristics so prized within the qualitative pre-1914 European armies and chronicled by Tim Travers for the British army—"manly courage," virility, and discipline—flowed away along the banks of the

Somme and sank into the mud at Passchendaele. However, the mythic view of aerial warfare persisted as part and parcel of the aviation thought style. The vestigial linkage to medieval combat as God's justice and the pilot-warrior's mastery over technology, rather than subservience to a complex system, returned some glory to war and appealed to younger naval officers for obvious reasons.

As with the battleship turboelectric drive, industries wishing to do business with the aviation navy had to fit through the relevant paradigmatic filter. In their early push for the single-seat F-18A strike fighter, McDonnell-Douglas's advertising theme supported the individual warrior schema, touting "One Man, One Plane." Senior naval aviators also tried to "spin" popular culture. The 1986 movie *Top Gun* followed hormone-super-charged fighter pilots, led by the charismatic actor Tom Cruise, through their training at the navy's Fighter Weapons School. Once knighted at graduation, these select warriors were sent immediately to an aircraft carrier in the Indian Ocean to handle a "crisis situation" caused by a botched American intelligence operation. After arriving, the "top guns" successfully dueled Communists flying the latest Soviet fighter aircraft — painted black to clarify their moral standing.

There was a lot of *Top Gun* in the Maritime Strategy and, in his autobiography, Secretary Lehman wrote with pride about his official aid in producing the movie. In the Maritime Strategy, as in the movie, superior U.S. naval technology, based mostly in aircraft carrier battle groups and used by warrior elites, would prevail over the Soviet navy. Nothing much had changed from Secretary of the Navy Charles Thomas's 1955 lecture to the Naval War College. There, Thomas issued a liturgical refrain from the qualitative Edwardian army. He identified the need for a naval offensive spirit in which U.S. naval officers refused "to be awed by mere numbers, the greater size of the enemy force, or dispirited by any atomic equation."[45]

The Maritime Strategy was ambitious. Given that it required 25 percent of the navy's carriers to attack Libya in 1986, the ability to prevail in attacks on the Soviet Union was problematic. The dissolution of the Soviet Union put an end to the Maritime Strategy, but not to the aviation technological paradigm.

In 1990 President George Bush called for a defense policy to address regional threats in a changing post–cold war world. In April 1991 the secretary of the navy, chief of naval operations, and Marine Corps commandant responded. Their white paper, "The Way Ahead," was followed by "From

the Sea" (1992) and "Forward . . . From the Sea" (1994). Each further refined U.S. naval strategy in terms reminiscent of the days of gunboat diplomacy. Instead of gunboats, the United States would employ power projection ashore from ships, typically aircraft carrier battle groups, stationed in
littoral waters. It was a mission that would be understood by a captain serving in the East India Squadron 150 years earlier.[46]

Local sea control was a key assumption in this war against the shore. This
was reflected in the introduction to "The Way Ahead": "Since the end of
World War II, the United States has been the world's preeminent military
power — especially at sea, where we enjoy clear maritime superiority. In
achieving and maintaining this preeminence, U.S. naval forces have sailed
the high seas virtually unchallenged for nearly half a century."

The white paper claim that the U.S. Navy had not been challenged in
combat since 1945 was true enough — there was little to fear from hostile
navies in Korea, Vietnam, Grenada, Panama, the Persian Gulf, and the
Adriatic Sea. The absence of any naval challenge in earlier combat operations reinforced the perception of the continued ability of the navy to
achieve local sea control in important littoral areas. The post–cold war period has been marked by the proliferation of sophisticated weapons — the
purchase of modern Russian submarines by Iran and antiship missiles there
and elsewhere. The U.S. Navy may counter these weapons through a continuation of the cold war strike-at-source doctrine. However, such an offensive act could be easily circumscribed by political and even legal considerations. Rules of engagement might require U.S. forces to act only in
self-defense, increasing their vulnerability to surprise strikes. This was one
of the reasons Admiral Zumwalt wanted to patrol the littorals with the expendable sea-control carrier.

A New Presumptive Anomaly

The modern cruise missile presents the clearest technological challenge
to the naval aviation technological paradigm. The cruise missile, along
with other uncrewed combat aerial vehicles (UCAVs), portend twenty-first-
century naval warfare increasingly dominated by machines with less direct
human involvement — at its darkest, future warfare presented in the movie
Terminator. These technologies also contribute to presumptions that the
aviation paradigm is in decline.

Kenneth Werrell has chronicled the postwar history of the cruise missile within the air force, navy, and Department of Defense.[47] Aviators in both services generally resented the missile, denigrated its capabilities, and bemoaned the lack of a pilot able to adjust to the dynamics of combat. Naval aviators perceived the Tomahawk sea-launched cruise missile (SLCM) as infringing on their antiship and land-attack missions. One aviator chief of naval operations berated an officer providing a brief on the early cruise missile: "We already have a cruise missile, it's an A-7 [a single-seat attack jet]. We don't need your cruise missile." In testimony before Congress in 1973, Secretary of the Navy John Chafee minimized the need for cruise missiles and parroted the aviation thought style by testifying that "carrier aircraft are essentially 'manned cruise missiles.'"[48]

The Tomahawk SLCM finally had the opportunity to demonstrate its effectiveness during the Persian Gulf War. Important Iraqi facilities in Baghdad were protected by an antiaircraft network judged "more formidable than those of any Eastern European target at the height of the Cold War, and seven times as dense as Hanoi's during the full-scale U.S. air offensive (Linebacker) in 1972."[49] During initial strikes to break the Iraqi air defense and command networks, Allied planners used 282 cruise missiles and F-117A stealth fighters with laser-guided bombs. The stealth fighters were vulnerable during the day and on moonlit nights and required clear weather to deliver their ordnance accurately. Conversely, time of day or overcast skies did not affect cruise missiles. When adverse weather forced a four-day delay in an aircraft strike against an Iraqi MiG-21 drone jet fighter equipped with chemical weapons, five Tomahawk SLCMs destroyed the airplane and its hangar.[50]

Navy assessments after the Gulf War defended the aviation technological paradigm built on the aircraft carrier and crewed aircraft. Like submarines and carrier aviation during the battleship era, the 1991 cruise missile was presented as a subordinate technology that allowed the technological exemplar — the aircraft carrier with its air wing — to act decisively in destroying the main enemy force.[51]

Both air force and navy aviators used the Gulf War as justification for newer, stealthy crewed bomber aircraft such as the B-2 (air force) and the proposed A-12 (navy). The B-2 has been subject to criticism over the degree of its stealthiness since it entered service, and the secretary of defense canceled the A-12 at the navy's request because of price overruns.[52] The new hope for crewed tactical air power for the twenty-first century is the Joint

Strike Fighter. The air force's new F-22 fighter (able to carry only six radar-guided antiair missiles in its internal weapons bay) was priced at close to $180 million per plane in the spring of 1998. Financially, this equates to approximately 450 SM-2 antiaircraft missiles used on *Aegis* system-equipped navy surface ships. This poses the question, especially for a navy wedded to increasingly expensive crewed aircraft, "What price air defense?"

The Tomahawk had other chances to demonstrate its capabilities when the United States launched two air strikes against Iraq during 1993. Iraq's failure to comply with United Nations' resolutions in January led President George Bush, a former navy carrier pilot, to approve part of a plan proposed by the chairman of the Joint Chiefs of Staff, General Colin Powell. Bush ordered aircraft strikes on Iraq, but ignored Powell's proposal to use cruise missiles as well. The aircraft attacked from above a 10,000-foot altitude to avoid antiaircraft gunfire and cloud cover prevented the use of laser-guided bombs. The aircraft hit only sixteen of thirty-three "aiming points," and the Iraqi diplomatic position did not change.[53]

Bush then had the navy launch forty-five Tomahawks from four ships against the Zaafaraniyah nuclear facility near Baghdad. Thirty-seven of the missiles destroyed four of seven targeted buildings and damaged the remaining three. Surprisingly, the missiles were sent into Baghdad via the same routes used during the Gulf War.[54] One of the missiles, apparently hit by ground fire, crashed into a hotel. This tactical misjudgment was reminiscent of aviator indictments of the misuse of the new technology of naval aviation by battleship officers before and during World War II.

When Iraq was implicated in an assassination plot against former President Bush in June 1993, no aircraft carrier was available for a punitive strike. The Clinton administration launched twenty-three missiles against an intelligence center in Baghdad and sixteen hit their target. The success rate was lower than the January attack. Defenders of the SLCM pointed out that these were older missiles (Block II), not the newer Block III versions capable of precise navigation using Global Positioning System (GPS) satellites.[55]

Cruise missile advocates believe these 1993 strikes marked a new policy consensus to use cruise missiles to prevent political fallout from killed or captured aircrew. This was not the case during early peacekeeping operations in Bosnia, where crewed reconnaissance/combat flights were the norm. The loss of one air force fighter to ground fire forced a successful but dangerous combat extraction of the pilot by a marine rescue team in June 1995. Three months later, military operations shifted to thirteen SLCMs launched from the cruiser *Normandy* in the Adriatic. During the spring 1999 NATO air war against Serbia, cruise missiles played a critical, early role.

Like interwar aviation advocates, supporters of the cruise missile attribute the reliance on crewed aircraft in Bosnia not to a "dearth of technology" but to the military's "failure to see the trend" toward standoff capability inherent in modern cruise missiles.[56] The trend became more visible in August 1998, when the Clinton administration relied solely on SLCMs to strike terrorist facilities in Afghanistan and Sudan. It is unclear if newspaper headlines — "Navy Fires Cruise Missiles from Arabian, Red Seas . . . *no aircraft were used* [emphasis added]" — expressed surprise or merely reported a new era of naval power.[57]

Recent studies within the chief of naval operation's Strategic Studies Group pose additional potential anomalies to the aviation technological paradigm. New technologies under consideration include electric rail guns with the projected capability to deliver hypervelocity projectiles out to 250 miles to 300 miles (the domain of carrier-based tactical aircraft), uncrewed combat aerial vehicles (UCAVs), and more advanced, less expensive SLCMs. Uninhabited combat aircraft, being tested by the air force, promise vehicles that will be recoverable and smaller, as well as able to exceed the physiological limits of crewed aircraft and retain a person in the decision process. The fact that the "pilot" could control the UCAV from a ship or even from a computer in the Pentagon poses a serious threat to the aviation warrior ethos and strikes at the heart of the defining relationship between aviators and their technology.[58]

In June 1998, *Navy Times* reported that the navy had placed its new carrier design, the CVX, on hold because of budgetary problems. Expenditure of $3.2 billion on CVX research, development, and design threatened acquisition of the last *Nimitz*-class supercarrier, CVN 77, scheduled to begin in 2001.[59] The CVX design team had promised an innovative new carrier embodying an "architecture for change" to build what chief of naval operations Admiral Jay L. Johnson characterized as "the right ship for the second half of the 21st Century."[60] However, the first CVX design criterion, according to the Naval Sea Systems Command — successor to the Bureau of Construction & Repair and the Bureau of Ships — was a carrier that "maintains [the] core capabilities of Naval Aviation." These core capabilities were not specified, but continued reliance on crewed aircraft to perpetuate the core *values* of *Top Gun* is a reasonable assumption.[61]

One reason that the battleship hierarchy retained its preeminent position was the fact that other nations also measured naval power in terms of battleships. During the cold war, naval aviation thought stylists exaggerated any Soviet interest in naval aviation. This American desire to see a mirror image ranged from the hybrid, antisubmarine cruiser-aircraft carriers of the *Kiev* class in the late 1970s through the *Kuznetsov*-class carriers constructed during the 1980s. The American aviation hierarchy tried to transform these ships into aircraft carriers designed for power projection based on the U.S. model. However, both were designed to assist in maintaining local sea control in defense of the Soviet submarine bastions.[62]

Like the battleship, the aircraft carrier is an extremely expensive expression of offensive naval power. Peter Karsten, writing amid the debate over intervention in Vietnam, questioned why the United States built large supercarriers as opposed to inherently defensive technologies, such as coastal defense submarines.[63] The answer is complex, and part of a debate extending back to the Early Republic, but rests in part in the aviation technological paradigm. A recent proposal to construct cheap "arsenal ships," carrying approximately five hundred cruise missiles, met widespread opposition within the navy. After Admiral Mike Boorda's suicide in 1996, fighter pilot Admiral Jay Johnson became chief of naval operations. Johnson defended the aviation technological paradigm and its extensive technological base. He dismissed the arsenal ship concept he inherited as a supporting ("complementary") technology to the aircraft carrier battle group, stating emphatically that "Arsenal Ship IS NOT a replacement for an aircraft carrier [original emphasis]."[64]

A decade ago, Paul Kennedy argued that the staggering expense necessitated by the United States's worldwide strategic commitments, and the requisite military and naval force structures, presented more of a threat to American power than any military opponent.[65] Every nation has a legitimate need to maintain military forces with which to defend itself, but as Kennedy pointed out, determining just what is a reasonable level of defense has been historically very difficult. Justifying expensive supercarriers promises to be very difficult in a post–cold war world populated with new technologies that pose viable presumptive anomalies, such as cruise missiles and UCAVs, and old ones like the submarine.

FRAMEWORKS FOR CHANGE

The history of the battleship supports the position that the momentum of established technologies is difficult to overcome. Proponents of the 1980s Strategic Defense Initiative (SDI) combined a broad coalition of political and economic interests with an organizational bureaucracy to produce significant momentum for the program. SDI opponents overcame this momentum only when the Soviet Union no longer provided a military threat to justify SDI.[66] The parallel with the battleship is clear: with no enemy battle fleet extant in late 1945, its existence could no longer be justified. The aircraft carrier faced the same irrelevance in an early postwar era defined by atomic weapons and no appreciable maritime challenge.

Technological trajectory is also related to technological momentum and thus to technological change. Henk van den Belt and Arie Rip consider technological trajectory a "further articulation" of a technological paradigm, "partly influenced by the selection environment." In relating "societal evolution involving technological development" — certainly evident in the American naval profession — they see a technological "selection environment" that is dependent on technological trajectory. The American naval profession's desire to acquire aircraft carriers to augment the battleship immediately after World War I was a clear example of this. Senior naval officers encouraged and protected aviation development and also provided funding because of its potential to support the existing battleship technological paradigm. Van den Belt and Rip would consider post–World War I encouragement of naval aviation as providing a "niche" that protected this budding technology from the natural selection process.[67]

Edward Constant considered the process of change, based in the concept of presumptive anomaly, "abstruse." It was recognized initially by "those very close to the intellectual foundations of their respective fields."[68] For nascent naval aviation in the early 1920s, this was true of the inventor-theoretician Rear Admiral Bradley A. Fiske and theoretician and Naval War College president Admiral William S. Sims. What made these two rare in their support for an aviation presumptive anomaly was that they were not typical paradigm shifters, that is, "young outsiders . . . not fully committed to the conventional technology." Both were lifelong champions of large battleships.

Constant contended that for a presumptive anomaly to be credible, such as one based on naval aviation, it must be "expressible quantitatively." For a military presumptive anomaly, the ultimate audit is war. In peacetime, quantitative evaluation typically occurs in war games and operational exercises. Robert L. O'Connell argued that late-nineteenth-century war games at the Naval War College were self-referencing and stacked to favor the battleship status quo. An operational exercise or fleet exercise, akin to a field experiment, often had similar biases built in.[69] Experimental rigor was lacking during most of the interwar fleet exercises, which led to results based on selective data or arbitrary assessments of damage. This certainly seemed the case in my own experience with the supercarrier *Enterprise* during the mid-1970s.

In the military arts, a presumptive anomaly has usually required a wartime framework to demonstrate its value. This is due to technological momentum, technological trajectory, and inherent paradigmatic inertia. During the late 1930s, there was increasing professional awareness of the presumptive anomaly posed by naval aviation. However, the shift to an aviation-based technological paradigm required clear demonstrations of naval aviation's superiority in combat during World War II.

In addition to technological paradigms, technological change has also been analyzed within social constructivist, system, and network models.[70] During a recent seminar on the history of technology, my students minimized the conceptual and semantic differences of these approaches. In contrast to my students, I prefer to echo Calvin Coolidge's last will and testament. I remain "mindful" of these frameworks, their contributions to the history of technology, and their applicability to this study.

Social constructivists, such as Wiebe Bijker, have been interested in the artifacts that social groups both identify with and champion. Social constructivists refer to ascendancy of one artifact over its rivals, or "stabilization" of a hierarchy of artifacts, as "closure." This results in relative social tranquility regarding that area of technology and society. For social constructivists, social issues, which tend to be stable rather than revolutionary, are most important. Bijker's concept of normal sociotechnology, in which a dominant group is able "to insist upon its definition of both problems and appropriate solutions," is especially resonant with the naval profession.[71]

The systems approach, typified by the work of Thomas Hughes, relates technology to "social, economic, political and scientific factors." Technological innovators juggle various problems relating to these factors as they build a functioning technological system.[72] For system builders, social values are variable but no more important than other factors. Network proponents, such as Michel Callon and John Law, perceive the "stability and form of artifacts" as "a function of the interaction of heterogeneous elements as these are shaped and assimilated into a network." Unlike the systems approach, network advocates emphasize conflict.[73]

An analysis of the history of U.S. naval technology since the Civil War indicates periods during which each of these models, as well as "soft" technological determinism based in military utility, has been more valid. The ascension of steam propulsion beginning during the Civil War, the replacement of turboelectric propulsion by gear reduction during the 1930s, and the aircraft carrier's suppression of the battleship all typify technologies that overcame the social forces trying to shape them. Conversely, the development and adoption of the turboelectric drive embodies the power of contextual, or social forces, to shape technology. Additional examples can be found in the case of the interwar submarine, forced to serve guerre d'escadre, the airplane and its service to the battleship technological paradigm, and the postwar suppression of cruise missiles.

In a technology-based society such as the navy, new technologies have often been trouble. As Merritt Roe Smith observed, "No matter how innocuous new technologies seem to be, they challenge old values."[74] Innovation leads to questioning of the social, professional, and technological status quo and may pose a valid anomaly that threatens the dominant theoretical paradigm. To preserve "normal sociotechnics," to borrow from Bijker, technological innovation must be strictly controlled.[75] This can be as simple as ignoring new technologies. This was easier for the naval profession before World War II since most naval technologies, such as turboelectric propulsion, were developed in the private sector. Their incorporation into the navy was subject to paradigm-based filters.

Retarding or managing technological innovation often involved control of "engineer-sociologists" within the navy, those engineers capable of upsetting the technology-based status quo.[76] The long, post–Civil War campaign by the line to control naval engineers culminated in the facade of the warrior-engineer amalgamated line in 1899. The warrior-engineer became more of a reality with the accession of a younger generation of officers to the middle ranks of the profession prior to World War I.

For a competitive technological society such as the navy, repression of a new technology is a path fraught with hazards if an anomalous technology is not adopted or co-opted and finds sponsors elsewhere. Co-optation of new technology can be part of the "barrier" Howard Margolis based in the "habits of mind." This is akin to the "sociotechnical ensemble"which supports Bijker's "obdurate boundary artifact" — a perfect characterization of the battleship — and also the trajectories related to technological paradigms.[77]

These concepts are relevant to the introduction and evolution of the airplane and submarine in the early twentieth century. "Normal" practitioners of the battleship technological paradigm viewed aviation as a technology that supported, rather than challenged, the battleship. In fact, the number one acquisition priority of the staff of the chief of naval operations in 1920 was aircraft carriers. In keeping with the battleship thought style, airplanes were "a new form of projectile."[78] The limited capabilities of emergent 1920s aviation technology supported this commensurate view of aviation. This occurred despite the claims of air power advocates, such as Billy Mitchell, who advocated a new national strategic paradigm based on military aircraft.

Although aviation technology advanced during the early 1930s, its proponents were unable to pose a viable presumptive anomaly to challenge the existing battleship paradigm. By the mid-1930s, naval aviators formed a strong thought collective within the naval profession. Nevertheless, they were an entrenched but weaker "sociotechnical frame" — to use Bijker's term. Bijker correctly observed that arguments carry "little weight" in other sociotechnical frames, and this was largely true in many respects for the mainstream interaction between aviation and battleship officers.[79]

Merritt Roe Smith maintained that "a review of the relevant historical literature" suggested that "technical creativity" within the military is dependent upon "able personnel and institutional flexibility, coupled with an infectious enthusiasm for technological change."[80] In the navy, untrammeled enthusiasm for technological change has been rare. Battleship advocates were not flexible — nor did they need to be as long as the battleship technological paradigm was truly preeminent within the international naval framework.

In contrast to aviators, interwar submarine officers were not agitated by externalist arguments similar to Mitchell's for aviation. Submariners' professional survival lay within the battleship paradigm. The development of the "fleet" submarine in the interwar years reflected its subordinate role within the technical hierarchy of the battle fleet and the strategy of guerre d'escadre. It also bore the imprint of the technological trajectory and momentum of the battleship technological paradigm. Ironically, the commencement of hostilities in the Pacific in December 1941 resulted in fleet submarines being used in the strategy of guerre de course, resurrected as the secondary task-oriented U.S. naval strategic mission in 1940.

Other people's wars have been touchstones for modern military professionals. The Russo-Japanese War of 1904–5 provided ammunition for competing refinements of the battleship paradigm. Later, the European War between September 1939 and December 1941 provided the U.S. naval profession with empirical reinforcement of the battleship technological paradigm. Britain's pursuit of the German battleship *Bismarck* in May 1941 dovetailed well with the belief that lesser technologies located and even harassed enemy battleships. The coup de grace could only occur in a battleship versus battleship slugfest. Apparent anomalies, such as the British torpedo bomber attack on Italian battleships in Taranto Harbor in 1940, were dismissed by ethnic prejudice.[81]

It was little wonder that the disastrous fate of the battleship force at Pearl Harbor in December 1941 provoked such angst among most American naval officers. War, in the form of attacking Japanese naval aircraft, seemed to many to be a ruthless agent of natural selection. The Pearl Harbor attack shook the battleship technological paradigm by dramatically destroying much of its technological foundation.

Popular culture holds Pearl Harbor as the demarcation between the old battleship navy and the new, aviation-based navy. However, the beginning of a paradigmatic shift cannot be placed earlier than July 1943, when Admiral Ernest J. King, chief of naval operations and commander-in-chief, U.S. Fleet, canceled the *Montana*-class superbattleships. The industrial and material resources they required could no longer be justified. Nevertheless, King's action only occurred in the wake of the assessment that completion of four of the *Iowa*-class battleships would provide sufficient fast battleships to ensure victory.

During World War II, the fast aircraft carrier, developed at the end of the 1930s, joined with the post-1937 fast battleships to dominate three-dimensional naval warfare — above, on, and under the sea. One must remember that the extremely capable carrier task forces fielded by the navy during the war were ineffective at night and in bad weather. This was not true for the battleship. The fast task force of battleship *and* carriers was an interim step in the transition to the postwar naval aviation technological paradigm capable of day and night combat operations and domination of war at sea.

World War II provided the venue for naval aviation to prove the superiority of the aircraft carrier over the battleship. As Rear Admiral Fiske predicted in 1917, naval aviation could concentrate "mechanical power" at a greater distance than the battleship.[82] Carrier aircraft also successfully countered the submarine, something the battleship never did. The ascension of the aircraft carrier technological paradigm after World War II reordered the naval profession. As a result, supporters of the naval aviation technological paradigm — not all aviators — have dominated the naval profession for much of the past fifty years.

"History Is Bunk"

Post-1945 American naval strategy has focused on conventional (nonnuclear) war at sea, which maintains sea control in order to project power ashore. The aircraft carrier dominated this version of naval warfare in the same way the battleship ruled naval warfare before World War II. Like the battleship officers in the late 1930s, the naval aviators of the 1970–80s wistfully minimized the vulnerability of the modern supercarrier. Nevertheless, that vulnerability was rooted in the increased capabilities of nuclear-powered submarines. The great difficulty in detecting and countering them was a technological discontinuity of the cold war. Carrier aviators minimized the submarine and considered the submarine problem solved by strike-at-source operations. This echoed the post–World War I opinion of Sims and many of his colleagues that the submarine was no longer a threat thanks to the tactics and acoustic detection devices developed during World War I.

The naval aviation technological paradigm remained unchallenged until the 1991 Persian Gulf War. Then, senior military officers, faced with a sophisticated and difficult air defense system within Iraq, decided to employ cruise missiles. This decision was critical in demonstrating and quantifying the cruise missile presumptive anomaly — a technology carried by surface ships and submarines controlled by nonaviators. The selection of cruise missiles again in 1993, 1995, 1998, and 1999 has marked the crumbling of a significant Margolian intellectual barrier among national security officials. Henry Kissinger's earlier query amid international crises, "Where are the carriers?" has been replaced by "Where are the cruise missile ships?" This shift has enfranchised SLCM-equipped surface ships and submarines and empowered their officer-operators within the naval profession.

The mid-1992 reorganization of the office of the chief of naval operations weakened the institutional status quo that supported the aviation technological paradigm. The technology-specific deputy chiefs of naval operations — the DCNOs for air, surface, and submarine warfare — ceased to exist. According to former vice chairman of the Joint Chiefs of Staff, Admiral William A. Owens (a submariner), the change occurred because of the "awareness that if the Navy was going to go beyond the rhetoric of change and actually incorporate the tenets of its new operational concept into the size and structure of naval forces, the influence of those institutions within the staff that had become the bastions of the older Maritime Strategy had to be altered."[83] Since 1992, the principal officers of each technology-defined warfare specialty (aviation, surface ships, and submarines) now report to the vice admiral serving as deputy chief of staff for resources, warfare requirements, and assessments (in staff terms, "N8"). Admiral Owens, the first N8, was optimistic about the ability to engender change. Nevertheless, change can only occur if the admiral serving as N8, and the N8 bureaucracy, is able to escape the influence and intellectual boundaries of past thought styles.

A subtle indication of paradigmatic uncertainty and a move toward the flexible "three-plane" navy Admiral William Fullam advocated in 1924 can be found in the navy's recent rhetoric. In 1997 the naval hierarchy started talking about minimizing the importance of individual technologies (aircraft carriers, cruisers, aircraft, or submarines) — termed "platform centric warfare" — in favor of technologically diverse "network centric warfare."[84]

For the naval profession, in many ways, past is prologue. Just as World War II was necessary for a paradigm shift from the battleship, the Persian Gulf War provided a catalyst for a shift away from singular emphasis on the aircraft carrier. The end of the 1990s was also much like the end of the 1930s, replete with a shifting strategic focus and a viable presumptive anomaly pressuring the navy's dominant technological paradigm. In 1937 Admiral William D. Leahy, chief of naval operations and battleship sailor, characterized the battleship as "the backbone of naval power."[85] Sixty years later, Admiral Jay L. Johnson, chief of naval operations and fighter pilot, adamantly defended the "obvious" relevance of the aircraft carrier and its role as "the right ship for the second half of the 21st Century."[86] Such comments are the late-term affirmations of ancien regimes struggling to control technological change. Admiral Leahy's opinion was overtaken by the dynamics of World War II. Admiral Johnson's prediction may well fall victim to the legacy of the Persian Gulf War, to the unacceptable political fallout from losing aircrews while projecting naval power ashore in operations short of large-scale war, to the presumptive anomalies posed by new technologies, to an increasing political consensus that the aircraft carrier has had its day, or to a genuine movement toward "network centric" warfare with a deemphasis on the aircraft carrier.

American naval capabilities have come a long way since 1804, when Commodore Edward Preble borrowed six gunboats and two bomb ketches from the Neapolitan government to lob a few rounds into the walled fortress at Tripoli.[87] The questions the American naval profession must answer in the year 2000 are similar to those it faced in 1900: What type of navy does the United States require? What is the proper measure of naval power? What is the nature of war at sea? What is heroism and who is a warrior in a modern, high-technology navy? How, and to what extent, can technological change be controlled? The answer to each of these questions has been linked inextricably to the navy's technological paradigms.

The intellectual, social, military, and political dynamics surrounding these questions during the battleship era were complex. The dynamics of twenty-first-century American naval power will be just as convoluted and bedeviling. However, a better understanding of the battleship era is instructive in the dynamics of technology selection, in general, and the specific decisions being made for the future of the U.S. Navy and the nation it serves.

This study cautions against the use of social constructivism or technological determinism as monolithic explanatory frameworks for technological change. Perhaps most important, this work illustrates an essential aspect of the nature of technological innovation and the evolution of the modern military arts. Although this study focuses on the battleship, it is applicable to the postwar navy and, as a theoretical model, to other military services.[88]

I. B. Holley concluded his 1953 seminal study, *Ideas and Weapons*, by observing that "to exist in a warring world the nation must pick winning weapons; if military analysts will distill every possible lesson from the history of two world wars such weapons will be easier to find and the odds on national survival will go up."[89] Holley assumed that senior military officers are receptive to potentially destabilizing intellectual analyses. The irony is that the modern U.S. Army and U.S. Air Force cultivate intellectual growth within their officer corps, especially in history, more than the U.S. Navy. Yet the army and air force seem more restrictive in their internal dialectics. The U.S. Navy has tolerated broad debate within its officer corps, yet today does little to encourage historical thinking.[90] Almost fifty years after Holley, it seems a broad spectrum of dynamics — from thought styles and professional rivalry to the defense of the dominant technological paradigm from competing technologies — have clouded naval perceptions of which technology is a "winner."

A century after the heyday of its historical prophet, Alfred Thayer Mahan, the American naval profession, always technically oriented, has become increasingly a technocracy in which historical analysis has been pushed to the far periphery. Chief of Naval Operations Johnson has stated repeatedly that the navy cannot steer by its wake into the next century. One can compare Johnson's pronouncement with the progressivism of automobile pioneer Charles Kettering, who often said, "You never get anywhere looking in your rearview mirror." Kettering, however, was a technological innovator, not the guardian of an established technological paradigm. Many officers of the modern naval profession confuse technical competence with an understanding of technology, embodying Henry Ford's dictum "history is bunk." Modern naval leaders routinely tout "thinking outside the box," but the absence of historical understanding of the dynamics of technology and change merely reinforces their self-referential thinking and their allegiance to their current castles of steel.[91]

During the twenty-first century many developing countries will acquire sophisticated weapons. They will be least able to afford them and most likely to use them in conflicts exacerbated by their hypernationalism. The U.S. Navy will often be involved either as peacekeeper or enforcer, operating "from the sea." If the American naval profession refuses to pay attention to its wake — as any competent deck officer does — its members may find the U.S. Navy steaming full circle, like the doomed battleship *Bismarck*, and falling victim to someone else's ability to perceive and exploit an effective presumptive anomaly.

Notes

INTRODUCTION

1. When surprised by a superior Japanese force of battleships and cruisers during the Battle of Leyte Gulf, Rear Admiral C. A. F. Sprague ordered his escorting destroyers and destroyer escorts to attack. The Japanese sank two destroyers and one destroyer escort, and 526 American destroyer sailors died before noon. See Theodore Roscoe, *United States Destroyer Operations in World War II* (Annapolis: Naval Institute Press, 1953), chap. 33.

2. British admiral's quote from Robert L. O'Connell, *Sacred Vessels: The Cult of the Battleship and the Rise of the U.S. Navy* (Boulder, Colo.: Westview Press, 1991), 149.

3. "2. Of a discarded type; out of date; *an obsolete battleship." The American College Dictionary* (New York: Random House, 1965), 837.

4. H. Bruce Franklin, *War Stars: The Superweapon and the American Imagination* (New York: Oxford University Press, 1988).

5. Martin van Creveld, *Technology and War: From 2000 B.C. to the Present* (New York: The Free Press, 1989). Conversely, a solid attempt to escape technological determinism and to address the social, intellectual, and institutional context of the evolution of war and strategic thought is *Tools of War: Instruments, Ideas, and Institutions of Warfare, 1445–1871*, ed. John A. Lynn (Urbana: University of Illinois Press, 1990). Lynn's volume ends at 1871, prior to the explosion of scientific engineering, the rapid increase in military-industrial relations, and the creation of the modern command economies described by William H. McNeill in *The Pursuit of Power: Technology, Armed Force, and Society since A.D. 1000* (Chicago: University of Chicago Press, 1982). A technologically determinist treatment of naval history is Karl Lautenschlager, "Technology and the Evolution of Naval Warfare," in *Naval Strategy and National Security*, ed. Steven E. Miller and Stephen Van Evera

(Princeton, N.J.: Princeton University Press, 1988), 173–221. Advocates of linear technological development would benefit from a careful reading of George Basalla, *The Evolution of Technology* (Cambridge: Cambridge University Press, 1988).

6. Merritt Roe Smith, "Introduction," in *Military Enterprise and Technological Change: Perspectives on the American Experience,* ed. Merritt Roe Smith (Cambridge, Mass.: The MIT Press, 1985), 4.

7. Edward W. Constant II, *The Origins of the Turbojet Revolution* (Baltimore: Johns Hopkins University Press, 1980), chap. 1.

8. For an introduction to technological momentum, see Thomas Parke Hughes, "Technological Momentum," in *Does Technology Drive History?: The Dilemma of Technological Determinism,* ed. Merritt Roe Smith and Leo Marx (Cambridge, Mass.: The MIT Press, 1996), 101–14. For technological trajectories, see Henk van den Belt and Arie Rip, "The Nelson-Winter-Dosi Model and Synthetic Dye Chemistry," in *The Social Construction of Technological Systems: New Directions in the Sociology and History of Technology,* ed. Wiebe J. Bijker, Thomas P. Hughes, and Trevor Pinch (Cambridge, Mass.: The MIT Press, 1994), 135–58. See their account of Neo-Schumpeterian economic theories' treatment of "technological trajectories" as by-products of technological paradigms.

9. See Thomas Kuhn, *The Structure of Scientific Revolutions,* 2nd ed. (Chicago: University of Chicago Press, 1970). For an application of Kuhn to the military, see Tim Travers's study of technology and doctrine in the pre-1914 Edwardian army: *The Killing Ground: The British Army, the Western Front and the Emergence of Modern Warfare, 1900–1918* (London: Allen & Unwin, 1987).

10. Kenneth J. Hagan, *This People's Navy: The Making of American Sea Power* (New York: The Free Press, 1991), 228.

11. I am referring to the concept of thought collective (*Denkkollektiv*) and thought style (*Denkstil*) elucidated by Ludwik Fleck, *Genesis and Development of a Scientific Fact,* ed. Thaddeus K. Trenn and Robert K. Merton (Chicago: University of Chicago Press, 1979), 39.

12. It is important to distinguish between paradigms. The strategic paradigm based on guerre d'escadre theoretically could be served by different technological paradigms, for example, either battleship-based or aircraft carrier-based.

13. Constant, *Origins,* 12–15.

14. Ibid., 16.

15. Ibid.

16. Introduction, "Third Report of Operations of the United States Navy in World War II, 1 March 1945–1 October 1945," in Fleet Admiral Ernest J. King,

USN, *U.S. Navy at War, 1941–1945: Official Reports to the Secretary of the Navy* (Washington, D.C.: U.S. Government Printing Office, 1946), 169.

17. Ibid., 167.

18. Peter Paret, "Introduction," in *Makers of Modern Strategy: From Machiavelli to the Nuclear Age,* ed. Peter Paret (Princeton, N.J.: Princeton University Press, 1986), 8.

CHAPTER 1.   THE POSTBELLUM NAVAL PROFESSION

1. On warriors and new technologies, see Robert L. O'Connell, *Of Arms and Men: A History of War, Weapons, and Aggression* (New York: Oxford University Press, 1989), chap. 1.

2. Ibid., especially 7–11.

3. See Carlo M. Cipolla, *Guns, Sails, and Empires: Technological Innovation and the Early Phases of European Expansion, 1400–1700* (New York: Pantheon, 1966), chap. 2, especially 101–2; and John Law, "Technology and Heterogeneous Engineering: The Case of Portuguese Expansion," in *The Social Construction of Technological Systems,* ed. Wiebe E. Bijker, Thomas P. Hughes, and Trevor Pinch (Cambridge, Mass.: The MIT Press, 1987), 111–34. For a detailed analysis of the social, economic, and cultural dynamics surrounding the interaction of two technologically discontinuous systems, see John F. Guilmartin, *Gunpowder and Galleys: Changing Technology and Mediterranean Warfare at Sea in the Sixteenth Century* (London: Cambridge University Press, 1974).

4. The bureau system and Engineering Corps date from congressional legislation approved in 1842. Five bureaus replaced the three-captain Board of Navy Commissioners in advising the secretary of the navy. The legislation specified that a navy captain should head the Bureau of Navy Yards and Docks and Bureau of Ordnance and Hydrography. The chief of the Bureau of Medicine and Surgery was a navy surgeon. The legislation placed no restriction on who could be chief of the Bureau of Provisions and Clothing. The Bureau of Construction, Equipment and Repairs, in which steam engineering was placed, was to be commanded by a "skillful naval constructor"; *History of the Construction Corps of the United States Navy* (Washington, D.C.: U.S. Government Printing Office, 1937), 16.

5. The Russian navy employed shell guns at Sinope in 1853, sinking nine of ten wooden Turkish warships and killing four thousand sailors. For an overview of shell guns and ironcladding, see O'Connell, *Of Arms and Men,* 193–95.

6. For the definition of a presumptive anomaly, see Edward W. Constant II, *The Origins of the Turbojet Revolution* (Baltimore: Johns Hopkins University Press, 1980), 15–16.

7. An overview of the steam frigate *Demologos* is in Passed Assistant Engineer Frank M. Bennett, U.S. Navy, *The Steam Navy of the United States: A History of the Growth of the Steam Vessel of War in the U.S. Navy, and of the Naval Engineer Corps* (Pittsburgh, Pa.: Warren and Company, 1896), chap. 2, especially 8–15.

8. Constant quoted in John M. Staudenmaier, *Technology's Storytellers: Reweaving the Human Fabric* (Cambridge, Mass.: The MIT Press, 1989), 109.

9. Bennett, *Steam Navy*, 16–19; quote from Captain John Rodgers, Board of Navy Commissioners, to the secretary of the navy, 30 December 1835, reprinted in Bennett, 18.

10. For the U.S. Navy's early, far-flung deployments in support of the American "empire of commerce," see Kenneth J. Hagan, *This People's Navy: The Making of American Sea Power* (New York: The Free Press, 1991), chap. 4.

11. See Elting Morison, *Men, Machines, and Modern Times* (Cambridge, Mass.: The MIT Press, 1966); Monte A. Calvert, *The Mechanical Engineer in America, 1830–1910: Professional Cultures in Conflict* (Baltimore: Johns Hopkins University Press, 1967), chap. 13; Edward William Sloan III, *Benjamin Franklin Isherwood, Naval Engineer: The Years as Engineer in Chief, 1861–1869* (Annapolis: Naval Institute Press, 1965) especially chaps. 9–11; Robert G. Albion, *Makers of Naval Policy 1798–1947* (Annapolis: Naval Institute Press, 1980); and Lance C. Buhl, "Mariners and Machines: Resistance to Technological Change in the American Navy, 1865–1869," *Journal of American History* 59 (1974): 703–27.

12. Herman Melville, *White-Jacket: The World in a Man-of-War* (New York: Oxford University Press, 1990).

13. Hagan, *This People's Navy*, 108–9; and Bennett, *Steam Navy*, chap. 3.

14. On American steam propulsion in the Pacific, see Hagan, *This People's Navy*, 109.

15. A useful summary of steamship operations during the war with Mexico is in Bennett, *Steam Navy*, chap. 7.

16. Craig L. Symonds, *Historical Atlas of the U.S. Navy* (Annapolis: Naval Institute Press, 1995), 72.

17. William H. McNeill, *The Pursuit of Power: Technology, Armed Force, and Society since A.D. 1000* (Chicago: University of Chicago Press, 1982), chap. 5.

18. For details on the Engineering Corps during the 1850s, see Bennett, *Steam Navy*, chap. 11.

19. Because of their work within a bureaucracy, Calvert argued that naval engineers identified with "Engineers" and wished "to avoid identification with the mechanic-machinist role"; Calvert, *Mechanical Engineer*, 260. The term "naval engineer" commonly refers to propulsion engineers who were more numerous but

can include, in a broad sense, naval constructors, who were relatively few in number and relegated to shore duty in Washington and at shipyards. The Engineering Corps originated with the 1842 administrative reorganization of the navy while the Construction Corps was created by an act of Congress in July 1866; see *History of the Construction Corps*, 1.

20. Hasseltino was ordered to duty in Rear Admiral David D. Porter's flagship, *Essex*, where his great fear was "that the *Essex* would be ordered to get underway and to go somewhere, and he would consequently be called upon to do something with the machinery"; Bennett, drawing on Hasseltino's diary, *Steam Navy*, 213. For other accounts of engineer amateurs, see Bennett, chap. 13.

21. See Peter Karsten, *The Naval Aristocracy: The Golden Age of Annapolis and the Emergence of Modern American Navalism* (New York: The Free Press, 1972), especially chap. 3.

22. Commander Alfred Thayer Mahan, USN, to Samuel Ashe, 27 January 1876, quoted by Karsten, *Naval Aristocracy*, 65.

23. Du Pont blamed failure of his attack on Fort Sumter on the poor quality of his ironclads, a view disputed in Stimers's report. Stimers was vindicated by a court of inquiry. Assistant Secretary of the Navy Gustavus Fox wrote of Du Pont: "He is of a wooden age, eminent in that, but in an engineering age, behind the time." The Du Pont–Stimers affairs is the subject of Bennett, *Steam Navy*, chap. 23; Fox quote from a letter to John Ericsson on 419. Wartime commendations for engineers are sprinkled throughout Bennett's chapters on the Civil War (chaps. 12–27).

24. Bennett, *Steam Navy*, 401–2.

25. *Annual Report of the Secretary of the Navy 1863*, pertinent sections reprinted in Bennett, *Steam Navy*, 654–56. According to Bennett, the relevant section in the annual report was most likely written by the pro-engineer Fox.

26. Quote from *Annual Report of the Secretary of the Navy 1864* reprinted in Bennett, *Steam Navy*, 658–59.

27. Bennett, *Steam Navy*, 660–63.

28. Quote from *Annual Report of the Secretary of the Navy 1866* reprinted in Bennett, *Steam Navy*, 664.

29. Bennett, *Steam Navy*, 668.

30. Ibid., 603.

31. Assistant Secretary of the Navy Gustavus Fox made a point of crossing the Atlantic in 1866 in a monitor, *Miantonomah*, when carrying a note to the Russian tsar; William J. Sullivan, "Gustavus Fox and Naval Administration, 1861–66" (Ph.D. diss., The Catholic University of America, 1977), 297–98.

32. On Porter and Borie, see Albion, *Naval Policy*, 56.

33. Morison, *Men, Machines, and Modern Times*, chap. 3.

34. Buhl, "Mariners and Machines," passim.

35. Albion, *Naval Policy*, 201.

36. For an introduction to the literature of technology transfer and technological cultural imperatives, see John Staudenmaier, *Technology's Storytellers: Reweaving the Human Fabric* (Cambridge, Mass.: The MIT Press, 1989), chap. 4 and passim. On the machine and the craft tradition at Harpers Ferry, see Merritt Roe Smith, *Harpers Ferry Armory and the New Technology: The Challenge of Change* (Ithaca, N.Y.: Cornell University Press, 1977).

37. Hagan, *This People's Navy*, 228.

38. See the statement of Sir Edward Reed and letter by Sir Thomas Brassey regarding the alarm *Wampanoag* caused in British naval circles; Sloan, *Isherwood*, 188.

39. Sloan, *Isherwood*, passim, especially chap. 10.

40. Porter quoted in Karsten, *Naval Aristocracy*, 66; the lobbying fund is mentioned on 67.

41. See "Report of Admiral D.D. Porter to the Secretary of the Navy, November 7, 1874," in *Annual Report of the Secretary of the Navy on the Operation of the Department with Accompanying Documents for the year 1874* (Washington, D.C.: U.S. Government Printing Office, 1875), 198–222.

42. On the Mamelukes' cavalry society and their defeat by the gunpowder army of the Ottomans, see John Keegan, *A History of Warfare* (New York: Knopf, 1993), 36–37.

43. For the transformation of the guild of ship construction into the engineering discipline of naval architecture, see *The Institution of Naval Architects, 1860–1960: An Historical Survey of the Institution's Transactions and Activities over 100 Years*, ed. Kenneth C. Barnaby (London: Royal Institution of Naval Architects in association with Allen & Unwin, 1960). The first armored iron warship was HMS *Warrior* (1860). On her development, see Oscar Parkes, OBE, *British Battleships from* Warrior *to* Vanguard: *A History of Design, Construction and Armament* (London: Seeley Service, 1970).

44. France, which embarked on a divergent naval strategy in 1694 (privateering), continued that trend with the Jeune École of the latter nineteenth century, which favored the torpedo boat as a technologically discontinuous response to Britain's strategy of guerre d'escadre based in the battleship. On French naval divergence late in the reign of Louis XIV, see Geoffrey Symcox, *The Crisis of French Sea Power, 1688–1697: From the Guerre d'Escadre to the Guerre de Course* (The Hague: Nijhoff, 1974); and McNeill, *Pursuit of Power*, 178–79. For the Jeune École, see

O'Connell, *Of Arms and Men*, 221. Also see Theodore Ropp, *The Development of a Modern Navy: French Naval Policy, 1871–1904* (Cambridge, Mass.: Harvard University Press, 1937).

45. Naval engineers also formed professional societies; the American Society of Naval Engineers was created in 1888. For parallel moves to achieve professional status among civilian engineers, see Edwin T. Layton Jr., *The Revolt of the Engineers: Social Responsibility and the American Engineering Profession* (Cleveland, Ohio: Press of the Case-Western Reserve University, 1971), chap. 2

46. Karsten, *Naval Aristocracy*, chap. 6.

47. The naval public relations effort centered upon the importance of a battle fleet to the maintenance of American economic power; see Karsten, *Naval Aristocracy*, 301–2.

48. For the difficulties technologically naïve senior naval officers experienced when faced with selection of competing technologies and the rise of technocrat officers, see McNeill, *Pursuit of Power*, 274.

49. Calvert, *Mechanical Engineer*, 257–58.

50. For a discussion of science in the nineteenth-century navy, see A. Hunter Dupree, *Science in the Federal Government: A History of Policies and Activities* (Baltimore: Johns Hopkins University Press, 1986), chap. 9; and Calvert, *Mechanical Engineer*, 255.

51. Calvert, *Mechanical Engineer*, 255; and Albion, *Naval Policy*, 201.

52. Jack Sweetman, *The U.S. Naval Academy: An Illustrated History* (Annapolis: Naval Institute Press, 1979), 108.

53. Calvert, *Mechanical Engineer*, 255.

54. Graduation data from United States Naval Academy Alumni Association, Inc., *Register of Alumni: Graduates and former Naval Cadets and Midshipmen* (Annapolis: Naval Academy Alumni Association, 1994 edition), 158–60.

55. The pertinent section of the Act is reprinted in Bennett, *Steam Navy*, 675.

56. Sweetman, *U.S. Naval Academy*, 116–17.

57. The legal battles of the Naval Academy cadet engineers is in Bennett, *Steam Navy*, 752–56.

58. Naval Cadet George F. Zinnel et al. to Secretary of the Navy William E. Chandler, 2 October 1882, Record Group 405, Records of the U.S. Naval Academy and Correspondence of the Superintendent, U.S. Naval Academy Archives, Nimitz Library, Annapolis, Md., formerly Record Group 405, National Archives (hereafter cited as RG 405, USNA), Entry 25, Box 22.

59. Ramsay's capability to act the martinet was demonstrated throughout his tenure. During graduation in June 1883, Ramsay interrupted the ceremony to con-

fine twenty cadets, including some graduates, to the prison ship *Santee,* for cheering the graduation of the first member of the class. See Sweetman, *U.S. Naval Academy,* 129.

60. William E. Chandler to Captain F. M. Ramsay, USN, 17 October 1882, Entry 25, Box 22, RG 405, USNA.

61. Secretary Chandler to Naval Cadet Zinnell et al., 20 October 1882, Entry 25, Box 22, RG 405, USNA.

62. Secretary Chandler to Superintendent Ramsay, 3 May 1883, Entry 25, Box 22, RG 405, USNA.

63. Ibid.

64. Ibid.

65. This is based upon material in RG 405, USNA, and upon inference from the imprecise information provided for the Class of 1883 in the U.S. Naval Academy Alumni Association's *Register of Alumni, 1845–1986* (Annapolis: Naval Academy Alumni Association, 1986), 160.

66. Secretary Chandler to Naval Cadet George F. Zinnell, 20 October 1882 (n. 61 above).

67. Naval Constructor John D. Tawresay, USN, to Secretary of the Navy John D. Long, 31 October 1898, General Records of the Secretary of the Navy, Record Group 80, National Archives, Washington, D.C. (hereafter RG 80), 1897–1915. The British director of naval construction, a civilian, Mr. William White, was a strong proponent of having foreign students at the school. He had been successful in blocking exclusionary rulings before, but was absent when the Admiralty Board brought the question up this last time.

68. *Report Concerning Certain Alleged Defects in Vessels of the United States Navy, by Washington Lee Capps, Chief Constructor, U.S. Navy, and Chief of the Bureau of Construction and Repair, Senate Document No. 297, 60th Congress, 1st Session* (Washington, D.C.: U.S. Government Printing Office, 1908), 1–2. The navy sent its first student abroad in 1879 when a naval cadet requested leave to study naval architecture at the Royal Naval College; see Sweetman, *U.S. Naval Academy,* 111.

69. Memorandum for the Superintendent, U.S. Naval Academy from Assistant Naval Constructor R. P. Hobson, USN, 17 February 1897, RG 405, USNA, Box 1.

70. Ibid.

71. Ibid.

72. William A. Baker, *A History of the Department of Naval Architecture and Marine Engineering, Massachusetts Institute of Technology, Department of Naval Architecture and Marine Engineering, Report No. 69–3* (Cambridge, Mass.: Massachusetts Institute of Technology, 1969), (hereafter *MIT Report 69–3*), 9.

73. *MIT Report* 69–3, 2–7.

74. Ibid., 10.

75. Assistant Naval Constructor Hobson to Superintendent, U.S. Naval Academy, 23 September 1897, RG 405, USNA.

76. "Memorandum for the Superintendent on the Necessity for *Three Years* for Post-Graduate Course," Hobson to Superintendent, U.S. Naval Academy, 2 October 1897, RG 405, USNA. Hobson's course of study was comparable to those attended by American naval constructors abroad. A British student spent three years at the Royal Naval College, Greenwich, after five years in the dockyard schools. In France, students spent two years at the École Polytechnique before their final two years of study at the École du Genie Maritime. The course at Glasgow lasted two years, but presupposed extensive knowledge on shipbuilding obtained in the shipyard schools. The course at Glasgow was less desirable and only used by American students when the Greenwich and École du Genie Maritime classes were filled. All the U.S. Navy students had spent three years at Greenwich and between three and four years at Paris.

77. Ibid.

78. A. S. Crowninshield, Acting Secretary, to Rear Admiral John A. Howell, USN, 25 October 1898, RG 405, USNA.

79. Academicians at the Naval Academy were assigned naval professorial rank to clarify disciplinary problems arising from interactions with the student body.

80. "Report of the Board on the Course in Naval Architecture, 9 January 1899," RG 405, USNA.

81. Ibid. The other subjects to be studied during the first two years were thermodynamics, differential and integral calculus, analytic geometry, analytical mechanics, statics of structures, strength of materials, kinematics and dynamics of mechanics, and hydromechanics.

82. Ibid.

83. Ibid.

84. Ibid.

85. Navy Secretary John D. Long to the Secretary of the Treasury, 13 January 1899, and John D. Long to Senator Eugene Hale, Chairman, Committee on Naval Affairs, 13 January 1899, RG 405, USNA.

86. Bureau of Construction & Repair to Superintendent, US Naval Academy, 1 February 1899, and William A. Fairburn to Commodore Hichborn, 28 January 1899, in RG 405, USNA, Entry 29.

87. Navy Secretary John D. Long's Endorsement on the "Report of the Board on the Course in Naval Architecture," 3 February 1899, RG 405, USNA, Entry 29.

88. Naval Constructor Lawrence Spear, USN, to Superintendent, US Naval Academy, 10 February 1899, RG 405, USNA, Entry 29.

89. Ibid.

90. Spear's letter of 7 March 1899 and the superintendent's endorsement that led to the cancellation of the postgraduate program at Annapolis have not survived.

91. See, for example, Naval Constructor Spear, USN, to Superintendent, US Naval Academy, 10 December 1898, RG 405, USNA, Entry 29.

92. Rear Admiral Crowninshield, Chief, Bureau of Navigation to Superintendent, US Naval Academy, 16 March 1899, RG 405, USNA, Entry 29.

93. Captain Sigsbee (ONI) to the President of the Boston School of Technology, 21 September 1900, and MIT President H. S. Pritchett to Sigsbee, 28 September 1900, Institute Archives and Special Collections, Massachusetts Institute of Technology, AC 13, Box 21, Folder 604. Sigsbee may have been selected to contact MIT because he was a scientific veteran of the Coast Survey.

94. *MIT Report 69–3*, 9.

95. Commander Richard Wainwright, Superintendent, US Naval Academy, to Professor Hendrickson et al. of the Academy Academic Board, 10 February 1902, RG 405, USNA, Entry 29.

96. Special Board (Professor Hendrickson) to Superintendent, U.S. Naval Academy, 25 March 1902, RG 405, USNA, Entry 29.

97. See Lieutenant Commander F. H. Eldridge, Head, Department of Marine Engineering and Naval Construction, to Superintendent, 8 April 1902, and Professor N. M. Terry, Head, Department of Physics and Chemistry, to Superintendent, 29 March 1902, RG 405, USNA, Entry 29.

98. Superintendent Wainwright to Chief of the Bureau of Navigation, 18 April 1902, and Superintendent Wainwright to Chief, Bureau of Navigation, second endorsement, 22 April 1902, RG 405, USNA, Entry 29.

99. Engineer-in-Chief George Melville, USN, to Chief, Bureau of Navigation, 28 April 1902, RG 405, USNA, Entry 29.

100. Ibid.

101. See Hagan, *This People's Navy*, 232 and 207.

102. Admiral George Dewey, USN, President, General Board, to Chief, Bureau of Navigation, sixth endorsement to USNA letter No. 168, 20 June 1902, RG 405, USNA, Entry 29.

103. Engineer-in-Chief Melville to Chief, Bureau of Navigation, eighth endorsement to USNA letter No. 168, 1 July 1902, RG 405, USNA, Entry 29.

104. The report of the bureau chiefs is reprinted in Commander Reginald R.

Belknap, USN, "The Postgraduate Department of the Naval Academy," *U.S. Naval Institute Proceedings* (hereafter *USNIP*) 39 (1913): 135–53, 141–42.

105. Passed midshipmen had completed their undergraduate requirements, but had not yet completed the postgraduation cruises required prior to commissioning.

106. Belknap, "The Postgraduate Department of the Naval Academy," 141–42.

107. Ibid., 143–44.

108. Secretary of the Navy George von L. Meyer to Commanders Afloat and Ashore, 3 March 1909, RG 80, 1897–1915, File 2091.

109. General Order No. 233 cited by Superintendent, US Naval Academy, in his letter to the Chief, Bureau of Navigation, 19 April 1913, RG 405, USNA, File 468B.

110. Head of Post Graduate Department to Superintendent, US Naval Academy, 4 February 1913, and Head of Post Graduate Department to Superintendent, 12 February 1913, RG 405, USNA, File 468B.

111. Head of Post Graduate Department to Superintendent, US Naval Academy, 20 February 1913, RG 405, USNA, File 468B.

112. Lieutenant Commander John Halligan Jr., USN, "Post Graduate Education in Naval Engineering," *Journal of the American Society of Naval Engineers* (hereafter *ASNE Journal*) 28 (1916): 215–29; 220.

113. Halligan, "Post Graduate Education in Naval Engineering," 228.

114. Opposition to postgraduate education among older line officers remained strong after World War I. Naval Academy Superintendent Henry B. Wilson, who actively worked to have the Annapolis Post Graduate School discontinued, or at least moved away from Annapolis, was representative. See the interesting record of Wilson's activities in RG 405, USNA, Box 1, File 6, as well as Fleet Admiral Ernest J. King, USN, and Walter Muir Whitehill, *Fleet Admiral King: A Naval Record* (New York: Norton, 1952), chap. 12.

115. Roosevelt was quoted by retired Engineer-in-Chief George Melville, USN, in a lecture delivered before the Stevens Institute, 9 September 1909; the text is contained in the Papers of George H. Melville, Manuscript Division, Library of Congress (hereafter cited as the Melville Papers), Washington, D.C.

116. See Lieutenant Commander John L. Gow, USN, "Discussion — Is Amalgamation a Failure?" *USNIP* 32 (1906): 307–8.

117. Melville's Stevens Institute Speech, Melville Papers.

118. Ibid.

119. Ibid.

120. Ibid. The type of sailing rig for the new training ship was more important than postgraduate education among Academy line officers and sparked quite a debate. See Commander C. T. Hutchins, USN, Commandant of Cadets to Super-

intendent, US Naval Academy, 24 October 1898, Entry 29, RG 405, USNA, and Rear Admiral McNair to Bureau of Navigation, 26 October 1898, Entry 29, RG 405, USNA.

121. Lieutenant Commander L. H. Chandler, "Is Amalgamation a Failure? — Being an Examination of the So-Called Proofs that such is the Case, and a Defense of Our Present Engineering Organization of the Commissioned Personnel of the Navy," *USNIP* 31 (1905): 823–943.

122. Ibid., 829.

123. Ibid., 830.

124. *Chief of the Bureau of Steam Engineering Annual Report 1904*, quoted in Chandler, "Is Amalgamation a Failure?" 833.

125. Rear Admiral G. W. Baird, USN, "Discussion — Is Amalgamation a Failure?" *USNIP* 32 (1906): 311.

126. Ibid.

127. Ibid., 315.

128. Ibid., 314.

129. Chief, Bureau of Construction & Repair to Chairman, House Naval Affairs Committee and to Secretary of the Navy, 18 February 1913, letter serial 575.A490, 249A, 821-A, in Record Group 10, Box 10, C&R 1913 Folder, Papers of Franklin Delano Roosevelt as Assistant Secretary of the Navy, Franklin D. Roosevelt Library, Hyde Park, N.Y.

130. "The four men who resigned did so to accept positions in the line of their profession with shipbuilding corporations where pay was materially greater and the prospects better than those offered in the Navy at the present time to members of the Construction Corps. Three of the men stood five, ten, and eleven on the [seniority] list of Naval Constructors, with the rank of lieutenant-commander, with prospective promotion to the rank of commander in 1926, 1931, and 1931 respectively. The classmates in the Line, of the officer who could count on promotion in 1926, were promoted to the rank of commander in 1911. Those of the two officers who could count on promotion in 1931 will probably be promoted to the rank of commander in 1914 and 1915 respectively. This inequity in promotion was not without a material influence in the resignations of these efficient officers." Ibid.

131. The "tougher" old days is a constant in military societies.

132. Michel Callon, "Society in the Making: The Study of Technology as a Tool for Sociological Analysis," in *The Social Construction of Technological Systems*, ed. Wiebe E. Bijker, Thomas P. Hughes, and Trevor Pinch (Cambridge, Mass.: The MIT Press, 1994), 83.

133. On the size of the Construction Corps, see *History of the Construction Corps*, 36–37 and 55.

CHAPTER 2.   COMPETING FOR CONTROL

1. Mahan's follow-on work, *The Influence of Sea Power upon the French Revolution and Empire, 1793–1812*, was published in 1892. On Jomini's interpretation of Napoleonic warfare, see John Shy, "Jomini," in *Makers of Modern Strategy: From Machiavelli to the Nuclear Age* (Princeton, N.J.: Princeton University Press, 1986), 143–85; and Russell Weigley, *The American Way of War: A History of United States Military Strategy and Policy* (New York: Macmillan, 1977), chap. 9. A contrary opinion of Mahan and his strategic framework is in Jon Tetsuro Sumida, *Inventing Grand Strategy and Teaching Command: The Classic Works of Alfred Thayer Mahan Reconsidered* (Washington, D.C.: Woodrow Wilson Center Press, 1997).

2. The battleship, like the ironclad warship — despite the strong popular legacy of the *Monitor* and *Virginia* (ex-*Merrimac*) — was also a European development. See Stanley Sandler, *The Emergence of the Modern Capital Ship* (Newark: University of Delaware Press, 1979). The two-power standard mandated a British Royal Navy equal to the next two navies in the world combined.

3. See Lance Buhl, "Maintaining an 'American Navy,' 1865–1889," in *In Peace and War: Interpretations of American Naval History, 1775–1984*, ed. Kenneth J. Hagan (Westport, Conn.: Greenwood, 1984), 145–85. Also see Mark Russell Shulman, *Navalism and the Emergence of American Sea Power, 1882–1893* (Annapolis: Naval Institute Press, 1995), especially chap. 7, for a detailed discussion of congressional politics and the emergence of the battleship.

4. Naval budget figure from Roger Dingman, *Power in the Pacific* (Chicago: University of Chicago Press, 1979), 3.

5. Fred A. Hobbs to Admiral of the Navy Dewey, 5 June 1905, Admiral of the Navy George Dewey Papers, Manuscript Division, Library of Congress, Washington, D.C. (hereafter Dewey Papers), Box 21.

6. Lieutenant Commander John H. Gibbons, USN, "The Need of a Building Program for Our Navy," *U.S. Naval Institute Proceedings* (hereafter *USNIP*) 29 (1903): 331.

7. For shared intellectual paradigms, shared rules as derivatives of shared paradigms, and the normal practice of science as the pursuit of efficiency, see Thomas Kuhn, *The Structure of a Scientific Revolution* (Chicago: University of Chicago Press, 1970), chaps. 3–5, especially 36, 42, and 46.

8. Margaret Sprout, "Mahan: Evangelist of Sea Power," in *Makers of Modern Strategy*, ed. Edward M. Earle (Princeton, N.J.: Princeton University Press, 1943),

cited in Philip A. Crowl, "Alfred Thayer Mahan: The Naval Historian," in *Makers of Modern Strategy: From Machiavelli to the Nuclear Age*, ed. Peter Paret (Princeton, N.J.: Princeton University Press, 1986), 469; Peter Karsten, *The Naval Aristocracy: The Golden Age of Annapolis and the Emergence of Modern American Navalism* (New York: The Free Press, 1972), 310–17. See Shulman, *Navalism*, chaps. 7 and 6.

9. See Karsten, *The Naval Aristocracy*, 300–317.

10. Ibid., 316–17.

11. Alex Roland, *Underwater Warfare in the Age of Sail* (Bloomington: Indiana University Press, 1978), 180–81.

12. Ensign Roger Welles, USN, to his mother, 4 February 1886, quoted by Karsten, *The Naval Aristocracy*, 301.

13. Parker cited in Kenneth J. Hagan, "Alfred Thayer Mahan: Turning America Back to the Sea," in *Makers of American Diplomacy: From Benjamin Franklin to Alfred Thayer Mahan*, 2 vols., ed. Frank J. Merli and Theodore A. Wilson (New York: Charles Scribner's Sons, 1974), vol. I, 289.

14. Rear Admiral Stephen B. Luce, USN, "Our Future Navy," *The North American Review*, July 1884, cited in Benjamin F. Cooling, *Benjamin Franklin Tracy: Father of the American Fighting Navy* (Hamden, Conn.: Archon Books, 1973), 73. Also see Hagan, "Alfred Thayer Mahan," 291.

15. Cooling, *Benjamin Franklin Tracy*, passim. For a very useful treatment of the offensive-defensive strategic debate during the 1880s, see Shulman, *Navalism*. Shulman dates the modern battleship from 1886 and considers armor unnecessary in a cruiser designed for commerce raiding. This ignores the American naval tradition of designing for strength. Armor would be useless against merchant ships but might provide the margin for victory against an enemy warship. Shulman's genealogical search for the battleship's near ancestors would benefit from exposure to an evolutionary model of technological development. See Shulman, *Navalism*, chap. 7.

16. The *Atlanta, Boston, Chicago*, and the dispatch ship *Dolphin* authorized in March 1883 at the cost of $1.3 million and finished five years later; Shulman, *Navalism*, 118.

17. Cooling, *Benjamin Franklin Tracy*, 75.

18. Quotes from Tracy, *Annual Report of the Secretary of the Navy, 1889, in Two Parts* (Washington, D.C.: U.S. Government Printing Office, 1890), Part I, 4.

19. Walter R. Herrick, *The American Naval Revolution* (Baton Rouge: Louisiana State University Press, 1966), 58.

20. *A Compilation of the Messages and Papers of the Presidents*, 11 vols., com-

piled by James D. Richardson (Washington, D.C.: U.S. Government Printing Office, 1890–1912), vol. IX, 10 and 44–45, cited by Herrick, *The American Naval Revolution*, 59.

21. Herrick, *The American Naval Revolution*, 62–63.

22. Shulman views the grandiose Policy Board report as useful in making three battleships palatable; Shulman, *Navalism*, 128–29.

23. Deriving their generic name from the Civil War USS *Monitor*, these ships usually mounted a few large guns, had a low freeboard, and were unsuited for sailing the high seas. Senator Chandler's point is in a letter to Secretary Tracy, 10 January 1890, Tracy Papers, Library of Congress, cited in Herrick, *The American Naval Revolution*, 71.

24. Herrick, *The American Naval Revolution*, 74.

25. Ibid., 143–44.

26. Crowl, "Mahan: Naval Historian," 471. Also see Herrick, *The American Naval Revolution*, 158–59.

27. For one interpretation of the blockade of the Confederacy, see William N. Still Jr., "A Naval Sieve: The Union Blockade in the Civil War," in *Readings in American Naval Heritage* (New York: American Heritage, 1997), chap. 7.

28. Herrick, *The American Naval Revolution*, 160.

29. See William H. McNeill, *The Pursuit of Power: Technology, Armed Force and Society since A.D. 1000* (Chicago: University of Chicago Press, 1982), 269–70.

30. For naval policy vis-à-vis Japan, see William Reynolds Braisted, *The United States Navy in the Pacific, 1897–1909* (Austin: University of Texas Press, 1958).

31. See Herrick, *The American Naval Revolution*, 173–75.

32. Ibid., 177–78. *Iowa* mounted a 12-inch main gun battery because of imbalances in the 13-inch gun turrets on the three *Indiana*-class battleships. The navy employed newly designed 13-inch guns in the *Kearsarge* class. See John C. Reilly Jr. and Robert L. Scheina, *American Battleships 1886–1923: Predreadnought Design and Construction* (Annapolis: Naval Institute Press, 1980), 72 and 86.

33. Sprout, *The Rise of American Naval Power*, 224; also see Herrick, *The American Naval Revolution*, 194–95.

34. Roosevelt also pushed naval preparedness as the way to avoid war. See "Washington's Forgotten Maxim," Address of Hon. Theodore Roosevelt, Assistant Secretary of the Navy, before the Class at the U.S. Naval War College, Newport, R.I., 2 June 1897, in *USNIP* 23 (1897): 447–61. For the need for an "offensive" naval strategy based in the battleship, see especially 457–58.

35. Roosevelt quoted in Howard K. Beale, *Theodore Roosevelt and the Rise of America to World Power* (Baltimore: Johns Hopkins University Press, 1956), 61, also

see 61–63; and Herrick, *The American Naval Revolution*, 219–20. Roosevelt's theatrics notwithstanding, Long remained in firm control of the department and its war efforts.

36. A summary of the features of these ships may be found in Reilly and Scheina, *American Battleships 1886–1923*, 161–73.

37. Navy General Order 544 of 13 March 1900, cited in Robert G. Albion, *Makers of Naval Policy, 1798–1947* (Annapolis: Naval Institute Press, 1980), 78.

38. Albion, *Makers of Naval Policy*, 79.

39. Ibid., 81.

40. "Duties of the General Board" (March 1900 [?]), Dewey Papers, Box 56.

41. Although most bureaus dealt with some aspects of naval technology, not all were headed by engineering specialists. The Bureaus of Ordnance and Equipment usually were commanded by line officers. Dewey had been chief of the Bureau of Equipment.

42. In 1901 the chief of the Bureau of Ordnance served as president of the Board on Construction. Even so, the coordination of all ordnance, equipment, and propulsion technology and material was performed in the Bureau of Construction & Repair, giving it, and its chief, a preeminent position in all Board on Construction affairs.

43. *Board on Construction Report*, 12 July 1901, General Records of the Secretary of the Navy, Record Group 80, National Archives, Washington, D.C. (hereafter RG 80), 1897–1915, File 8557–28.

44. Sims quoted in Elting Morison, *Admiral Sims and the Modern American Navy* (Boston: Houghton Mifflin Co., 1942), 80 and n. 5.

45. Sims's testimony in hearings before the House of Representatives, 1925, quoted by Morison, *Admiral Sims*, 80–81, n. 5.

46. See Reilly and Scheina, *American Battleships*, 84–86.

47. *Board on Construction Report*, 12 July 1901, 8. A myriad of factors could have affected the volume of fire from these guns besides the method of mounting.

48. Ibid.

49. Ibid., 9.

50. Secretary of the Navy Long to Dewey, 26 July 1901, Dewey Papers, Box 13.

51. Dewey to Lieutenant Commander N. Sargent, USN, 3 August 1901, Dewey Papers, Box 13.

52. Sims would have approved of Dewey's rejection of open casements, as they were a holdover from the gun-deck frigate with muzzle-loading guns; Lieutenant William S. Sims, USN, "The Board of Construction and the Design of Battleships: Comments on our Method of Determining the Designs of Men-of-War, Guns,

Mounts, Turrets, details of Protection, Etc.; The Methods Employed in France, and a Comparison of the Results Accomplished," 11 December 1901, essay forwarded to the Secretary of the Navy; RG 80, 1897–1915, File 13529–38 (hereafter Sims, "Board on Construction and the Design of Battleships"), 9.

53. Admiral of the Navy Dewey to the Secretary of the Navy, 30 September 1901, Dewey Papers, Box 13.

54. *Board on Construction Report, 26 November 1901*, cited by Reilly and Scheina, *American Battleships*, 163.

55. Lieutenant Sims to Rear Admiral Harold T. Bowles, USN, Chief Constructor, Bureau of Construction & Repair, 24 September 1901, Admiral William S. Sims Papers, Manuscript Division, Library of Congress, Washington, D.C. (hereafter Sims Papers), Box 12. Bowles had graduated from Annapolis in 1879, one year before Sims, and was entitled to temporary rank as a rear admiral as chief of a bureau.

56. Sims, "Board on Construction and the Design of Battleships," 6.

57. Ibid., 13.

58. Ibid.

59. Ibid., 15.

60. Ibid.

61. Ibid., 15–16.

62. Such was the case when HMS *Monmouth*, Rear Admiral Craddock's flagship, was sunk by the German force under Admiral Graf von Spee off the Coronelles in 1914. *Monmouth*'s low freeboard resulted in its windward guns being unusable since the gun ports were shipping too much water.

63. Sims, "Board on Construction and the Design of Battleships," 57.

64. See Bowles to Sims, 9 November and 2 December 1901, Sims Papers, Box 12.

65. Bowles to Sims, 14 January 1902, Sims Papers, Box 12.

66. Lieutenant Commander A. P. Niblack, USN, "The Tactics of the Gun," reprinted in *USNIP* 38 (1902): 925–36.

67. Ibid., 925. Underwater bow rams, similar to those on ancient galleys, were in vogue in late-nineteenth-century warships; see O'Connell, *Sacred Vessels*, 41.

68. Board on Construction to Secretary of the Navy, 27 December 1902, RG 80, 1897–1915, File 15365.

69. Ibid.

70. Ibid.

71. See Roland, *Underwater Warfare*, passim; and O'Connell, *Sacred Vessels*, chap. 6.

72. *Board on Construction Minority Report,* 12 *February* 1902, RG 80, 1897–1915, File 15365.

73. Board on Construction (Majority members) to Secretary of the Navy, 13 February 1902, RG 80, 1897–1915, File 15365.

74. Commander J. B. Murdock, USN, "Torpedo Tubes in Battleships," *USNIP* 29 (1903): 547–51.

75. Murdock, "Torpedo Tubes in Battleships," 548.

76. American antiship missiles, such as Harpoon, which entered the fleet in the 1970s, were designed against U.S. warships on the premise that the damage they would afflict on U.S. ships would be repeated, if not increased, against a foreign ship design. This proved an optimistic assumption with the discovery that the Soviet navy, unlike the U.S. Navy, was incorporating armor in their ships.

77. See the Discussion section on Murdock's article featuring comments by Rear Admiral Bradford and eleven other senior line officers (and one naval constructor) in *USNIP* 29 (1903): 551–68.

78. See Lieutenant F. K. Hill, USN, *Minority Report and Endorsement, Board on Construction,* 25 *July* 1903, RG 80, 1897–1915, File 15365–2.

79. Admiral of the Navy Dewey, President of the General Board, to the Secretary of the Navy, 26 September 1903, RG 80, 1897–1915, File 15365–3.

80. *Board on Construction Report [3rd Endorsement],* 27 *February* 1904, RG 80, 1897–1915, File 15365–3.

81. So-called because American warships had white hulls. Roosevelt sent the modern U.S. Fleet around the world as a demonstration of America's new naval power and especially to inform the Japanese that "there were fleets of the white races which were totally different from the fleet of poor Rodjestvensky [which the Japanese destroyed in 1905]." Roosevelt quoted in Beale, *Theodore Roosevelt,* 328, note b. Also see James R. Reckner, *Teddy Roosevelt's Great White Fleet* (Annapolis: Naval Institute Press, 1988).

82. See Reilly and Scheina, *American Battleships,* 167.

83. Henk van den Belt and Arie Rip, "The Nelson-Winter-Dosi Model and Synthetic Dye Chemistry," in *The Social Construction of Technological Systems: New Directions in the Sociology and History of Technology,* ed. Wiebe J. Bijker, Thomas P. Hughes, and Trevor Pinch (Cambridge, Mass.: The MIT Press, 1994), 140–41.

84. For the 1842 reorganization, see Geoffrey S. Smith, "An Uncertain Passage: The Bureaus Run the Navy, 1842–1861," in *In Peace and War,* 79–106.

85. "Statement of Hon. William H. Moody, Secretary of the Navy," *Hearings Before the Committee on Naval Affairs, House of Representatives on Appropriation Bill for 1905 Subjects and on H.R. 15403, for General Board, Fifty-eighth Congress,*

Second Session (hereafter *HNAC Testimony for 1905*) (Washington, D.C.: U.S. Government Printing Office, 1904), 909.

86. Secretary Moody's testimony on 11 April 1904, *HNAC Testimony for 1905*, 920.

87. Ibid.

88. Memorandum on Present Duties of the General Board, Dewey to Secretary Moody, 19 April 1904, reprinted in *HNAC Testimony for 1905*, 936.

89. Testimony of Admiral Dewey, *HNAC Testimony for 1905*, 946.

90. Rear Admiral Capps's testimony, *HNAC Testimony for 1905*, 958–59.

91. Rear Admiral Rae's Testimony, *HNAC Testimony for 1905*, 965–66. For a description of protected and armored cruisers, see Naval Constructor David W. Taylor, USN, "The Present Status of the Protected Cruiser," *USNIP* 30 (1904): 145–49.

92. Testimony of Assistant Navy Secretary Darling, *HNAC Testimony for 1905*, 931.

93. Ibid.

94. President of the General Board to the Secretary of the Navy, G.B. No. 420–2, 28 October 1905, RG 80, 1897–1915, File 8557.

95. "Board on Construction 2nd Endorsement on Recommendations of the General Board to the Numbers and Types of Ships," 23 November 1905, RG 80, 1897–1915, File 8557–51.

96. President of the General Board to the Secretary of the Navy, 29 December 1905, RG 80, 1897–1915, File 21122.

97. President of the General Board to the Secretary of the Navy, 30 March 1906, quoting from the Board of Inspection and Survey report on the trial of the battleship *Tennessee*, RG 80, 1897–1915, File 21670.

98. President of the Board of Inspection and Survey to the Secretary of the Navy, 25 April 1906, RG 80, 1897–1915, File 21670.

99. The relations between the bureaus and the navy outside Washington, D.C., became more of an "us against them" affair and replaced the line-engineer rivalry. An example was the alliance between line Captain Yates Stirling and New York navy yard naval constructor Captain H. T. Wright. These officers combined to force the Bureau of Construction & Repair to cancel a large contract for "pillow" lifesavers during World War I. As captain of a troop transport, Stirling had seen many crewmen from the torpedoed *Antilles* drown with their pillow vests on. The pillows, on each side of the head, actually forced the head forward and into the water. Stirling and Wright prevailed on the bureau to cancel a large contract for pillow life vests in favor of the more expensive, and more effective, vest-style life preservers. See Yates Stirling, Rear Admiral, USN (Ret.), *Sea Duty: Memoirs of a Fighting Admiral* (New York: G. P. Putnam's Sons, 1939), 179.

CHAPTER 3.   REFINING THE TECHNOLOGICAL IDEAL

1. A mixture of Kuhn's "normal science as puzzle solving" and "anomaly and the emergence of scientific discoveries"; see Thomas Kuhn, *The Structure of Scientific Revolutions*, 2nd ed. (Chicago: University of Chicago Press, 1970), chaps. 4 and 6.

2. Roosevelt quoted by Howard K. Beale, *Theodore Roosevelt and the Rise of America to World Power* (Baltimore: Johns Hopkins University Press, 1956), 38. For more on Roosevelt and the navy, see William Reynolds Braisted's thorough study, *The United States Navy in the Pacific, 1897–1909* (Austin: University of Texas Press, 1958), as well as his *The United States Navy in the Pacific, 1909–1922* (Austin: University of Texas Press, 1968), chaps. 1 and 2.

3. See Robert L. O'Connell, *Sacred Vessels: The Cult of the Battleship and the Rise of the U.S. Navy* (Boulder, Colo.: Westview Press, 1991), 134–35, and caption on 136.

4. Braisted, *The United States Navy in the Pacific, 1909–1922*, 3.

5. For the Pacific focus, see Kenneth J. Hagan, *This People's Navy: The Making of American Sea Power* (New York: The Free Press, 1991), chap. 4.

6. Captain Asa Walker, USN, "With Reference to the Size of Fighting Ships," *U.S. Naval Institute Proceedings* (hereafter *USNIP*) 26 (1900): 515–22.

7. Until the *Arleigh Burke*-class of guided missile destroyers designed in the early 1980s, U.S. warships continue to be designed for long-range cruising rather than speed. This was done to maximize the navy's operating radius throughout the large expanse of the Pacific Ocean and is contrary to most other nations' warship design philosophies. The Italian navy's Mediterranean focus has led them to design their ships for high speed. This emphasis on high speed can be seen in Soviet naval designs as well — a legacy of the influence of Italian naval architects on their Soviet colleagues.

8. Lieutenant Matt Signor, USN, "A New Type of Battleship," *USNIP* 28 (1902): 1–20. Signor called for a 16,000-ton battleship, with a speed of 18 knots and an endurance of 3,000 miles at 16 knots. The main battery would be arranged in two turrets, each containing three 13-inch guns. An intermediate battery of 10-inch guns would be located in triple-gun turrets on each side of the main deck amidship. Six 5-inch broadside guns were retained on the port and starboard beams. Signor discarded the superposed turrets of the *Kearsarge* class, but retained the smaller-caliber broadside guns of the secondary battery. The first six warships of the U.S. Navy, authorized in 1794, included superfrigates equipped with more numerous, heavier cannons than contemporary frigates.

9. Lieutenant Homer Poundstone, USN, "Size of Battleships for U.S. Navy," *USNIP* 29 (1903): 161–74.

10. Ibid., 161.

11. Roosevelt quoted by Beale, *Theodore Roosevelt*, 39.

12. Winston S. Churchill, *The World Crisis*, vol. I (New York: Charles Scribner's Sons, 1928), 225.

13. Francesco Maurizio di Giorgio Martini (1439–1502) revamped fortification architecture to counter the effect of siege cannon.

14. See Stephen W. Roskill, *The War at Sea, 1939–1945*, vol. I (London: H.M. Stationery Office, 1954), 398. He described this sort of dilemma during the *Hood–Bismarck* engagement in 1941: "Admiral Holland must also have considered whether it would be to his advantage to fight the enemy at long or short ranges. He had no information regarding the ranges at which the *Bismarck* would be most vulnerable to the gunfire of his own ships, but he did know that the *Prince of Wales* [battleship] should be safe from vital hits by heavy shells from maximum gun range down to about 13,000 yards, and that the *Hood* should become progressively more immune from such hits as the range approached 12,000 yards and the enemy shell trajectories flattened. At long ranges, the *Hood*, which lacked heavy horizontal armour, would be very vulnerable to plunging fire by heavy shells."

15. To their dismay, the British found German battleship armor design relatively impervious to punishment at Jutland in 1916, and in 1941, HMS *Hood*'s armor design failed catastrophically during its six-minute engagement with *Bismarck* and the cruiser *Prinz Eugen*, and 1,415 British sailors died very quickly. On the *Hood*, see the battle log in Ulrich Elfrath and Bodo Herzog, *Schlachtschiff Bismarck: Ein Bericht in Bildern und Dokumenten* (Dorheim: Podzun-Verlag, 1975), 107–9; and excerpts from the Admiralty Court of Inquiry reprinted in Alan Raven and John Roberts, *British Battleships of World War II* (Annapolis: Naval Institute Press, 1976), 350–51.

16. Robert G. Albion, *Makers of Naval Policy, 1798–1947* (Annapolis: Naval Institute Press, 1980), 86.

17. Poundstone, "Size of Battleships," 172–73.

18. Lieutenant Commander John H. Gibbons, USN, "Discussion on 'Size of Battleships for U.S. Navy,'" *USNIP* 29 (1903): 435.

19. Assistant Naval Constructor T. G. Roberts, USN, "Discussion on 'Size of Battleships for the U.S. Navy,'" *USNIP*, 29 (1903): 440. For the progression in battleship size, see O'Connell, *Sacred Vessels*, 133.

20. Naval Constructor Horatio G. Gillmor, USN, "Discussion on 'Size of Battleships for U.S. Navy,'" *USNIP* 29 (1903): 441.

21. Representative James A. Tawney, Chairman, House Appropriations Committee, quoted by Richard Hough, *Dreadnought* (New York: Macmillan, 1964), 37.

22. The First Sino-Japanese War (1894–95) featured armored ships but not modern battleships such as those that fought in the Russian and Japanese navies in 1904–5. The initial shot of the First Sino-Japanese War was fired by the Japanese protected cruiser *Naniwa* at the Chinese cruiser *Chi' Yuan*. The first Japanese battleship, *Fuji*, was laid down at the Thames Ironworks in 1894 and not completed until after the war with China. See Hansgeorg Jentschura, Dieter Jung, and Peter Mickel, *Warships of the Imperial Japanese Navy, 1869–1945* (Annapolis: Naval Institute Press, 1977), 16 and 95–96.

23. "Preliminary Statement, Committee on Designs, 3 January 1905," in *The Papers of Admiral Sir John Fisher*, vol. I, ed. Lieutenant Commander P. K. Kemp, RN (London: Navy Records Society, 1960) (hereafter cited as *Fisher Papers, I*), 215–16.

24. Committee on Designs Statement "The Battleship," *Fisher Papers I*, 217.

25. According to one Japanese official, "By the time the 6-inch guns were coming into play the action was already going against Russia"; quotation by Special Committee on Design, *Fisher Papers, I*, 218.

26. A nautical mile contains 2,000 yards as compared to the 1,760 yards in a statute mile.

27. Special Committee on Designs, "The Battleship," *Fisher Papers, I*, 218–19.

28. Ibid., 219.

29. Ibid.

30. Jon Tetsuro Sumida, *In Defence of Naval Supremacy: Finance, Technology, and British Naval Policy, 1889–1914* (London: Unwin Hyman, 1989), especially chap. 2.

31. Admiral of the Navy George Dewey, President of the General Board, to the Secretary of the Navy, 26 September 1903, General Records of the Secretary of the Navy, Record Group 80, National Archives, Washington, D.C. (hereafter RG 80), 1897–1915, File 15365–3.

32. Commander Bradley A. Fiske, USN, "American Naval Policy," *USNIP* 31 (1905): 1–80.

33. See, for example, the discussion of Fiske's essay by Rear Admiral C. F. Goodrich, *USNIP* 31 Part I (1905): 693–98.

34. Commander Bradley A. Fiske, USN, "Compromiseless Ships," *USNIP* 31 (1905): 549–53.

35. Ibid., 550.

36. Ibid., 550–51.

37. Captain Alfred Thayer Mahan, USN, *Collier's Weekly*, 17 June 1905, quoted by Fiske, "Compromiseless Ships," 551–52.

38. Fiske, "Compromiseless Ships," 552.

39. Ibid.

40. Fiske's proposal was analyzed and judged technically sound; see Assistant Naval Constructor Richard D. Gatewood, USN, "Approximate Dimensions for a 'Compromiseless Ship,' " *USNIP* 32 (1906): 571–83.

41. Captain A. T. Mahan, USN, "Reflections, Historic and Other, Suggested by the Battle of the Japan Sea," *USNIP* 32 (1906): 447–71.

42. Ibid., 452.

43. Ibid., 466.

44. Ibid., 458.

45. Ibid., 460.

46. Ibid., 461.

47. Lieutenant Commander William S. Sims, USN, "The Inherent Tactical Qualities of All-Big-Gun, One-Caliber Battleships of High Speed, Large Displacement and Gun-Power," *USNIP* 32 (1906): 1337–66.

48. Lieutenant R. D. White, USN, "With the Baltic Fleet at Tsushima," *USNIP* 32 (1906): 597ff.

49. Sims, "Inherent Tactical Qualities of All-Big-Gun Battleships," 1341.

50. Ibid., 1342.

51. For surface ships, the horsepower to speed curve becomes very steep above cruising speed—typically around 15 to 17 knots. A one-knot increase in speed will require the addition of extremely large amounts of horsepower (discussed in the case of the 1916 battlecruisers in the next chapter). The strategic focus of the British and Germans, for example, was the North Sea. They could afford to convert hull volume allocated to fuel (range) to additional propulsion machinery (speed). Limited by the vast reaches of the Pacific, such a trade-off was impossible for U.S. warships. Speed could only come from additional propulsion machinery—which of course required additional fuel in an increasing volumetric (size) design spiral.

52. Sims, "Inherent Tactical Qualities of All-Big-Gun Battleships," 1347–48.

53. Ibid.

54. Ibid., 1365–66.

55. See Elting Morison, *Admiral Sims and the Modern American Navy* (Boston: Houghton Mifflin, 1942), 174.

56. Park Benjamin, "The Shout for 'Big Ships,'" *The Independent*, 31 January 1909, clipping in the Papers of Admiral William S. Sims, Manuscript Division,

Library of Congress, Washington, D.C. (hereafter Sims Papers), Mahan Folder, Box 71.

57. For the genesis of the dreadnought-type battleship, see Oscar Parkes, OBE, *British Battleships*, Warrior *to* Vanguard: *A History of Design, Construction, and Armament* (London: Seeley Service, 1957), 466–76.

58. See Hough, *Dreadnought*, 47.

59. Much like one aspect of the military view of the atomic bomb — an efficient replacement for 1,000-plane bomber raids.

60. See Morison, *Admiral Sims*, chap. 8.

61. Sims to Lieutenant Commander A. P. Niblack, USN, 14 March 1902, quoted in Morison, *Admiral Sims*, 130.

62. Elting Morison describes this article and the subsequent Senate investigation and Newport Conference in detail, focusing, of course, on Sims's role; see Morison, *Admiral Sims*, chaps. 12–13.

63. See Morison, *Admiral Sims*, 184.

64. See Secretary of the Navy to Lieutenant Commander Frank Hill, 15 February 1908, RG 80, 1897–1915, File 26000–3.

65. Lieutenant Commander Hill to Secretary of the Navy, 24 February 1908, RG 80, 1897–1915, File 26000–3.

66. Telephone message memorandum, White House usher, Mr. Hackett (for the President) to the Secretary of the Navy, 19 February 1908, 4 P.M., RG 80, 1897–1915, File 26000.

67. See Albion, *Makers of Naval Policy*, 214.

68. Ibid., 213.

69. Senator Hale in the *Congressional Record, Fifty-eighth Congress, 2nd Session*, 2727, quoted by Albion, *Makers of Naval Policy*, 214. The two smaller battleships, known in the navy as the *"Hale"* class, were sold to Greece during the 1914 Greco-Turkish dispute.

70. In addition to his testimony and written statement to the Senate Naval Affairs Committee, Rear Admiral Converse also prepared a lengthy report on the Reuterdahl article for the secretary of the navy. See "Statement Regarding Criticism of the Navy," 3 March 1908, and "Exhibit 'A' to accompany Report Dated March 3rd, 1908 of Rear Admiral Converse," in RG 80, 1897–1915, Files 26000 and 26000–11.

71. Originally commissioned as a line officer, Capps was sent to Glasgow for naval architectural training and transferred to the Construction Corps in 1889. Although naval constructors held commissions with naval rank after 1899, for example, "commander," naval usage — determined by line officers — did not allow them

to use these titles. Instead they were referred to as "assistant naval constructor" or "naval constructor" until this practice was ended by Secretary of the Navy Josephus Daniels in 1918. After that, naval constructors were referred to by naval rank, that is, "Commander, Construction Corps, U.S. Navy"; see *History of the Construction Corps of the United States Navy* (Washington, D.C.: U.S. Government Printing Office, 1937), 42–43.

72. Capps, *Report Concerning Alleged Defects in Vessels of the United States Navy, by Washington Lee Capps, Chief Constructor, U.S. Navy, and Chief of the Bureau of Construction and Repair, Senate Document No. 297, 60th Congress, 1st Session* (Washington, D.C.: U.S. Government Printing Office, 1908), (hereafter cited as Capps, *Alleged Defects*), 1–2.

73. See the information relative to foreign armor belt design in Office of Naval Intelligence, "Memorandum on the Distribution of Water-Line Armor," 28 February 1908, reprinted in *Hearings Before the Committee on Naval Affairs, United States Senate on the Bill (S.3335) to Increase the Efficiency of the Personnel of the Navy and Marine Corps of the United States*, (hereafter cited as *Senate Hearings [S.3335]*), 110–11.

74. Testimony of Rear Admiral Capps, *Senate Hearings (S.3335)*, 73.

75. Capps in Testimony of Sims, *Senate Hearings (S.3335)*, 170–71.

76. *Senate Hearings (S.3335)*, 62.

77. Testimony of Rear Admiral Converse, President of the Board on Construction, *Senate Hearings (S.3335)*, 23.

78. Testimony of Sims, *Senate Hearings (S.3335)*, 163.

79. Ibid., 167.

80. See Rear Admiral Capps's testimony, *Senate Hearings (S.3335)*, 52–53.

81. The solicitation of private-sector designs was in response to a clause in the naval appropriation bill approved on 29 June 1906.

82. Secretary of the Navy V. H. Metcalf to Senator Eugene Hale, 9 March 1908, RG 80, 1897–1915, File 26000–4; also printed in *Senate Hearings (S.3335)*, 253–57.

83. Rear Admiral Caspar Goodrich, commandant of the New York Navy Yard, had also responded to General Order 49 with some pointed criticism regarding the location of the armor belt: "I have the honor to suggest that placing the waterline armor so low in our original designs was a pardonable error; that retaining it so low in succeeding vessels was, to say the least, obstinate; that to retain it so low in future ships, if this does not border on the criminal, certainly involves a responsibility which none should care to accept. . . . It is a faulty system [technical bureaus and Board on Construction] at which my remarks were directed, and not at any individual past or present." See Goodrich to Secretary of the Navy, 12 August 1907,

and Secretary of the Navy to Goodrich, 18 February 1908, RG 80, 1897–1915, File 26000–1; also Rear Admiral Goodrich to Secretary Metcalf, 27 February 1908, RG 80, 1897–1915, File 26000, Key Folder.

84. *Senate Hearings* (S.3335), 253–57.

85. Ibid.

86. Ibid.

87. Rear Admiral Robley D. Evans to Secretary of the Navy, 6 March 1908, quoted in Board on Construction memorandum to Secretary of the Navy Metcalf, 26 March 1908, and in 2nd endorsement by Board on Construction on letter of Senator Hale, 9 April 1908, included with Secretary of the Navy to Senator Hale, 13 April 1908, RG 80, 1897–1915, Files 26000 and 26000–7.

88. Secretary Metcalf to Senator Hale, 13 April 1908, RG 80, 1897–1915, File 26000–7.

89. See testimony of Lieutenant Commander Sims in *Senate Hearings* (S.3335), 155–98; and Morison, *Admiral Sims*, 192–95.

90. Morison, *Admiral Sims*, 200.

91. Secretary Metcalf to Senator Hale, 19 May 1908, RG 80, 1897–1915, File 26000–5.

92. Ibid.

93. For an overview, see Morison, *Admiral Sims*, 201–14.

94. Commander Key to the Secretary of the Navy, 9 June 1908, Records of the 1908 Battleship Conference, Naval Historical Collection, Naval War College, Newport, R.I.

95. Ibid.

96. Ibid.

97. Sims to William Loeb (Presidential Secretary), quoted in Morison, *Admiral Sims*, 203.

98. A ship hull can be thought of as a beam subjected to alternating peak loads (buoyant force of the water) in its middle and then at each end. These loads, coupled with the downward weight of the ship (including components such as a 12-inch gun mount), exert significant forces on a ship's structure.

99. The General Board was quite concerned with German dreadnought construction, and viewed the Germans as a more ready yardstick for American naval power.

100. Key to President Roosevelt, 6 January 1909, Sims Papers, Box 64.

101. Admiral of the Navy Dewey to Congressman William Alden Smith, 16 June 1906, Dewey Papers, Box 23.

102. Key to Sims, 30 April 1917, Sims Papers, Box 68.

Chapter 4.   Technological Trajectory

1. For a discussion of Neo-Schumpeterian models of technological innovation, technological trajectories, and technological paradigms, see Henk van den Belt and Arie Rip, "The Nelson-Winter-Dosi Model and Synthetic Dye Chemistry," in *The Social Construction of Technological Systems: New Directions in the Sociology and History of Technology*, ed. Wiebe E. Bijker, Thomas P. Hughes, and Trevor Pinch (Cambridge, Mass.: The MIT Press, 1994), 135–218; especially 136–39.

2. As discussed earlier, the question was moot as Congress, fearing imperialistic adventurism, restricted the steaming range of the navy's first three battleships appropriated in 1890. The navy's fourth battleship, *Iowa*, authorized in 1892, was not subject to any range limitations in response to President Harrison's call for a modern fleet to protect American property throughout the world. See Walter R. Herrick, *The American Naval Revolution* (Baton Rouge: Louisiana State University Press, 1966), 74 and 143–44.

3. See John C. Reilly Jr. and Robert L. Scheina, *American Battleships, 1886–1923: Predreadnought Design and Construction* (Annapolis: Naval Institute Press, 1980), 22.

4. Captain C. W. Dyson, USN, "The Development of Machinery in the U.S. Navy During the Past Ten Years," *Journal of the American Society of Naval Engineers* (hereafter *ASNE Journal*) 29 (1917): 217. Also see Engineer-in-Chief Hutch I. Cone, USN, "Naval Engineering Progress," a lecture to the Naval War College, 9 August 1910, reprinted in *ASNE Journal* 22 (1910): 1019.

5. One comparison placed the steam turbine at one-half the length and height and two-thirds the weight of a reciprocating steam engine delivering a comparable power output; see "Comparison of a Turbine and a Reciprocating Engine for the United States Navy," *Scientific American*, reprinted in *U.S. Naval Institute Proceedings* (hereafter cited as *USNIP*) 32 (1906): 1615–16.

6. The high speed of the turbine caused the propeller to rotate at a speed above its optimum efficiency. Cavitation, with attendant propeller blade erosion, was the result. See Rear Admiral C. W. Dyson, USN, "The Passing of the Direct-Connected Turbine for the Propulsion of Ships," *ASNE Journal* 31 (1919): 555–76.

7. The concept of a reverse technological salient originated with Thomas Hughes, who defined it as uneven growth of the components of a technological system which requires the attention of engineers and technicians so that the system can be made to function optimally; Thomas P. Hughes, *Networks of Power: Electrification in Western Society, 1880–1930* (Baltimore: Johns Hopkins University Press, 1983), 14.

8. William H. McNeill, *The Pursuit of Power: Technology, Armed Force and So-*

ciety since A.D. 1000 (Chicago: University of Chicago Press, 1982), chap. 8; and Alex Roland, "Science and War," *Osiris* 1 (1985): 258.

9. For example, in 1910, the Engineering Experimental Station was evaluating steam traps, steam nozzles, and an air compressor offered for naval service by private industry. See *Annual Report of the Experimental Station, 1910,* in Record Group 405, Records of the U.S. Naval Academy and Correspondence of the Superintendent, U.S. Naval Academy Archives, Nimitz Library, Annapolis, Md. (formerly Record Group 405, National Archives, Washington, D.C.); (hereafter RG 405, USNA).

10. Melville's statement was quoted in *Hearings before the Committee on Naval Affairs, House of Representatives, Sixty-fourth Congress, First Session, On Estimates Submitted by the Secretary of the Navy, 1916* (Washington: D.C.: U.S. Government Printing Office, 1916), 3384.

11. Bradley A. Fiske, *From Midshipman to Rear-Admiral* (New York: The Century Company, 1919), 580.

12. Testimony of Captain Bradley Fiske, *Hearings Before the Committee on Naval Affairs, United States Senate on the Bill (S.3335) to Increase the Efficiency of the Personnel of the Navy and Marine Corps of the United States* (Washington, D.C.: U.S. Government Printing Office, 1908) (hereafter *Senate Hearings [S.3335]*), 147.

13. Ibid., 148.

14. For histories of the Consulting Board, see Daniel J. Kevles, *The Physicists: The History of a Scientific Community in Modern America* (New York: Knopf, 1979), chaps. 8 and 9; Thomas P. Hughes, *Elmer Sperry: Inventor and Engineer* (Baltimore: Johns Hopkins University Press, 1971), chap. 9; David K. Allison, *New Eye for the Navy: The Origin of Radar at the Naval Research Laboratory,* NRL Report 8466 (Washington, D.C.: U.S. Government Printing Office, 1981); idem, "The Origins of the Naval Research Laboratory," *USNIP* 15 (July 1979): 63–69; and William M. McBride, "The 'Greatest Patron of Science'?: The Navy-Academia Alliance and U.S. Naval Research, 1896–1923," *Journal of Military History* 56 (1992): 7–33.

15. A history of the early development of the Parsons turbine and its applications to marine propulsion can be found in Edward W. Constant II, *The Origins of the Turbojet Revolution* (Baltimore: Johns Hopkins University Press, 1980), 69–79.

16. The disparaging comments of Mark Robinson at the 1897 meeting of the Institution of Naval Architects were typical; see *The Institution of Naval Architects, 1860–1960: An Historical Survey of the Institution's Transactions and Activities Over 100 Years,* ed. Kenneth C. Barnaby (London: Royal Institution of Naval Architects

in association with Allen & Unwin, 1960), 197; and Charles Parsons, "The Application of the Compound Steam Turbine to the Purpose of Marine Transportation," *Transactions, Institution of Naval Architects* (hereafter *INA Transactions*) 38 (1897): 232–42.

17. For an account of the *Wampanoag*, see Elting Morison, *Men, Machines, and Modern Times* (Cambridge, Mass.: The MIT Press, 1966), chap. 6.

18. Parsons had more success selling his turbine in the merchant marine market. By 1903, two passenger ships on the Clyde and ferries on the Dover-Calais and Newhaven-Dieppe routes employed Parsons steam turbines. See Charles Parsons, "The Steam Turbine and its Application to the Propulsion of Vessels," *INA Transactions* 85 (1903): 284–311.

19. See William Le Roy Emmet, *The Autobiography of an Engineer* (New York: American Society of Mechanical Engineers, 1940), 135–36; and John W. Hammond, *Men and Volts* (New York: J. B. Lippincott Co., 1941), 275–78.

20. Emmet, *Autobiography of an Engineer*, 175–76.

21. Ibid., 107–15; and Hammond, *Men and Volts*, 237.

22. See Hughes, *Networks of Power*, 211, for a description of the Insull–Consolidated Edison agreement with General Electric for the development of the 5,000 kw turbogenerator.

23. AEG entered into a similar agreement with the Swiss firm of Brown, Boveri & Company, who held the European rights to the Parsons turbine; see Hughes, *Networks of Power*, 179.

24. Ibid., 232–33.

25. The Rateau turbine fell victim to Parsons's reputation when the French government selected Parsons turbines, manufactured under license, for the six battleships of the 1906 Program. See "France Adopts Turbines," translated from *Le Yacht* and reprinted in *USNIP* 33 (1907): 878; and Siegfried Breyer, *Schlachtshiffe und Schlachtkreuzer, 1905–1970* (Munich: J. F. Lehmann, 1970), 442–43.

26. *Scientific American*, "Comparison of a Turbine and a Reciprocating Steam Engine for the United States Navy," reprinted in *USNIP* 32 (1906): 1615–16. Also see Cone, "Naval Engineering Progress."

27. In 1907, the director of naval construction in Germany announced that, as a result of cruiser tests involving reciprocating and turbine propulsion, all future warships of the Imperial German Navy would be equipped with turbines. See "The Turbine Warship *Luebeck*," *The Engineer*, reprinted in *USNIP* 31 (1905): 990; and "Germany Adopts Turbines," *International Marine Engineering*, reprinted in *USNIP* 33 (1907): 878.

28. Testimony of Rear Admiral Converse, *Senate Hearings (S.3335)*, 42.

29. See Rear Admiral C. W. Dyson, USN, "A Fifty Year Retrospect of Naval Marine Engineering," *ASNE Journal* 30 (1918): 289–91, for a discussion of the Curtis turbine problems in *North Dakota*.

30. "A Marine Steam Turbine Reducing Gear," *Iron Age*, reprinted in *USNIP* 35 (1909): 1301.

31. For a brief description of each system, see William Hovgaard, *Modern History of Warships: Comprising a Discussion of Present Standpoint and Recent War Experiences for the Use of Students of Naval Construction, Naval Constructors, Naval Officers, and Others Interested in Naval Matters* (London: E. & F. N. Spon, Ltd., 1920), 378–81.

32. See Dyson, "Development of Machinery in the U.S. Navy," 224–26; Francis Hodgkinson, "Progress in Turbine Ship Propulsion," a paper presented to the (U.S.) Society of Naval Architects and Marine Engineers, 14 November 1918, and reprinted in *ASNE Journal* 31 (1919): 193–94; and George Westinghouse, *Broadening the Field of the Marine Steam Turbine, The Problem and its Solution (The Melville-MacAlpine Reduction Gear)* (Pittsburgh, Pa.: "Printed for Geo. Westinghouse," 1909).

33. C. A. Parsons and R. J. Walker, "Twelve Months' Experience with Geared Turbines in the Cargo Steamer *Vespasian*," *INA Transactions* 53 (1911): 29–36; and "Speed Reduction Gears for Marine Turbines," *The Engineer*, reprinted in *USNIP* 36 (1910): 305.

34. "A Hydraulic Transmission Gear for Marine Turbines," *Engineering Magazine*, reprinted in *USNIP* 36 (1910): 293–96.

35. Fessenden claimed to have offered his plans for an electric-driven ship to the navy as early as 1899, and again in 1902 and 1904. In 1908, the Navy Department told him to take his idea to the "electric companies." See Fessenden to the Secretary of the Navy, 21 August 1911, Record Group 80, General Records of the Secretary of the Navy, National Archives, Washington, D.C. (hereafter RG 80), 1897–1915, File 24666–22.

36. American naval expansion had absorbed an average of approximately 17 percent of the annual federal budget during Theodore Roosevelt's presidency; Roger Dingman, *Power in the Pacific* (Chicago: University of Chicago Press, 1979), 3.

37. See Fessenden to Secretary of the Navy, 14 August 1908, RG 80, 1897–1915, File 26272–13a.

38. In September 1908, the U.S. naval attaché in Berlin reported on the claim for reduced fuel consumption in the new turboelectrically propelled submarine mother ship, *Vulkan*. This report was in the possession of the Bureau of Steam Engineering when they reviewed Fessenden's turboelectric propulsion proposal. See *Board on Construction Endorsement on National Electric Signal Company Pro-*

pelling Plant for Battleships Nos. 30 and 31, 25 January 1909, RG 80, 1897–1915, File 26272–13.

39. Acting Secretary Newberry to Fessenden, 19 August 1908, RG 80, 1897–1915, File 26272–13a; also see n. 37 above.

40. Emmet to Navy Secretary George von L. Meyer, 4 October 1909, RG 80, 1897–1915, File 26840–208.

41. Memorandum from the Secretary of the Navy to the Bureau of Steam Engineering, 8 October 1909, RG 80, 1897–1915, File 26840–208.

42. Lieutenant Commander J. McNamee, USN, to the Secretary of the Navy, 10 November 1909, RG 80, 1897–1915, File 26840–208.

43. Emmet to Congressman Weeks, 26 January 1910, RG 80, 1897–1915, File 26840–208.

44. Emmet to Arthur Sherwood (his brother-in-law), 27 January 1910, and Sherwood to Secretary Meyer, 29 January 1910, RG 80, 1897–1915, File 26840–208.

45. See Elting E. Morison, *Admiral Sims and the Modern American Navy* (Boston: Houghton Mifflin, 1942), chaps. 12 and 13.

46. Memorandum from Engineer-in-Chief Cone to the Secretary of the Navy, 3 February 1910, RG 80, 1897–1915, File 26840–208.

47. Memorandum from the Secretary of the Navy to the General Board, 10 April 1911, RG 80, 1897–1915, File 24666–14.

48. Dyson, "A Fifty Year Retrospect of Naval Marine Engineering," 297.

49. Secretary Meyer to Congressman Knowland, 21 April 1911, RG 80, SecNav, 1897–1915, File 26466–15.

50. The loss of this auxiliary machinery work did not sit well with the civilian machinists at the Mare Island Naval Shipyard, where the electrically powered USS *Jupiter* was being constructed. See congressional complaints lodged by the Vallejo (California) Chamber of Commerce and Vallejo Trades and Labor Council in RG 80, SecNav, 1897–1915, Files 24666–14, 24666–15, and 24666–22.

51. Emmet to Assistant Secretary of the Navy Beekman Winthrop, 29 August 1912, RG 80, 1897–1915, File 26840–181; the battleships were most likely *Nevada* and *Oklahoma*.

52. Memorandum from the Aide for Material to the Bureaus, 3 September 1912, and Memorandum from the Chief, Bureau of Steam Engineering to the Aide for Material, 17 September 1912, RG 80, 1897–1915, File 26840–181.

53. "Electric Propulsion of Ships," *Electrical World*, reprinted in *USNIP* 38 (1912): 367–68.

54. Lieutenant Commander P. W. Foote, USN, "Turbine Electric Propulsion of a Battleship Compared with Other Means," *ASNE Journal* 27 (1915): 54–82.

55. Emmet to Secretary Daniels, 14 July 1913; Roosevelt to Emmet, 22 July 1913; Endorsement on Emmet's 29 July letter by Rear Admiral Griffin, RG 80, 1897–1915, File 29840–181.

56. Emmet to Roosevelt, 29 July 1913, RG 80, 1897–1915, File 29840–181.

57. Lieutenant S. M. Robinson, USN, "Operation and Trials of the U.S. Fleet Collier *Jupiter*," *ASNE Journal* 26 (1914): 339–53.

58. See Lieutenant Commander H. C. Dinger, USN, "Gearing and the Electric Drive," *ASNE Journal* 26 (1914): 563–74; 563.

59. The Westinghouse Electric Manufacturing Company complained to Secretary Daniels about the noncompetitive awarding of the *California* contract to GE. Daniels called for new bids over the bitter objections of Emmet. Westinghouse had access to all the bid submissions, and Emmet believed that they would tender a "me too, twenty-five percent less" bid. Although GE did reduce their price, the Westinghouse bid was lower. Daniels awarded the contract to GE based upon their experience in the collier *Jupiter*. See Emmet, *Autobiography of an Engineer*, 167–68.

60. These ships replaced the small battleships of the same name (the so-called *Hale* class sold to Greece). *California* entered service under the name *New Mexico*.

61. *New York Times*, 27 April 1915, 4. The *California* was to be built in the New York Navy Yard, which had submitted a bid of $631,000 for steam turbine propulsion. The General Electric bid for the electric drive was $431,000. See *New York Times*, 7 March 1915, Section III, 2, for cost comparisons between the turboelectrically powered collier *Jupiter*, built at the Mare Island Naval Shipyard, and the colliers *Jason* and *Orion*, built in private yards at three-quarters the cost.

62. See *New York Times*, 26 October 1916, 14, and 16 November 1916, 10.

63. *New York Times*, 6 December 1916, and RG 80, SecNav, 1916–1926, File 28662, Memorandum from the Bureau of Steam Engineering to the Secretary of the Navy, 27 December 1916.

64. *New York Times*, 7 December 1917, 14. Newport News Shipbuilding and Wm. Cramp & Sons (Philadelphia) set their profits at 10 percent. Union Iron Works (San Francisco) and Fore River Shipbuilding (Quincy, Mass.) bid at a profit margin of 15 percent. New York Shipbuilding did not bid, as "their plant facility was taxed with navy and commercial contracts."

65. *New York Times*, 3 February 1917, 12, and 16 February 1917, 8. Besides Pupin, other electrical experts said to express doubts in the electric drive were Charles G. Curtis and Luther Lovekin, president of New York Shipbuilding.

66. William S. Sims to J. Bernard Walker, 19 February 1917, Admiral William

S. Sims Papers, Manuscript Division, Library of Congress, Washington, D.C., Box 83.

67. See *History of the Bureau of Steam Engineering During the Great War*, Navy Department Publication #5, Office of Naval Records and Library (Washington, D.C.: U.S. Government Printing Office, 1922), 21.

68. *Army and Navy Register*, 17 February 1917, 193.

69. See Rear Admiral David Taylor's statement before the House Naval Affairs Committee, reprinted in *USNIP* 43 (1917): 801.

70. Ibid.

71. Claiming to be an impartial electrical engineer, Wheeler revealed his close association with the private shipyards when he discussed confidential design details known only to shipyard management and the navy. See "Professional Notes," *USNIP* 43 (1917): 390–94, for a reprint of Griffin's letter of 5 January 1917 to Congressman Padgett, which discusses the Wheeler interview in detail.

72. See Tesla's statement in the *New York Herald*, reprinted in *USNIP* 43 (1917): 798–800; and Frank J. Sprague's letter to the *New York Times*, 21 February 1917.

73. Articles from *Shipping Illustrated* and *Marine Engineering*, reprinted in *USNIP* 43 (1917): 803–4.

74. Charles G. Curtis, Letter to the Editor, *Scientific American*, 17 February 1917: 173.

75. Letters from the Secretary of the Navy to the Compensation Board, RG 80, 1916–1926, File 28645–41:1.

76. Dyson, "Passing of the Directly Connected Turbine," 558–62.

77. See the address by Captain Sir Edward Chatfield, RN, Admiral Beatty's flag captain at Jutland, to the Institution of Naval Architects, "Current Tendencies in Naval Designs," *The Engineer*, reprinted in *ASNE Journal* 32 (1920): 539–40.

78. See Breyer, *Schlachtshiffe*, 122.

79. Article reprinted in *ASNE Journal* 32 (1920): 211.

80. See Hansgeorg Jentschura, Dieter Jung, and Peter Mickel, *Warships of the Imperial Japanese Navy, 1869–1945* (Annapolis: Naval Institute Press, 1977), 64.

81. William H. Garzke and Robert O. Dulin, *Battleships: Axis and Neutral Battleships in World War II* (Annapolis: Naval Institute Press, 1986), 291.

82. Ibid., 377–81.

83. *First Progress Report, Committee on Designs, February 1905*, reprinted in *The Papers of Admiral Sir John Fisher*, vol. I, ed. Lieutenant Commander P. K. Kemp, RN (London: Allen & Unwin, 1964), 201–7.

84. Marine Baumeister Gustav Berling quoted in Foote, "Turbine Electric Propulsion of a Battleship Compared with Other Means," 76.

85. See "Germany Adopts Turbines" (n. 27 above).

86. *Annual Report of the Secretary of the Navy, 1931* (Washington, D.C.: U.S. Government Printing Office, 1932), 313.

87. The increased speed was believed necessary to counter the 27-knot Japanese battlecruisers of the *Kongo* class; see Norman Friedman, *U.S. Battleships: An Illustrated Design History* (Annapolis: Naval Institute Press, 1985), 251.

88. "Prepared Statement by Engineer-in-Chief (BUENG), General Board Hearing, 8 October 1935," and "Statement by Rear Admiral H. G. Bowen, Engineer-in-Chief, USN, to the Advisory Board on Battleship Plans, 20 October 1937," Papers of Vice Admiral Harold G. Bowen, Seeley G. Mudd Manuscript Library, Princeton University, Princeton, N.J., Box 2, Battleship File.

89. In January 2000, the U.S. Navy announced that a new version of electric propulsion was being developed for DD 21, its new twenty-first-century "land-attack" destroyer. See " 'Integrated Power Systems & Electric Drive for DD 21,' Special Defense Department Briefing on U.S. Navy Announcement Subject: Embrace and Application of New Technologies to Navy Destroyers," 6 January 2000, available at http://dd21.crane.navy.mil/Library/library.htm and from the author's files.

Chapter 5. Anomalous Technologies of the Great War

1. Refer to the concepts of thought collective (*Denkkollektiv*) and thought style (*Denkstil*) elucidated by Ludwik Fleck, *Genesis and Development of a Scientific Fact*, ed. Thaddeus K. Trenn and Robert K. Merton (Chicago: University of Chicago Press, 1979), 39.

2. Ibid.

3. The *Amagi*-class ships were to carry ten 16-inch/45-caliber guns to match the U.S. capital ships of the 1916 Program and, at 30 knots, were considerably faster. In the wake of the Five-Power Treaty, two ships were canceled and *Amagi* and *Akagi* were to be converted into aircraft carriers. *Amagi* was heavily damaged by an earthquake in 1923 while on the building way, and she was scrapped. *Akagi* was completed as an aircraft carrier, took part in the attack on Pearl Harbor, and was sunk at Midway (1942); see Hansgeorg Jentschura, Dieter Jung, and Peter Mickel, *Warships of the Imperial Japanese Navy, 1869–1945* (Annapolis: Naval Institute Press, 1977), 36.

4. See the write-up on the *Invincible* Class in Oscar Parkes, OBE, *British Battleships, Warrior 1860 to Vanguard 1950: A History of Design, Construction and Armament* (London: Seeley Service, 1970), chap. 84.

5. Jon Tetsuro Sumida, *In Defence of Naval Supremacy: Finance, Technology, and British Naval Policy, 1889–1914* (London: Unwin Hyman, 1989).

6. Fisher, "The Function of Battleships and Cruisers," in *The Papers of Admiral Sir John Fisher*, vol. II, ed. Lieutenant Commander P. K. Kemp, RN (London: Allen & Unwin, 1960) (hereafter *Fisher Papers, II*), 260.

7. Testimony of Rear Admiral Badger, 8 December 1914, *Hearings before the Committee on Naval Affairs of the House of Representatives on Estimates Submitted by the Secretary of the Navy, 1915,* (hereafter *HNAC Hearings for 1915*) (Washington, D.C.: U.S. Government Printing Office, 1915), 479.

8. Meyer's speech quoted in the *Army and Navy Journal*, 13 February 1915, reprinted in *U.S. Naval Institute Proceedings* (hereafter *USNIP*) 41 (1915): 549.

9. Ibid.

10. See table C-16, Robert Dulin and William Garzke, *Battleships: United States Battleships in World War II*(Annapolis: Naval Institute Press, 1976), 248; and table 4, Alan Raven and John Roberts, *British Battleships of World War II: The Development and Technical History of the Royal Navy's Battleships and Battlecruisers from 1911 to 1946* (Annapolis: Naval Institute Press, 1976),17.

11. Assistant Naval Constructor B. S. Bullard, USN, "A Plea for the Battle-Cruiser," *USNIP* 41 (1915): 1916.

12. Commander Yates Stirling, USN, "The Arrival of the Battle-Cruiser," *USNIP* 41 (1915): 1920.

13. Ibid., 1924.

14. *General Board Report, 30 July 1915*, reprinted in *USNIP* 42 (1916): 588–89.

15. *Naval Lessons from the European War, 1915*, Bureau of Construction & Repair Report, Record Group 80, General Records of the Secretary of the Navy, National Archives, Washington, D.C. (hereafter RG80), 1897–1915, 2.

16. "The Naval Appropriation Bill of 1916," Professional Notes, *USNIP* 42 (1916): 1637.

17. *Scientific American*, "The Big Gun in the Big Ship," 18 November 1917, reprinted in *USNIP* 43 (1917): 184–85.

18. Ibid.

19. Daniels's report to Congress in "Limitation on Size of Battleships," *Army and Navy Journal*, April 1917, reprinted in *USNIP* 43 (1917): 790.

20. See Park Benjamin, "The Shout for 'Big Ships,'" *The Independent*, 31 January 1909, clipping in Admiral William S. Sims Papers, Manuscript Division, Library of Congress, Washington, D.C. (hereafter Sims Papers), Mahan Folder, Box 71.

21. *Scientific American*, "Great Defensive Strengths of Battleships," 29 July 1916, reprinted in *USNIP* 42 (1916): 1682–83.

22. See Capps, *Alleged Defects*, 24.

23. See, for example, the account in Parkes, *British Battleships*, chap. 111.

24. The U.S. Navy was aware of the loss of the battlecruisers at Jutland. However, the naval hierarchy remained ignorant of the exact circumstances until after the United States entered the war. For a treatment of the damage inflicted to ships during the Battle of Jutland, see N. J. M. Campbell, *Jutland: An Analysis of the Fighting* (Annapolis: Naval Institute Press, 1986).

25. Fisher marginalia on Admiral Sir John Jellicoe to Lord Fisher, 15 June 1916, letter 315, *Fear God and Dread Nought: The Correspondence of Admiral of the Fleet Lord Fisher of Kilverstone*, vol. III: *Restoration, Abdication, and Last Years, 1914–1920*, ed. Arthur J. Marder (London: J. Cape, 1959) (hereafter *Fisher Correspondence, III*), 356.

26. See Parkes, *British Battleships*, chap. 112. *Hood* was still relatively weakly armored compared to contemporary battleships.

27. See Sims to Benson, 14 November 1918; Sims to Benson, 7 December 1918; and Planning Committee Memorandum (Secret) to Chief of Naval Operations, 7 October 1918, Sims Papers, Box 42.

28. See Dean C. Allard, "Anglo-American Naval Differences During World War I," in *In Defense of the Republic: Readings in American Military History*, ed. David Curtis Skaggs and Robert S. Browning III (Belmont, Calif.: Wadsworth, 1991), 239–51.

29. Admiral Benson's cable number 185 to Sims, cited in Sims to Benson, 13 January 1919, Sims Papers, Box 49.

30. Ibid.

31. *New York Times*, 16 March 1919, reprinted in *USNIP* 45 (1919): 660.

32. An account of the attack by *U-9*'s Kommandant may be found in *HNAC Hearings for 1915*, 997–99.

33. See Robert L. O'Connell, *Sacred Vessels: The Cult of the Battleship and the Rise of the U.S. Navy* (Boulder, Colo.: Westview Press, 1991), 162–63.

34. *Shipping Illustrated*, "Torpedo Nets for U.S. Ships," 10 October 1914, reprinted in *USNIP* 40 (1914): 1821; and *The Navy*, "Protection Against Torpedoes," November 1914, reprinted in *USNIP* 40 (1914).

35. Fisher to Beatty, 27 January 1915, letter 117, *Fisher Correspondence, III*, 146–47. Fisher to Beatty, 3 February 1915, letter 123, *Fisher Correspondence, III*, 152. Fisher to Beatty, 25 January 1915, letter 116, *Fisher Correspondence, III*, 146.

36. Testimony of Rear Admiral Fletcher, 9 December 1914, *HNAC Hearings for 1915*, 514–15.

37. Ibid., 514.

38. Ibid.

39. Ibid.

40. Ibid., 515.

41. See Ernest Andrade, "Submarine Policy in the United States Navy, 1919–1941," *Military Affairs* 35 (April 1971): 50–56.

42. Testimony of Rear Admiral Fletcher, *HNAC Hearings for 1915*, 521.

43. Testimony of Rear Admiral Badger, *HNAC Hearings for 1915*, 510–11.

44. General Board letter and memorandum of 25 September 1912, recommending the composition of the building program for the fiscal year ending 30 June 1914, quoted in *HNAC Hearings for 1915*, 478.

45. *Naval Lessons from the European War*, Bureau of Construction & Repair Report (1915), RG 80, SecNav, 1897–1915.

46. For the Naval Consulting Board, see William M. McBride, "The 'Greatest Patron of Science'?: The Navy-Academia Alliance and U.S. Naval Research, 1896–1923," *Journal of Military History* 56 (1992): 7–33, especially n. 4.

47. See Robert L. O'Connell, *Of Arms and Men: A History of War, Weapons and Aggression* (New York: Oxford University Press, 1989), 7–11.

48. Ensign V. N. Bieg, USN, "The Submarine and the Future," *USNIP* 41 (1915): 152. Bieg apparently continued to move in unconventional circles, being killed as a lieutenant commander in the crash of airship ZR2 on 24 August 1921.

49. Yates Stirling, Rear Admiral, USN (Ret.), *Sea Duty: The Memoirs of a Fighting Admiral* (New York: G. P. Putnam's Sons, 1939), 153–54. Also see John B. Lundstrom, "Chester W. Nimitz: Victory in the Pacific," in *Quarterdeck and Bridge: Two Centuries of American Naval Leaders*, ed. James C. Bradford (Annapolis: Naval Institute Press, 1997), 330.

50. Stirling, *Sea Duty*, 155–56.

51. Benson to Sims, 15 November 1915, Sims Papers, Box 48.

52. Sims to Benson, 18 November 1915, Sims Papers, Box 48.

53. Lieutenant Commander H. E. Yarnell, USN, "Notes on Naval Tactics," *USNIP* 42 (1916): 77.

54. See John Ellis, *The Social History of the Machine Gun* (Baltimore: Johns Hopkins University Press, 1975), chap. 5, especially 118–20.

55. *Shipping Illustrated*, 15 May 1916, reprinted in *USNIP* 42 (1916): 946.

56. Lieutenant (j.g.) F. A. Daubin, USN, "The Fleet Submarine," *USNIP* 42 (1916): 1815–1923.

57. Ibid., 1816.

58. Ibid., 1822.

59. Ibid., 1818.

60. Roland, *Underwater Warfare in the Age of Sail* (Bloomington: Indiana University Press, 1978), 181.

61. Rear Admiral William S. Sims, USN, *The Victory at Sea* (Garden City, N.Y.: Doubleday, Page, 1919), 115.

62. Since the acoustic detection technologies were funded by the Bureau of [Steam] Engineering, it is no wonder that Sims was loath to credit them with winning the war. See McBride, "Greatest Patron," 32–33.

63. Ronald H. Spector, *Professors of War: The Naval War College and the Development of the Naval Profession* (Newport, R.I.: Naval War College Press, 1977), 148.

64. Lieutenant R. C. Saufley, USN, "Naval Aviation: Its Values and Needs," *USNIP* 40 (1914): 1459–72.

65. Ibid., 1471.

66. Commander Ralph Earle, USN, "Naval Scouts: Their Necessity, Utility, and Best Type," *USNIP* 41 (1915): 1099–1120; see 1119.

67. Commander Thomas D. Parker, USN, "An Air Fleet: Our Most Pressing Naval Want," *USNIP* 41 (1915): 709–63.

68. Ibid., 710.

69. Ibid.

70. Ibid., 761–62.

71. Ibid., 732–33.

72. See a summary of the Board's recommendations in "Dirigibles for the U.S. Navy," *Professional Notes, USNIP* 40 (1914): 261–62.

73. *Report of the Aeronautical Board*, reprinted in part in Parker, "An Air Fleet," 753–59.

74. See Paolo E. Coletta, *Admiral Bradley A. Fiske and the American Navy* (Lawrence: University Press of Kansas, 1979), 109.

75. Due to a shortage of office space, Bristol had to share Fiske's desk for eighteen months; Coletta, *Admiral Bradley A. Fiske*, 110.

76. Captain Bristol, Director of Aeronautics, testimony before the House Naval Affairs Committee cited by Parker, "An Air Fleet," 759.

77. Parker, "An Air Fleet," 760.

78. Testimony of Rear Admiral Victor Blue, 30 November 1914, in *HNAC Hearings for 1915*, 17.

79. Testimony of Secretary Daniels, 11 December 1914, *HNAC Hearings for 1915*, 709.

80. Testimony of Rear Admiral Fiske, 17 December 1914, *HNAC Hearings for 1915*, 1019.

81. This seed money had a precedent as Fiske explained: "Some years ago when we were trying to get shells in this country we could not get any. Nobody could

make them, and we had difficulty in getting them from abroad. We could not get any manufacturers in this country to put the money into a plant, because they did not know whether the Government would buy any shells or whether the Navy Department had the money to buy them. So the Secretary of the Navy—I think it was Mr. Whitney—got a large appropriation from Congress, the idea being he could say to the steel people, 'I have this much money; if you will put up works here and make shells, I have this much money with which to buy shells.' So the American manufacturers got the rights from abroad and put up their works and made the shells"; testimony of Rear Admiral Fiske, 17 December 1914, *HNAC Hearings for 1915*, 1022.

82. General Board *Report on Aviation*, reprinted in *Hearings before the Committee on Naval Affairs of the House of Representatives on Estimates Submitted by the Secretary of the Navy, 1916* (hereafter *HNAC Hearings for 1916*) (Washington, D.C.: U.S. Government Printing Office, 1916), 3597–3601.

83. See testimony of Navy Secretary Daniels, 31 March 1916, *HNAC Hearings for 1916*, 3601.

84. Secretary Daniels to *Aviation and Aeronautical Engineering*, reprinted in *USNIP* 43 (1917): 383.

85. *Army and Navy Register*, "The Navy's Air Policy," reprinted in *USNIP* 43 (1917): 382.

86. Fiske's speech before the Pan-American Aeronautical Exposition, 12 February 1917, excerpted in *Army and Navy Journal* and reprinted in *USNIP* 43 (1917): 855.

87. See Coletta, *Admiral Bradley A. Fiske*, chap. 16.

88. Rear Admiral Bradley Fiske, USN, "Air Power," *USNIP* 43 (1917): 1701–4.

89. Ibid., 1701.

90. Ibid., 1702; also see Fiske, "Naval Power," *USNIP* 37 (1911): 683–786.

91. Fiske, "Air Power," 1704.

92. See testimony of Lieutenant Colonel Porte, RFC, 18 June 1918, *Hearings Before the General Board of the Navy, 1917–1950* (microfilm), Naval Historical Library, Washington Navy Yard, Washington, D.C. (hereafter cited as *General Board Hearings*).

93. The aircraft carried on scouts and battleships were seaplanes and were launched by catapults and hoisted back aboard after they landed in the sea. For the separate-ship carrier discussion, see the testimony of Captain Irwin, 18 June 1918, *General Board Hearings*.

94. Statement by Admiral Badger, 4 September 1918, *General Board Hearings*.

95. Statement by Rear Admiral Winterhalter, 4 September 1918, *General Board Hearings*.

96. Ibid.

97. Testimony of Admiral H. T. Mayo, USN, 17 April 1919, *General Board Hearings*.

98. Commander in Chief, U.S. Fleet [Admiral Mayo] to General Board, 5 May 1919, contained in *General Board Hearings* (1919, vol. IV), 660.

99. Ibid., 671.

100. Ibid., 660.

101. Rear Admiral Taylor to Commander Emory S. Land, USN, 14 May 1920, Admiral Emory S. Land Papers, Manuscript Division, Library of Congress, Washington, D.C. (hereafter Land Papers).

102. Fiske to Sims, 30 November 1920, Sims Papers, Box 57.

103. Sims to Fiske, 2 December 1920, Sims Papers, Box 57.

104. For the 1920 Sims controversy, see Morison, *Admiral Sims*, chap. 23.

105. Taylor to Commander Land, 26 January 1920, Land Papers, Box 1.

106. Chief of Naval Operations (Planning Division) Secret Memorandum to the General Board, Serial Op-12-D, P.D.138–15, 30 October 1920, included in *General Board Hearings*, Reel 3, 926ff.

107. Ibid.

108. Ibid.

109. Henry Woodhouse, "The Torpedoplane: The New Weapon Which Promises to Revolutionize Naval Tactics," *USNIP* 45 (1919): 751.

110. Testimony of Captain Mustin, 26 June 1922, *General Board Hearings*.

111. Fiske to Sims, 8 January 1921, Sims Papers, Box 57.

112. Ibid.

113. Sims to Fiske, 10 January 1921, and Fiske to Sims, 12 January 1921, Sims Papers, Box 57.

114. Sims to Fiske, 12 March 1921, Sims Papers, Box 57.

115. Sims to Fiske, 25 January 1921, Sims Papers, Box 57.

116. For useful discussions of the establishment of the Bureau of Aeronautics, see William F. Trimble, *Admiral William A. Moffett: Architect of Naval Aviation* (Washington, D.C.: Smithsonian Institution Press, 1994), especially chap. 4; and Clark G. Reynolds, *Admiral John H. Towers: The Struggle for Air Supremacy* (Annapolis: Naval Institute Press, 1991), chap. 7. A recent book by Paolo Coletta, *Admiral William A. Moffett and U.S. Naval Aviation* (Lewiston, N.Y.: E. Mellen Press, 1997), offers nothing new and falls far short of his excellent treatment of Bradley A. Fiske.

117. Testimony of Rear Admiral Moffett, 26 June 1922, *General Board Hearings*.

118. Testimony of Rear Admiral McVay, 26 June 1922, *General Board Hearings*.

119. Testimony of Captain Stocker (CC), USN, 26 June 1922, *General Board Hearings.*

120. See the comment of Rear Admiral W. V. Pratt, 26 June 1922, *General Board Hearings.*

121. In 1920, the army was also defining the role of aviation in terms of its contemporary strategy and tactics, that is, "teamwork with ground troops"; U.S. Army Command and General Staff School textbook, E. L. Naiden, *Air Service*, cited by I. B. Holley Jr., *Ideas and Weapons: Exploitation of the Aerial Weapon by the United States During World War I; A Study in the Relationship of Technological Advance, Military Doctrine, and the Development of Weapons* (Washington, D.C.: U.S. Government Printing Office, 1983 [second printing]), 171.

122. Secretary of the Navy to Commander Yates Stirling, USN (orders to command the Submarine Flotilla, Atlantic Fleet), 2 April 1914, and Sims to Stirling, 9 April 1914, and Stirling to Sims, 15 April 1914, Sims Papers, Box 86.

123. Sims to John Callan O'Laughlin, 23 September 1925, Sims Papers, Box 46. Sims characterized the appointment of these admirals of the Old School as "a crime against the people of the United States"; see Sims to Lieutenant Commander Dinger, 14 February 1925, Sims Papers, Box 54. For an overview of Moffett's career, see Clark G. Reynolds, "William A. Moffett: Steward of the Air Revolution" in Bradford, *Quarterdeck and Bridge*, 291–306.

CHAPTER 6.   CONTROLLING AVIATION AFTER THE WORLD WAR

1. Fisher to the Editor, *The Times*, 2 September 1919, letter 552, *Fear God and Dread Nought: The Correspondence of Admiral of the Fleet Lord Fisher of Kilverstone*, vol. III: *Restoration, Abdication, and Last Years, 1914–1920*, ed. Arthur J. Marder (London: J. Cape, 1959), 590.

2. Rear Admiral David W. Taylor, USN, "The Design of Vessels as Affected by the World War," *Journal of the American Society of Naval Engineers* 32 (1920): 745.

3. Ibid., 749.

4. Ibid., 754. In light of the aerial torpedo's smaller size and warhead, Taylor's point was valid.

5. Testimony of Captain Mustin before the General Board, 26 June 1922, *Hearings of the General Board of the Navy, 1917–1950*, microfilm, Nimitz Library, U.S. Naval Academy, Annapolis, Md. (hereafter *General Board Hearings*). For another view on the future of the torpedo plane, see Henry Woodhouse, "The Torpedoplane: The New Weapon Which Promises to Revolutionize Naval Tactics," *U.S. Naval Institute Proceedings* (hereafter *USNIP*) 45 (1919): 751.

6. U.S. Senate, "Hearings: Building up the United States Navy to the Strength

Permitted by the Washington and London Naval Treaties," 72nd Congress, 1st Session (1932), cited in Robert L. O'Connell, *Sacred Vessels: The Cult of the Battleship and the Rise of the U.S. Navy* (Boulder, Colo.: Westview Press, 1991), 291.

7. Roger Dingman, *Power in the Pacific* (Chicago: University of Chicago Press, 1979), 152. Battleship numbers from Christopher Hall, *Britain, America, and Arms Control, 1921–1937* (New York: St. Martin's Press, 1987), table 1.1, 13.

8. The Five-Power Treaty did not limit aircraft. A committee of aviation experts from Britain, Japan, Italy, France, and the United States deferred discussion of the aircraft question, believing "that it is not practicable to impose any effective limitations upon the numbers or characteristics of aircraft, either commercial or military, except in the single case of lighter-than-air craft. The committee is of the opinion that the use of aircraft in warfare should be governed by the rules of warfare as adapted to aircraft by a further conference which should be held at a later date"; *Report of Special Committee on Limitation of Aircraft in Warfare*, cited in *Tenth Annual Report of the N.A.C.A.* in *Aviation*, 22 December 1924, reprinted in *USNIP* 51 (1925): 495.

9. Or, as Billy Mitchell put it: "more than 1000 airplanes can be built and maintained for the outlay required for a single [$10 million] battleship"; see William Mitchell, *Winged Defense: The Development and Possibilities of Modern Air Power Economic and Military* (New York: Dover, 1988; reprint of the 1925 edition published by G. P. Putnam's Sons), 120.

10. The older battleships that were expended in these tests had been designed to fight at relatively shorter ranges than contemporary battleships, for example, those of the U.S. 1916 Program. At shorter ranges more vertical armor was installed as the trajectory of incoming shells was relatively flat. Long-range naval guns delivered their shells at a steeper angle and as a result the later battleship designs installed heavier horizontal armor that resisted aerial bombs as well.

11. *Report of the Special Board on Policy with Reference to the Upkeep of the Navy in its Various Branches*, 17 January 1925, in *General Board Hearings* (hereafter *1925 Special Board Report*), 23.

12. For the aviation side of the bombing trials, see Mitchell, *Winged Defense*, chap. 3.

13. Rear Admiral Moffett's statement before the Joint Military and Naval Affairs Committee of the House of Representatives printed in full in *Aviation* and reprinted in *USNIP* 50 (1924): 1364–70.

14. Ibid. Army flyers had recently flown across the northern Pacific from Dutch Harbor, Alaska, to Japan in April–May 1924; see "Professional Notes," *USNIP* 50 (1924): 1370.

15. The naval officer corps deeply resented the intrusion of New Era efficiency experts, led by the administration's budget office, into naval affairs. As Admiral Rodgers told the General Board: "We are continually confronted by Treasury people who want a change, and reformers who are constantly making the point in which their business training causes them to follow a line of thought which is not in accordance with [naval] necessity. . . . [The New Era] Congress, taking over from civil life its business ideas, looks to the great corporations which have grown up by consolidation to become trusts, and says, 'Let us consolidate everything that looks alike and put them together.' They say, 'We will take those things that look alike and consolidate them in the interests of efficiency'"; Rodgers's 26 June 1922 statement to the General Board, *General Board Hearings*. For many naval officers, Republican Party New Era efficiency threatened the U.S. Navy with extinction.

16. Secretary of the Navy Curtis D. Wilbur to General Board, 23 September 1924, reprinted in *General Board Hearings*.

17. See *Army and Navy Register*, "The Naval Budget," 25 October 1924, reprinted in *USNIP* 51 (1925): 127–28.

18. 1925 *Special Board Report*, 1–3.

19. Ibid., 11.

20. Ibid. As yet, the navy had no aircraft carriers save the converted collier *Langley*.

21. Ibid.

22. Ibid.

23. Ibid.

24. Ibid., 43.

25. Ibid., 70.

26. Ibid., 70–71.

27. Ibid., 20.

28. Ibid. 28.

29. For accounts of Sims's experience in revamping U.S. naval gunnery practices, see Elting Morison, *Men, Machines, and Modern Times* (Cambridge, Mass.: The MIT Press, 1966), chap. 2.

30. 1925 *Special Board Report*, 20.

31. Ibid.

32. Ibid. For a critical assessment of the battleship, see Rodrigo Garcia y Robertson, "The Failure of the Heavy Gun at Sea, 1898–1922," *Technology and Culture* 28 (1987): 539–57.

33. 1925 *Special Board Report*, 20.

34. Testimony of Admiral Strauss, *Hearings Before the Special Board*, contained in *General Board Hearings* (hereafter *Special Board Hearings*), 132.

35. See testimony of Rear Admiral Moffett and Captains Johnson and Land of the Bureau of Aeronautics, 30 September and 1–3 October 1924, *Special Board Hearings*.

36. Navy Secretary Curtis D. Wilbur in [Washington] *Star*, "Naval Engineers Have in View Vastly Improved Battleships," 15 January 1925, reprinted in *USNIP* 51 (1925): 501.

37. *1925 Special Board Report*, 27.

38. Ibid.

39. Combat experience in World War II forced the navy to take waterborne shock from nearby explosions more seriously than before the war. See Francis Duncan, "Hyman G. Rickover: Technology and Naval Tradition," in *Quarterdeck and Bridge: Two Centuries of American Naval Leaders*, ed. James C. Bradford (Annapolis: Naval Institute Press, 1997), 398.

40. Testimony of Rear Admiral Moffett, 30 September 1924, *Special Board Hearings*.

41. Moffett had entered his "largest gun" battleship essay for the Naval Institute's Lippincott Prize. See Commander William A. Moffett, USN, to Captain William S. Sims, USN, 12 June 1916; Sims to Moffett, 16 June 1916; and Moffett to Sims, 27 June 1916, in the Papers of William S. Sims, Manuscript Division, Library of Congress, Washington, D.C. (hereafter Sims Papers), Box 47.

42. Testimony of Rear Admiral Moffett, *Special Board Hearings*, 170. For Moffett's medal-winning in Mexico, see William F. Trimble, *Admiral William A. Moffett: Architect of Naval Aviation* (Washington, D.C.: Smithsonian Institution Press, 1994), chap. 3. Moffett was an astute politician and his use of the gun metaphor may have been to placate the battleship hierarchy. At this stage in his career at the Bureau of Aeronautics, Moffett is most likely speaking sincerely as a member of the battleship thought collective. The developments in aviation technology in the late 1920s and introduction of the large carriers *Lexington* and *Saratoga* provided the framework which allowed him to move away from this "normal" view of naval aviation.

43. Testimony of Lieutenant Commander Mitscher, *Special Board Hearings*, 42 and 39.

44. Testimony of Rear Admiral M. M. Taylor, *Special Board Hearings*, 310. Taylor was basing his estimate on a carrier costing $16 million; the carrier *Saratoga*, still under construction, actually cost $45 million, which would have enhanced his argument on behalf of the submarine.

45. For an overview of the U.S. Navy's War Plan Orange against Japan, see Rus-

sell F. Weigley, *The American Way of War: A History of United States Military Strategy and Policy* (Bloomington: Indiana University Press, 1977), chap. 12.

46. Testimony of Rear Admiral Magruder, 10 October 1924, *Special Board Hearings*.

47. Ibid.

48. The Board heard testimony regarding the potential use of poison-gas aerial bombs against ships. See the testimony of Rear Admiral Moffett and Lieutenant McMurrain, 30 September[?] 1924, *Special Board Hearings*.

49. 1925 *Special Board Report*, 75–77.

50. Testimony of Admiral William F. Fullam, USN (Ret.), *Special Board Hearings*, 716.

51. Ibid.

52. Ibid.

53. Ibid., 719.

54. Ibid., 723.

55. Coontz was a part of the what Rear Admiral Sims condemned as the reactionary "Daniels Cabinet"; see Sims to John Callan O'Laughlin, 23 September 1925, Sims Papers, Box 46. Sims characterized the appointment of these admirals of the Old School as "a crime against the people of the United States"; Sims to Lieutenant Commander Dinger, 14 February 1925, Sims Papers, Box 54. For Admiral Coontz and the submarine program, see the testimony of Admiral Fullam, *Special Board Hearings*, 723.

56. Testimony of Admiral Fullam, *Special Board Hearings*, 717.

57. Ibid.

58. The General Board, fearing a foreign shift in naval competition from the battleship to the air, was particularly interested in British efforts in naval aviation. Lieutenant Colonel Porte, of the Royal Flying Corps, was questioned vigorously in 1918 in order to ascertain whether the British were considering any radical employment of aircraft at sea. See the testimony of Porte in *General Board Hearings*, 18 June 1918.

59. For the Morrow Board, see Archibald D. Turnbull and Clifford L. Lord, *History of United States Naval Aviation* (New Haven, Conn.: Yale University Press, 1949), 249–58.

60. The Morrow Board Report was reprinted in *USNIP* 52 (1926): 196–225.

61. See Michael A. West, "Laying the Legislative Foundation: The House Naval Affairs Committee and the Construction of the Treaty Navy, 1926–1934" (Ph.D. diss., The Ohio State University, 1980), 44–47; and Turnbull and Lord, *Naval Aviation*, 257 and chap. 24.

62. Admiral Samuel S. Robinson, commander-in-chief, U.S. Fleet, to Captain Emory S. Land, (CC), USN, Bureau of Aeronautics, 17 May 1926, Emory S. Land Papers, Manuscript Division, Library of Congress, Washington, D.C., Box 5.

63. Captain Yates Stirling, USN, "The Place for Aviation in the Organization for War," *USNIP* 52 (1926): 1103–4.

64. Captain Yates Stirling, USN, "Some Fundamentals of Sea Power," *USNIP* 51 (1925): 913–14. Stirling extended his gun metaphor to include the submarine, a projectile that plunges beneath the sea, but one, like the airplane, which must return to the surface; ibid., 915.

65. Ibid., 917. Stirling's description of future aerial warfare typified the naval hierarchy's extension of its battleship worldview to this alternate technology: "[We may imagine] an air fleet . . . the great bombing planes and torpedo planes representing capital ships; scouting planes and spotting planes being the light cruisers and fighting planes the swift, agile destroyers of the air; fighting planes on each side concentrating to bring a superior force upon the enemy air fleet at a superior moment." See Stirling, "Aviation in Organization for War," 1105.

66. Statement of Rear Admiral Williams, *Special Board Hearings*, 122.

67. The first purpose-built aircraft carrier, *Ranger*, did not enter service until 1934. Early fleet exercises tended to pit *Lexington* against *Saratoga*. On the aviation lessons of the exercises, see Thomas Wildenberg, *Destined for Glory: Dive Bombing, Midway, and the Evolution of Carrier Airpower* (Annapolis: Naval Institute Press, 1998), chap. 7.

68. *Tactical Orders and Doctrine for the U.S. Fleet, 1941* quoted in Wildenberg, *Destined for Glory*, v.

69. Rear Admiral M. M. Taylor, commander of the Control Force, *Special Board Hearings*, 263.

70. Ibid. Other relevant articles include Hector C. Bywater, "The Battleship and Its Uses," *USNIP* 52 (1926): 407–25; Lieutenant Commander C. A. Pownall, USN, "The Airphobia of 1925," *USNIP* 52 (1926): 459–63; Lieutenant Commander O. C. Badger, USN, "History Repeats *or* The Application of Lessons of History on a National Problem of Today," *USNIP* 51 (1925): 707–21.

71. For a contemporary account of the navy's Naval Aircraft Factory, see Lieutenant Commander S. J. Ziegler (CC), USN, "The Naval Aircraft Factory," *USNIP* 52 (1926): 83–94.

72. Rear Admiral W. A. Moffett, USN, "Recent Technical Development of Naval Aviation," *USNIP* 57 (1931): 1182–83.

73. Ibid., 1185. For the development of aviation technology during Moffett's tenure as chief of the Bureau of Aeronautics, see Trimble, *Admiral Moffett*, passim.

74. Admiral W. T. Mayo, USN, Commander-in-Chief, U.S. Fleet, to General Board, 5 May 1919, a copy of which is contained in *General Board Hearings*, 660–71; quote on 660.

75. Testimony of Rear Admiral Moffett, 30 September 1924, *Special Board Hearings*. At the time of his testimony, the navy had only one carrier, the converted collier *Langley*; the converted battlecruisers *Lexington* and *Saratoga* would not be ready until 1927. The effect that New Era efficiency exerted on the navy can be seen in the naval budgets from 1922 through 1932. During this period, the naval appropriations averaged $359 million per year and totaled almost $4 billion. The total naval appropriations for the dreadnought navy (1906 through 1916) was less (almost $2 billion), but on average, 27 percent of each year's appropriation went to naval expansion, that is, new ship construction and alterations to existing vessels, as opposed to an average of 15 percent ($54 million) per year during the New Era. Data drawn from West, "Laying the Legislative Foundation," table 3, 38.

76. See, for example, Naval War College Staff Lectures: "The Employment of Aviation in Naval Warfare," serial 3429–1487/9-9-37, September 1937; "Tactical Employment of the Fleet," 29 October 1937; and "The Employment of Submarines," 3 October 1938, in Record Group 14, Staff Presentations and Lectures, Naval Historical Collection, Naval War College, Newport, R.I.

77. *New York Times*, 26 February 1931, reprinted in *USNIP* 57 (1931): 538.

78. Ibid. Also see Clark G. Reynolds, *The Fast Carriers: The Forging of an Air Navy* (Huntington, N.Y.: Robert E. Krieger, 1978), chap. 1, especially 14–21.

79. Wildenberg, *Destined for Glory*, 156–57.

80. Ibid., 156.

81. Ibid.

82. See ibid., chaps. 18 and 19; speed data on 193.

83. Wildenberg places the initial squadron employment of the SBD-2 in December 1940 (ibid., 161). TBD squadron data from Roy A. Grossnick, *United States Naval Aviation, 1910–1945* (Washington, D.C.: Naval Historical Center, 1996), appendix 6, "Combat Aircraft Procured," 494.

Chapter 7.   Disarmament, Depression, and Politics

1. Commager and Morris, editors' introduction, in William E. Leuchtenburg, *Franklin D. Roosevelt and the New Deal, 1932–1940* (New York: Harper & Row, 1963), ix.

2. Only two (*Alaska* and *Guam*) of the six expensive, lightly armored, and relatively slow *Alaska*-class battlecruisers were completed. Work on the third, *Hawaii*, was suspended in December 1943. The remaining three were not laid down, and

their allocations were shifted to aircraft carrier production. Although the navy designated the *Alaskas* as large cruisers (CB), they were, in fact, battlecruisers. See Robert O. Dulin and William H. Garzke Jr., *Battleships: U.S. Battleships in World War II* (Annapolis: Naval Institute Press, 1976), chap. 6.

3. Selig Adler, "Hoover's Foreign Policy and the New Left," in *The Hoover Presidency: A Reappraisal*, ed. Martin L. Fausold and George T. Mazuzan (Albany: State University of New York Press, 1974), 153–54; Ellis W. Hawley, "Herbert Hoover and Modern American History: Sixty Years After," in *Hoover and the Historians* (West Branch, Iowa: Hoover Presidential Library Association, 1989), 5. A useful review of U.S. foreign policy vis-à-vis the Japanese is Norman A. Graebner, "Hoover, Roosevelt, and the Japanese," in *Pearl Harbor as History: Japanese-American Relations 1931–1941*, ed. Dorothy Borg and Shumpei Okamoto (New York: Columbia University Press, 1973), 25–52. Also see Alexander DeConde, "Herbert Hoover and Foreign Policy: A Retrospective Assessment," in *Herbert Hoover Reassessed: Essays Commemorating the Fiftieth Anniversary of the Inauguration of Our Thirty-First President*, Senate Document No. 96–63, 96th Congress, 2nd Session (Washington. D.C.: U.S. Government Printing Office, 1981), 313–34; and Joan Hoff Wilson, "A Reevaluation of Herbert Hoover's Foreign Policy," in Fausold and Mazuzan, 164–86 (quote on 171).

4. The one exception to Hoover's general policy was his grudging permission for the navy to transfer funding allocated for shore establishments, under the July 1932 Emergency Relief and Construction Act, to the construction of three new destroyers. See Robert H. Levine, *The Politics of American Naval Rearmament, 1930–1938* (New York: Garland, 1988), 41–61. Roosevelt in 1932 proposed a $100 million reduction in the annual naval budget and a $30 million cap on annual naval construction funding over the next five years. See Michael A. West, "Laying the Legislative Foundation: The House Naval Affairs Committee and the Construction of the Treaty Navy" (Ph.D. diss., The Ohio State University, 1980), 275–77.

5. Arthur M. Schlesinger does not mention naval construction as part of the $3.3 billion NIRA public works package in his *The Age of Roosevelt: The Coming of the New Deal* (Boston: Houghton Mifflin, 1958); see chap. 6. Nor is there any mention in Leuchtenburg, *Roosevelt*; see chap. 3, especially 55–61. Frank Freidel mentions NIRA naval construction in passing, characterizing it as "palatable" to Roosevelt; see Frank Freidel Jr., *Franklin D. Roosevelt: Launching the New Deal* (Boston: Little, Brown, 1973), 432. Freidel observed that Hoover allocated $15 million in relief funds for ships and shore bases (although he did not point out that the funds were for operation and maintenance and not for new construction); ibid. See William H. McNeill, *The Pursuit of Power: Technology and Armed Force in European Society*

since A.D. 1000 (Chicago: University of Chicago Press, 1982), chap. 8, especially 270. On Cleveland and the navy, see Walter R. Herrick, *The American Naval Revolution* (Baton Rouge: Louisiana State University Press, 1966), 160 and 173–75.

6. Frank Freidel Jr., "Hoover and Roosevelt and Historical Continuity," in *Hoover Reassessed*, 288. Admiral Land to Secretary Swanson, 13 June 1933, Record Group 19, Records of the Bureau of Construction & Repair, National Archives, Washington, D.C. (hereafter RG 19); and Swanson to Roosevelt, 15 June 1933, Record Group 80, Records of the Secretary of the Navy, National Archives, Washington, D.C. (hereafter RG 80), cited in West, "Legislative Foundation," 329. See also Waldo H. Heinrichs Jr., "The Role of the United States Navy," in *Pearl Harbor as History*, 197–224.

7. The "Basic Naval Policy" may be found in General Board to Secretary of the Navy, G.Bd.#438, Serial 1347, 21 April 1927, in Foreign Affairs — Disarmament File, Presidential Papers, Hoover Papers, Herbert Hoover Presidential Library, West Branch, Iowa. Trade data from Gerald E. Wheeler, *Prelude to Pearl Harbor: The United States Navy and the Far East, 1921–1931* (Columbia: University of Missouri Press, 1963), 190. The disagreement between Hoover and the General Board over the worldwide strategy and large navy inherent in the "Basic Naval Policy" was paralleled by late cold war tensions among national strategy, developed by the President/National Security Council; Joint Chiefs of Staff (JCS)/commander-in-chiefs' strategies formulated by the JCS; defense strategy developed within the Department of Defense; and service strategies created within each of the armed services and used for "setting their own institutional agendas, rationalizing their requirements, and arguing for a larger or protected slice of the budget." Carl H. Builder, *The Masks of War: American Military Styles in Strategy and Analysis* (Baltimore: Johns Hopkins University Press, 1989), chap. 5, quote on 57–58.

8. [Washington] *Star*, "Navy Up to Par, Coolidge Believes," 10 January 1925, reprinted in *U.S. Naval Institute Proceedings* (hereafter *USNIP*) 51 (1925): 481–82. Captain Dudley Knox, USN (Ret.), "New Naval Limitation Conference Predicted," *The Sun* [Baltimore], 7 January 1925, reprinted in *USNIP* 51 (1925): 498–500. For cruiser funding, see Christopher Hall, *Britain, America, and Arms Control: 1921–1937* (New York: St. Martin's Press, 1987), 39–40.

9. See Wheeler, *Prelude to Pearl Harbor*, chap. 6; and Hall, *Arms Control*, 58. See Robert L. O'Connell, *Sacred Vessels: The Cult of the Battleship and the Rise of the U.S. Navy* (Boulder, Colo.: Westview Press, 1991), 292. Norfolk *Virginian-Pilot*, December 1928, quoted in Raymond G. O'Connor, *Perilous Equilibrium: The United States and the London Naval Conference of 1930* (New York: Greenwood, 1969), 20.

10. See Wilson, "Reevaluation." As William Appleman Williams observed, Hoover believed that "[t]he armed forces of the United States had the one purpose of guaranteeing 'That no foreign soldier will land on American soil. To maintain forces less than that strength is to destroy national safety, to maintain greater forces is not only economic injury to our people but a threat against our neighbors and would be a righteous cause for ill will among them.'" William Appleman Williams, "What This Country Needs . . . The Shattered Dream: Herbert Hoover and the Great Depression," in *Hoover Reassessed*, 445.

11. Hoover in a statement to his cabinet, reprinted in Ray L. Wilbur and Arthur M. Hyde, *The Hoover Policies* (New York: Charles Scribner's Sons, 1937), 601. Hoover published J. Reuben Clark's memorandum, suppressed by Coolidge, which invalidated use of the Monroe Doctrine as a justification for U.S. intervention in Latin America; Williams, "What This Country Needs," 444.

12. Military historiography also generally lauds Roosevelt while castigating the New Era Republicans. In light of the Japanese attack on Pearl Harbor in 1941, Hoover, Calvin Coolidge, and Warren Harding at best are portrayed as misguided in their pursuit of international arms reductions. Dudley Knox, the navy's interwar court historian, characterized treaty limitations as "disappointing" and the reason for the "defeat" of the Open Door Policy. Knox claimed that the opposite would have occurred had New Era Republicans emulated the "strong naval policies" of Theodore Roosevelt or followed "the wisdom of President Wilson in striving, after 1916, to create a preeminent American Navy"; Dudley W. Knox, *A History of the United States Navy* (New York: G. P. Putnam's Sons, 1948), 431. Also see Philip T. Rosen, "The Treaty Navy, 1919–1937," in *In Peace and War: Interpretations of American Naval History, 1775–1984*, ed. Kenneth J. Hagan (Westport, Conn.: Greenwood, 1984), 221–36. More recently, Robert W. Love Jr. claimed Hoover had a "crabbed view of the utility of naval power" and characterized his defense policies as "eccentric" and Hoover as a "pacifist"; *History of the U.S. Navy, 1775–1941* (Harrisburg, Pa.: Stackpole, 1992), 558–59. Stephen E. Pelz, in his *Race to Pearl Harbor: The Failure of the Second London Naval Conference and the Onset of World War II* (Cambridge, Mass.: Harvard University Press, 1974), simply wrote off Hoover and attributed "a fateful delay in American naval building" not to the New Era Republicans but to Roosevelt's difficulties with Congress. For a more balanced, recent treatment, see Dean C. Allard, "Naval Rearmament, 1930–1941: An American Perspective," *Revue Internationale d'Histoire Militaire* 73 (1991): 35–54.

13. Herbert C. Hoover, *The Memoirs of Herbert Hoover: The Cabinet and Presidency, 1920–1933*, vol. 2 (New York: Macmillan, 1952), 340. In the fall of 1929, the navy had 331 ships in commission: 16 battleships, 15 cruisers, 3 aircraft carriers, 8

minelayers, 103 destroyers, 82 submarines, 29 patrol ships, and 75 noncombatant service ships; *Report of the Chief of Naval Operations* in *Annual Reports of the Navy Department for the Fiscal Year 1929 (Including Operations to November 15, 1929)* (Washington, D.C.: U.S. Government Printing Office, 1929), 79–80.

14. Sperry pointed out to Hoover that measuring naval strength by tonnage alone was inaccurate; telephone message from Sperry to Hoover, December 12, 1921, Limitation of Armaments Conference File, Commerce Papers, Hoover Papers, Herbert Hoover Presidential Library, West Branch, Iowa. Hoover was serving as a member of the Executive Committee of the U.S. Advisory Committee to the American Delegation.

15. General Board to Secretary of the Navy, G.B. No. 438–1, Serial 1427, 10 June 1929, Foreign Affairs — Disarmament File, Presidential Papers, Hoover Papers, Hoover Library. For the genesis of U.S. naval war gaming, see Ronald H. Spector, *Professors of War: The Naval War College and the Development of the Naval Profession* (Newport, R.I.: Naval War College Press, 1977), chap. 6, especially 74–82.

16. See the detailed formulas contained in General Board to Secretary of the Navy, G.B. No. 438–1, Serial 1430 (Secret), 1 August 1929, Foreign Affairs — Disarmament File, Presidential Papers, Hoover Papers, Hoover Library.

17. For an overview of the London Naval Conference, see Wheeler, *Prelude to Pearl Harbor*, chap. 7; and O'Connor, *Perilous Equilibrium*. No notes were taken during the Hoover–MacDonald meeting; Hall, *Arms Control*, 78–79.

18. For a view of the London conference from the Japanese side, see Kobiyashi Tatsuo, "The London Naval Treaty, 1930," in *Japan Erupts: The London Naval Conference and the Manchurian Incident, 1928–1932*, ed. James William Morley (New York: Columbia University Press , 1984), 11–117. The Japanese government of Prime Minister Hamaguchi Osachi saw Japan's future in a reordering of the domestic economy and expansion of foreign trade. Naval agreement at London was viewed as essential to buy Western support for a solution to Japan's economic problems; Arthur E. Tiedemann, "Introduction," *Japan Erupts*, 4–11; 9. Hall, *Arms Control*, 80.

19. Hoover to Secretary of the Navy, 2 November 1929, Foreign Affairs — Disarmament File, Disarmament Conference 1929, November 1–10 Folder, Presidential Papers, Hoover Papers, Hoover Library.

20. See Wheeler, *Prelude to Pearl Harbor*, chap. 7; and William F. Trimble, "Admiral Hilary P. Jones and the 1927 Geneva Naval Conference," *Military Affairs* 43 (1979): 1–4. For Pratt, see Gerald E. Wheeler, *Admiral William Veazie Pratt, U.S. Navy: A Sailor's Life* (Washington, D.C.: Naval Historical Division, Department of the Navy, 1974); and the useful contextual essay by Craig L. Symonds, "William

Veazie Pratt: 17 September 1930 – 30 June 1933," in *The Chiefs of Naval Operations*, ed. Robert W. Love Jr. (Annapolis: Naval Institute Press, 1980), 69–86.

21. Hall, *Arms Control*, 101.

22. Both cruiser and submarine tonnage ratios were critical to Japanese strategic and tactical planning regarding war with the United States. Both ship types were necessary to reduce the advancing U.S. fleet through attrition to a size comparable to the Japanese fleet; Tiedemann, Introduction, in *Japan Erupts*, 9. On submarines, see O'Connell, *Sacred Vessels*, 295–96. Also see Edward P. Stafford, *The Far and the Deep* (New York: Putnam, 1967).

23. Department of State press release, 12 April 1930, cited by Hall, *Arms Control*, 104. This mirrored the $1.17 billion Navy Department proposal for new ships submitted to the director of the budget the previous year. Secretary of State Stimson pointed out that one modern 10,000-ton cruiser cost more than double the original price of the Library of Congress; Secretary of State Press Release, 31 May 1929, Foreign Affairs — Disarmament File, Presidential Papers, Hoover Papers, Hoover Library. Also see Hall, *Arms Control*, 106–7.

24. Quotes from President Hoover's Message to the Senate reprinted in Hoover, *The Memoirs of Herbert Hoover: Cabinet and Presidency, 1920–1933* (New York: Macmillan, 1952), 349–52. Also see West, "Legislative Foundation," 32.

25. Hall, *Arms Control*, 116–17. Historical characterizations of the naval budget reductions as "meat-ax" cuts accept the contemporary navy's view of the worldwide scope of its mission and are influenced by post facto knowledge of the conflict with Japan. See, for example, Robert G. Albion, *Makers of Naval Policy, 1798–1947* (Annapolis: Naval Institute Press, 1980), chap. 12; and West, "Legislative Foundation," 49–50. They ignore the fact that the definition of the navy's mission had been drafted by the General Board, which sought to aggrandize the navy through the adoption of the most comprehensive strategic responsibilities.

26. William L. Neumann, "Franklin Delano Roosevelt: A Disciple of Admiral Mahan," *USNIP* 78 (1952): 718. Eleanor Roosevelt said that her husband had always characterized Mahan's histories as books that he found "most illuminating"; ibid., 719. Two of Roosevelt's appointees to serve as chief of naval operations during the 1930s were eminent members of the "Gun Club" — Admirals William D. Leahy and Harold Stark. The successful careers of these Bureau of Ordnance officers, or rather the popularity of ordnance officers with Congress and politicians in general, may be linked with the great number of ordnance facilities scattered around the country and the magnitude of the contracts for armor plate and weapons that the bureau handled. See Albion, *Makers of Naval Policy*, 170–71; for the "Gun Club," see 384. For useful treatments of these two prewar CNOs, see

John Major, "William Daniel Leahy, 2 January 1937 – 1 August 1939," 101–17; and B. Mitchell Simpson III, "Harold Raynsford Stark, 1 August 1939 – 26 March 1942," 119–35, both in Love, *Chiefs of Naval Operations.*

27. Albion, *Makers of Naval Policy,* 67. On Roosevelt as ship designer, see, for example, Roosevelt to Captain Wilson, 17 December 1934; Roosevelt to the Assistant Secretary of the Navy, 15 March 1938; and Roosevelt to Commander Callaghan, 29 August 1938, all in President's Official File, Franklin D. Roosevelt Library, Hyde Park, N.Y. (hereafter FDR Official File), 18s.

28. See the correspondence on this dating from fall 1938 in FDR Official File 18i. Also of interest is Douglas H. Robinson and Charles L. Keller, *"Up Ship!": A History of the U.S. Navy's Rigid Airships* (Annapolis: Naval Institute Press, 1982).

29. Standley quoted in Thaddeus V. Tuleja, *Statesmen and Admirals: Quest for a Far Eastern Naval Policy* (New York: Norton, 1963), 91. While Theodore Roosevelt was instrumental in the rise of the U.S. Navy, his relationship with the service involved policy rather than technology issues. Although he ordered the Newport Conference of 1908 to review the charges made by William Sims and others that U.S. capital ship designs were inferior, Roosevelt did not interfere and let the naval hierarchy, and its technical experts, sort things out.

30. Felix Herbert, Republican National Committee, to Lawrence Richey, Secretary to President Hoover, 20 October 1932, Navy Correspondence File, Presidential Papers – Cabinet, Hoover Papers, Hoover Library. Commander Whortley memorandum to all naval reserve officers in the Great Lakes area, 5 October 1932, Naval Reserve File, Presidential Papers – Subject File, Hoover Papers, Hoover Library.

31. Harold G. Brownson, "The Naval Policy of the United States," *USNIP* 59 (1933) and editor's comments, both cited in Neumann, "Roosevelt," 719.

32. A treatment of Stark is B. Mitchell Simpson, *Admiral Harold R. Stark: Architect of Victory, 1939–1945* (Columbia: University of Missouri Press, 1989).

33. Stark to Roosevelt, 21 March 1933, and Stark to Roosevelt, 22 June 1933, President's Personal File, Franklin D. Roosevelt Library, Hyde Park, N.Y. (hereafter FDR Personal File), Box 166.

34. For Vinson's experiences during his first term as chairman of the Naval Affairs Committee and his work for renewed naval construction, see West, "Legislative Foundation," chap. 4. The *New York Times,* 30 November 1932, cited in West, "Legislative Foundation," 276; also see 275–77.

35. See West, "Legislative Foundation," 299–305; Secretary Adams to Congressman Vinson, 22 December 1932, RG 80, quoted on 299–300. See *The United States Navy in Peace Time: The Navy in Its Relation to the Industrial, Scientific,*

Economic, and Political Development of the Nation (Washington, D.C.: U.S. Government Printing Office, 1931).

36. Secretary Swanson wrote: "The recent [NIRA shipbuilding] program is in close accord with the purposes of the National Recovery Act, since it will substantially aid unemployment and the restoration of commercial, industrial, and agricultural activity. Approximately 85 percent of the moneys spent on this naval construction will go directly into the pockets of labor; about half at the shipyards and the remainder scattered throughout the country among the producers and fabricators of raw materials. Every State will benefit"; *Annual Report of the Secretary of the Navy for the Fiscal Year 1933* (Washington, D.C.: U.S. Government Printing Office, 1933), 2.

37. The *New York Times*, 21 March 1933, cited in West, "Legislative Foundation," 304–5. Commander A. B. Court memorandum to Chief of Naval Operations, "History of development of current Naval Building Programs, and the effect of these programs on Treaty Strength," 8 July 1934, RG 80, cited in West, "Legislative Foundation," ibid.

38. See Ellis W. Hawley, *The New Deal and the Problem of Monopoly: A Study in Economic Ambivalence* (Princeton, N.J.: Princeton University Press, 1966), 24–25. See West, "Legislative Foundation," 295.

39. Roosevelt's message to foreign leaders, his message to Congress, and foreign replies are reprinted in Franklin D. Roosevelt, *Roosevelt's Foreign Policy, 1933–1941: Franklin D. Roosevelt's Unedited Speeches and Messages* (New York: W. Funk, 1942), 11–19. My quotations are from 12 and 14 respectively.

40. These letters are typical of those in FDR Official File 18.

41. Moley wrote: "if Franklin Roosevelt can be said to have had any philosophy at all, that philosophy rested on the fundamental belief that the success of concerted international action toward recovery presupposed the beginnings of recovery at home. He did not believe that our depression could be conquered by international measures"; Raymond Moley, *After Seven Years* (New York: Harper & Brothers, 1939), 88. This assessment is echoed by Arthur M. Schlesinger Jr. in *The Age of Roosevelt: The Crisis of the Old Order, 1919–1933* (Boston: Houghton Mifflin, 1957), 442.

42. Roosevelt to Assistant Secretary of the Navy H. L. Roosevelt, 18 June 1934, FDR Official File 18.

43. CNO to Secretary of the Navy, 24 March 1933, an enclosure to Secretary Swanson to Roosevelt, 5 April 1933, in FDR Official File 18, Box 12.

44. On the technological limits placed by Congress on the first battleships, see John C. Reilly and Robert L. Scheina, *American Battleships, 1886–1923: Predread-*

nought Design and Construction (Annapolis: Naval Institute Press, 1980), chaps. 3–6. See West, "Legislative Foundation," 326–28.

45. "Appendix: United States Naval Policy," *Annual Report of the Secretary of the Navy for 1933*, 34–36.

46. Admiral Land to Secretary Swanson, 13 June 1933, RG 19, and Swanson to Roosevelt, 15 June 1933, RG 80, cited in West, "Legislative Foundation," 329. The *New York Times*, 3 August 1933, also cited in West, "Legislative Foundation," 332.

47. Shipyards located in Democratic strongholds that received NIRA contracts were Newport News Shipbuilding in Virginia; Bethlehem Shipbuilding in Quincy, Massachusetts; New York Shipbuilding in Camden, New Jersey; Federal Shipbuilding in Kearny, New Jersey; United Dry Dock in New York City; and the Boston, Philadelphia, New York, Charleston, and Norfolk navy yards; Navy Press Release, 19 June 1933, RG 80, cited in West, "Legislative Foundation," n. 56, 332. On Newall's conversion, see Roger S. McGrath to Louis Howe, Presidential Secretary, 15 August 1933, FDR Official File 18.

48. Regarding the funding, see the *New York Times*, 17 June 1933, cited in West, "Legislative Foundation," 330. For the Morrow Board and the 1,000-plane navy, see Archibald D. Turnbull and Clifford L. Lord, *History of United States Naval Aviation* (New Haven, Conn.: Yale University Press, 1949), chaps. 23 and 24.

49. King to Secretary of the Navy, 11 August 1933, serial Aer-P-3-ML, FDR Official File 18. See Secretary of the Navy to Secretary Ickes, 10 August 1933, FDR Official File 18.

50. See West, "Legislative Foundation," 338–40.

51. See ibid., 348–64. See John C. Walter, "William Harrison Standley: 1 July 1933–1 January 1937," in Love, *Chiefs of Naval Operations*, 93; and West, "Legislative Foundation," 375. Secretary Cordell Hull to Roosevelt, 21 February 1934, FDR Personal File 5901.

52. Senator Park Trammell was an old warhorse who initiated similar legislation in the Senate. On enactment of the Vinson–Trammell Act, see West, "Legislative Foundation," chap. 6. See Roosevelt's statement in the *New York Herald*, 28 March 1934, reprinted in West, "Legislative Foundation," 435. On WPA funding, see West, "Legislative Foundation," 436.

53. Arthur J. Marder, *Old Friends, New Enemies: The Royal Navy and the Imperial Japanese Navy; Strategic Illusions, 1936–1941* (Oxford: Clarendon Press, 1981), 9.

54. Sadao Asada, "The Japanese Navy and the United States," in *Pearl Harbor as History*, 238. According to Asada, the 1930 London Naval Treaty aggravated "internal splits within the navy and a steady erosion of the top ministry leadership by

the high command and middle-echelon officers. . . . With the rise of violently anti-British, anti-American, and pro-German elements, the navy's policy was increasingly dominated by highly emotional modes of thinking and myopic strategic preoccupations. By the mid-1930s the 'moderate' naval leadership had been reduced to a decided minority" (225–26). For opposition to the Five-Power Treaty, see Sadao Asada, "Japanese Admirals and the Politics of Naval Limitation: Katō Tomosaburo and Katō Kanji," in *Naval Warfare in the Twentieth Century, 1900–1945: Essays in Honour of Arthur Marder*, ed. Gerald Jordan (London: Croom Helm, 1977), 141–66.

55. Pelz, *Race to Pearl Harbor*, 19 and 25; Tuleja, *Statesmen and Admirals*, 95; also see "Warship Building," *Japan Advertiser*, 25 December 1935, reprinted in *USNIP* 62 (1936): 435–37.

56. Asada, "Japanese Navy," 240–42. Resignation of the navy minister would cause a constitutional crisis. For the Japanese domestic decision to abrogate the ratio system, see Pelz, *Race to Pearl Harbor*, chaps. 1, 3, and 4.

57. Asada, "Japanese Navy," 235. The Japanese relied on their new Type 6 submarines and torpedo-carrying Mitsubishi Type 96 bombers stationed in the Marshall and Caroline Islands; Pelz, *Race to Pearl Harbor*. Also see Tuleja, *Statesmen and Admirals*, 88–89. Asada, "Japanese Navy," 243 and n. 81. As Asada observed, such a strategy assumed the United States would pursue a decisive fleet encounter early in the war rather than build up overwhelming strength before advancing westward. The Japanese strategy for war with the United States was consistently myopic. The 1936 Tactical Plan, for example, specified only initial stages of operations against the United States; later stages would be "expedient measures as the occasion may demand"; 236.

58. Ibid., 234. Yamamoto believed that Japan lacked the resources and industrial might to win a war with the United States and endorsed the ratio system: "The 5:5:3 ratio works just fine for us; it [the Five-Power Treaty] is a treaty to restrict the *other* parties [the United States and Britain]"; Yamamoto quoted by Asada, ibid.

59. See Tim Travers, *The Killing Ground: The British Army, The Western Front and the Emergence of Modern Warfare, 1900–1918* (London: Allen & Unwin, 1987), chaps. 2 and 3.

60. Asada, "Japanese Navy," 237.

61. Ibid., 242. See Pelz, *Race to Pearl Harbor*, 31–32, on the *Yamato* class. As early as 1931, three years before *Yamato*-class design began in October 1934, the Japanese successfully tested an 18.1-inch battleship gun that delivered "30 percent more power" on impact than a 16-inch gun, the largest size then in existence; Pelz, ibid. Gun "power" is an imprecise term. An analysis based on the ballistic data in tables

C-7 and C-11 in Dulin and Garzke, *Battleships*, 234–35 and 240–41, indicates that at a range of 30,000 yards, the striking energy of the *Yamato*-class 18.1-inch gun was approximately 52 percent greater than that of the U.S. Navy's Mark 5, 16-inch/45-caliber gun carried by the superdreadnought *Maryland* of the 1916 Program; at a range of 20,000 yards, the Japanese 18.1-inch gun delivered approximately 51 percent greater striking energy than the U.S. Mark 5 gun. Striking energy — a function of shell mass, velocity, and shape — contributed to kinetic penetration or damage to armor plate.

62. See Pelz, *Race to Pearl Harbor*, 33–39.

63. Roosevelt to Swanson, Navy Day, 1935, *London Times* reprinted in *USNIP* 62 (1936): 116. *Chicago Tribune*, "Address by Admiral Standley" reprinted in *USNIP* 62 (1936): 115–16. For a discussion of naval issues during Standley's tenure as CNO, see Walter, "William Harrison Standley."

64. State Senator Charles S. Bream to Roosevelt, 26 November 1935, FDR Official File 1818.

65. Transcript of Presidential Press Conference, 13 November 1935 (excerpt), reprinted in *Franklin D. Roosevelt and Foreign Affairs*, vol. III, ed. Edgar B. Nixon (Cambridge, Mass.: Belknap Press of Harvard University Press, 1969), 58–60. Presidential memorandum to Charles Swanson, 20 September 1935, FDR Official File 18s. Standley resented Roosevelt's "inflated opinion of his knowledge of naval strategy and tactics"; Standley quoted in Tuleja, *Statesmen and Admirals*, 91. Roosevelt to Secretary of the Navy, 17 December 1934, FDR Official File 18i.

66. Naval Aide Wilson Brown memo to Roosevelt, December 1934, FDR Official File 18s.

67. Between 1935 and 1937, a total of seventy-seven battleship designs were prepared by the Bureau of Construction & Repair; Dulin and Garzke, *Battleships*, 27. "The biggest is the best" approach to battleship design dates from Lieutenant Homer Poundstone, USN, "Size of Battleships for U.S. Navy," *USNIP* 29 (1903): 161–74.

68. Hall, *Arms Control*, 184.

69. Marder, *Old Friends, New Enemies*, 9–10. For the common upper limit, see Pelz, *Race to Pearl Harbor*, passim. Arthur Marder wrote that the Japanese official war history maintained that the Japanese delegation was "pretty sure" Nagano's proposal would be rejected, but that "it was not necessarily a cloak for the unwillingness of the Navy to continue the shipbuilding limitations" and the Naval General Staff was sincere in their support for the proposal as a way to save on naval expenditures; *Senshi soshō (War History Series)*, vol. xci, *Daihon'ei kaigunbu: Rengo kantai (I): Kaisen made (Imperial General Headquarters, Navy Department, Com-*

bined Fleet (I): Until the Outbreak of War) (Tokyo, 1975), 284, cited by Marder,
11–12.

70. Capital ship size was limited to 35,000 tons and, for the first time, battleship
guns were limited in size (to a 14-inch bore). The maximum size for new aircraft
carriers was set at 23,000 tons. Tonnage and maximum gun-size restrictions were
adopted for cruisers and submarines. Italy subscribed to the provisions of the Treaty
in 1938. See Marder, *Old Friends, New Enemies,* 12.

71. Admiralty memorandum for the Committee on Imperial Defence, 6 January 1938, ADM 116/3735, cited in Marder, *Old Friends, New Enemies,* 13. Vinson's
threat is in *The Sun* [Baltimore], "Vinson to Push for Naval Construction Program
if Japan Spurns Treaty," 28 November 1934, FDR Personal File 5901.

72. Marder, *Old Friends, New Enemies,* 13–14; Japanese Foreign Office statement cited on 13.

73. Ibid.

74. See Allard, "Naval Rearmament," 45–50. See the *New York Herald Tribune,*
14 January 1937, reprinted in *USNIP* 63 (1937): 418–19.

75. Presidential press conference, 8 January 1937, reprinted in *Roosevelt and Foreign Affairs,* 575.

76. See articles from the *New York Herald Tribune,* 18 May 1937, and the *Washington Herald,* 22 May 1937, in *USNIP* 63 (1937): 1034.

77. *New York Herald Tribune,* "Difficulties with Steel," 14 February 1937,
reprinted in *USNIP* 63 (1937): 572–73.

78. Brunner to Roosevelt, 5 January 1937; Burnison to Roosevelt, 9 September
1937; Roosevelt to Burnison, 13 September 1937, FDR Official File 1698.

79. Private shipbuilders traditionally complained that the cost and building
estimates of government shipyards bore no relation to reality. Recall the case of the
turboelectric drive.

80. Representative Powers to Roosevelt, 11 October 1937; Senator Moore to Roosevelt, 11 October 1937; Roosevelt to Senator Moore, 20 October 1937, FDR Official File 1698.

81. McIntyre to Secretary Edison, 27 November 1937, FDR Official File 1698.

82. Secretary Edison to McIntyre, 9 December 1937, FDR Official File 1698.
Burnison to Roosevelt, undated, in December 1937 folder, FDR Official File 99.

83. Roosevelt quoted in Charles A. Beard, *American Foreign Policy in the Making, 1932–1940* (New Haven, Conn.: Yale University Press, 1946), 151.

84. William McNeill pointed out the link between the Gladstone government's
broadening of the political franchise with the Reform Act of 1884 and the subsequent increase in the British naval estimates later that year. The income tax was

paid by a small percentage of the electorate while the majority of the new electorate would benefit from government contracts, that is, naval construction, and
voted for them; McNeill, *Pursuit of Power*, 269–70. The political mechanism for
the 1884–1914 naval race was in place. Paul Kennedy apparently did not consider
naval construction employment as a social issue. He emphasized the limitations
placed on late-nineteenth-century British naval spending by the "demands of a
mass democracy for social and economic improvements"; *The Rise and Fall of
British Naval Mastery* (London: Allen Lane, 1976), 194.

85. President George Bush's administration did push fighter aircraft production
for export to help its campaign standing in California and other areas of the aero-
space gun belt.

86. See Merritt Roe Smith, *Harpers Ferry Armory and the New Technology: The
Challenge of Change* (Ithaca, N.Y.: Cornell University Press, 1977), chap. 1 and
chap. 9, "Regional Interests and Military Needs." A recent case involved the navy's
F/A-18 Hornet aircraft, which owed its acquisition, in great measure, to Senator
Edward Kennedy, a long-standing opponent of defense expenditures. Engines for
the Hornet were manufactured in Massachusetts, and Kennedy, along with
Speaker of the House Tip O'Neill, were "mighty proponents of the F-18, and piv-
otal to its acceptance by Congress"; James P. Stevenson, *The Pentagon Paradox:
The Development of the F-18 Hornet* (Annapolis: Naval Institute Press, 1993), 219.
A recent useful anthology is *The Pentagon and the Cities*, ed. Andrew Kirby (New-
bury Park, Calif.: Sage Publications, 1992), especially Richard Barff, "Living by the
Sword and Dying by the Sword?: Defense Spending and New England's Econ-
omy," 77–99.

87. See Ann Markusen, Peter Hall, Scott Campbell, and Sabrina Dietrick, *The
Rise of the Gun Belt: The Military Remapping of Industrial America* (New York: Oxford University Press, 1991). On 6 September 1994, with its Tri-Service Standoff
Attack Missile (TSSAM) facing congressional budget cuts, the Pentagon, using tactics devoid of subtlety, released graphics which indicated that the missile's contractors and subcontractors were spread over twenty states.

88. For a recent treatment of the interwar defense economy, see Paul A. C.
Koistinen, *Planning War, Pursuing Peace: The Political Economy of American War-
fare, 1920–1939* (Lawrence: University of Kansas Press, 1998). On "classifying"
Hoover, see David Burner and Thomas R. West, "A Technocrat's Morality: Con-
servatism and Hoover the Engineer," in *The Hofstadter Aegis: A Memorial*, ed.
Stanley Elkins and Eric McKitrick (New York: Knopf, 1974), 235–56. While
Hoover was "moral" regarding arms expenditures, he did authorize a political bur-
glary by a naval officer from the Office of Naval Intelligence; see Jeffery M. Dor-

wart, *Conflict of Duty: The U.S. Navy's Intelligence Dilemma* (Annapolis: Naval Institute Press, 1983), 3–5. Roosevelt, on the other hand, availed himself of "private" spies: his "personal naval secret agent" was Vincent Astor, his neighbor at Hyde Park and a naval reserve intelligence officer. See Dorwart, ibid., 114–15 and chap. 15.

89. Navy Secretary Swanson to Roosevelt, 21 December 1937, FDR Personal File 1722.

90. Peter Paret, "Clausewitz," in *Makers of Modern Strategy from Machiavelli to the Nuclear Age*, ed. Peter Paret (Princeton, N.J.: Princeton University Press, 1986), 186–213; Clausewitz quoted on 199. For more detail, see Carl von Clausewitz, *On War*, ed. and trans. Michael Howard and Peter Paret (Princeton, N.J.: Princeton University Press, 1984), Book 1, chaps. 1–2; 75–99.

CHAPTER 8.   WAR AND A SHIFTING TECHNOLOGICAL PARADIGM

1. Commander Hugh Douglas, USN, lecture on aviation, 23 October 1933, Naval War College, quoted in Thomas Wildenberg, *Destined for Glory: Dive Bombing, Midway, and the Evolution of Carrier Air Power* (Annapolis: Naval Institute Press, 1998), 128.

2. Chief of Naval Operations Admiral Harold R. Stark originally wanted to transfer a carrier, four cruisers, and two squadrons of destroyers — along with three battleships. Secretary of State Cordell Hull opposed any weakening of American Pacific naval power. Stark reduced his request to the three battleships, and Roosevelt approved their transfer on 13 May 1940. B. Mitchell Simpson III, *Admiral Harold R. Stark: Architect of Victory, 1939–1945* (Columbia: University of South Carolina Press, 1989), 86.

3. A recent treatment of Nimitz's career is John B. Lundstrom, "Chester W. Nimitz: Victory in the Pacific," *Quarterdeck and Bridge: Two Centuries of American Naval Leaders*, ed. James C. Bradford (Annapolis: Naval Institute Press, 1997), 327–44. A more lengthy version is E. B. Potter, *Nimitz* (Annapolis: Naval Institute Press, 1976).

4. Thomas B. Buell, *The Quiet Warrior: A Biography of Admiral Raymond A. Spruance* (Boston: Little, Brown, 1974), 217.

5. The Japanese used planes launched from aircraft carriers to sink four U.S. battleships, severely damage another three, and cause one to be beached in order to avoid sinking. The first wave of attacking aircraft employed bombs converted from 16-inch armor-piercing battleship shells, as well as aerial torpedoes, to destroy the American battleship force. At Pearl Harbor, the Japanese lost a relatively high percentage of their attacking aircraft (8.5 percent — 30 planes), lending credence to prewar estimates of the effectiveness of antiaircraft fire. Japanese losses from Eric

Brown, *Duels in the Sky: World War II Naval Aircraft in Combat* (Annapolis: Naval Institute Press, 1988), 73.

6. Clark Reynolds, *The Fast Carriers: The Forging of an Air Navy* (Huntington, N.Y.: Robert E. Krieger, 1978), 39.

7. Captain Wayne P. Hughes, USN (Ret.), *Fleet Tactics: Theory and Practice* (Annapolis: Naval Institute Press, 1986), 92. According to Hughes, "a battleship or heavy cruiser formation could travel two hundred nautical miles at night . . . [and] air strikes were mounted at ranges of around two hundred nautical miles." This gives new meaning to Robert O'Connell's dismissal of hard-to-retire battleships as the "Vampires of Sea Power." During World War II, battleships ruled the night but were vulnerable during daylight—as long as the weather was good enough for flight operations. Carrier commanders needed to keep a safe buffer between themselves and enemy surface forces; Hughes, *Fleet Tactics*, 91.

8. Fleet Admiral Ernest J. King, USN, *U.S. Navy at War 1941–1945: Official Reports to the Secretary of the Navy by Fleet Admiral Ernest J. King, U.S. Navy, Commander-in-Chief, United States Fleet and Chief of Naval Operations* (Washington, D.C.: U.S. Government Printing Office, 1946), 15.

9. Aviators wore brown shoes with their khaki uniforms; all other officers wore black shoes. This made for a simple classification. Aviators also could wear a unique green service uniform.

10. Fleet Admiral Ernest J. King, USN, Introduction, "Third Report of Operations of the United States Navy in World War II, 1 March 1945–1 October 1945," in *U.S. Navy at War 1941–1945*, 169.

11. Ibid., 167.

12. Rear Admiral Moffett's testimony regarding H.R. 9690, "To Authorize the Construction and Procurement of Aircraft and Aircraft Equipment in the Navy and Marine Corps," quoted in Douglas H. Robinson and Charles L. Keller, *"Up Ship!": A History of the U.S. Navy's Rigid Airships, 1919–1935* (Annapolis: Naval Institute Press, 1982), 178.

13. Robinson and Keller, *"Up Ship!"* 193–94.

14. Ibid., 182–83.

15. The difficulty of locating ships from aircraft cannot be overemphasized. During Pacific Fleet exercises in 1976, the author's destroyer was never detected during numerous over flights by carrier search planes during a forty-eight-hour period of unlimited visibility.

16. Robinson and Keller, *"Up Ship!"* 183. *Akron* was assigned to the "Black" (enemy) force in Fleet Problem XIII and was to have assisted the weaker Black force resist seizure of an atoll by the Blue (United States) fleet. See Commander-in-

Chief, U.S. Fleet to Fleet, CinC File No. A16–3(2) 705 of 23 May 1932, in *Records Relating to United States Navy Fleet Problems I to XXII 1923–1941*, National Archives Microfilm Publication M964 (Washington, D.C.: National Archives [NARS/GSA], 1974).

17. Robinson and Keller, *"Up Ship!"* 184.

18. Ibid.

19. Ibid., 184–85. See the search diagram in Richard K. Smith, *The Airships Akron & Macon: Flying Aircraft Carriers of the United States Navy* (Annapolis: United States Naval Institute, 1965), 110.

20. The low pressure of the storm had caused the barometric altimeter to report an altitude that was between 320 and 600 feet too high. When the order was given to climb, the commander did not realize how close the tail was to the sea; see Robinson and Keller, *"Up Ship!"* 185–86.

21. The retiring Admiral Pratt, who had demonstrated a remarkable flexibility and breadth of vision during his term as CNO, informed Roosevelt that he personally would select Admiral Hepburn to replace him as CNO, but that Standley was the service's choice; Admiral Pratt to Roosevelt, 16 February 1933, FDR Personal File 586, Franklin D. Roosevelt Library, Hyde Park, N.Y.

22. Robinson and Keller, *"Up Ship!"* 186.

23. Ibid., 186–87. For a defense of the airships' operations with the fleet, and the differing opinions of fleet commanders as to their usefulness, see the testimony of Commander C. E. Rosendahl, USN, on "Lighter-than-Air Ship Policy," *Hearings Before the General Board of the Navy, 1917–1950*, Nimitz Library, U.S. Naval Academy, Annapolis, Md. (Microfilm) (hereafter cited as *General Board Hearings*), 1 February 1937, vol. 1, 1937, 18–19.

24. Sellers's report quoted in Robinson and Keller, *"Up Ship!"* 188.

25. Robinson and Keller, *"Up Ship!"* 189.

26. Ibid., 189–90.

27. The damaged member was scheduled for repair upon the *Macon*'s return to Lakehurst, New Jersey. Robinson and Keller, *"Up Ship!"* 191–92.

28. Statement of Admiral William D. Leahy, Chief of Naval Operations, 21 January 1937, "On Naval Operations," *Hearings Before the Subcommittee of the Committee on Appropriations, House of Representatives, Seventy-fifth Congress, First Session, on the Navy Department Appropriations Bill of 1938* (Washington, D.C.: U.S. Government Printing Office, 1937), 34.

29. Statement of Rear Admiral A. B. Cook, USN, chief of the Bureau of Aeronautics, *General Board Hearings*, 1 February 1937, 33.

30. Assistant Navy Secretary Edison to Roosevelt, 4 June 1938, FDR Official File 18i, FDR Library.

31. Ibid.

32. Roosevelt to Assistant Secretary Edison and Admiral Leahy, 10 September 1938, FDR Official File 18i.

33. CNO, Admiral Leahy, to Roosevelt, 16 September 1938, FDR Official File 18i. Rear Admiral Cook's recommendation for a training airship dovetailed with his proposal for government-subsidized merchant airships which would be nationalized during war and whose crews would be augmented by a small naval cadre.

34. "Statement of Admiral William D. Leahy, United States Navy, Chief of Naval Operations during Hearings on H.R. 9218 to Establish the Composition of the United States Navy, to Authorize the Construction of Certain Naval Vessels, and for Other Purposes," *Hearings Before Committee on Naval Affairs of the House of Representatives on Sundry Legislation Affecting the Naval Establishment 1937– 1938*, Seventy-fifth Congress, Second and Third Sessions (Washington, D.C.: U.S. Government Printing Office, 1938), 2090.

35. Statement of Dr. William F. Durand to the General Board of the Navy regarding "Lighter-than-Air Ship Policy," 1 February 1937, *General Board Hearings*, roll 10, 29.

36. Statement of Admiral Leahy (see n. 28).

37. Congressman Thom to Roosevelt, 12 October 1938, FDR Official File 18i.

38. Roosevelt to Congressman Thom et al., 4 November 1938, FDR Official File 18i.

39. Hoover to Julius H. Barnes, 17 July 1926, Disarmament 1921–28 File, Commerce Papers, Hoover Papers, Hoover Presidential Library, West Branch, Iowa.

40. For a thorough treatment of the development of U.S. naval aviation, see Clark G. Reynolds, *Admiral John H. Towers: The Struggle for Naval Air Supremacy* (Annapolis: Naval Institute Press, 1991).

41. For a brief account of the ascension of the dive bomber in the interwar navy, see Stephen McFarland, *Pursuit of Precision Bombing, 1910–1945* (Washington, D.C.: Smithsonian Institution Press, 1995), chap. 6. In 1931 the General Board was willing to remove the torpedo magazines from the carrier *Ranger* design as an economy measure since dive bombers were considered better offensive weapons. See Norman Friedman, *U.S. Aircraft Carriers: An Illustrated Design History* (Annapolis: Naval Institute Press, 1983), 69.

42. See Hughes, *Fleet Tactics*, 85.

43. Friedman, *U.S. Aircraft Carriers*, 62–63.

44. See statement of Rear Admiral Robert L. Ghormley, War Plans Division, reprinted in Friedman, *U.S. Carriers*, 114–16.

45. Rear Admiral J. D. Beuret quoted in Friedman, *U.S. Aircraft Carriers*, 63–64. Heavy seas prevented Craddock's squadron from using their weather guns against the German squadron under Graf von Spee.

46. Ibid.

47. See Friedman, *U.S. Aircraft Carriers*, chap. 4, especially 79–80.

48. Ibid., 13.

49. Analogous to the continued production of the inferior Sherman M4 tank during the war. Numbers were deemed more important than quality.

50. See Friedman, *U.S. Carriers*, chap. 9.

51. Figures attributed to Rear Admiral Cochrane, chief of the Bureau of Ships in 1944; Friedman, *U.S. Carriers*, 143.

52. For a representative squabble over who damaged whom, see Thomas B. Buell, *Master of Sea of Sea Power: A Biography of Fleet Admiral Ernest J. King* (Boston: Little, Brown, 1980), 116–17.

53. While senior officers may have seen the War College as an important institution, that apparently was not a view shared by many mid-level officers (lieutenant commanders and lieutenants), who viewed it as "a waste of time." Buell, *Quiet Warrior*, 71.

54. Ibid., 53.

55. Ibid., 72. Interestingly, Turner abandoned naval aviation, supposedly to return to the battleship world to obtain a major ship command in order to make admiral. Reynolds, *Admiral Towers*, 275.

56. Commander Ralph C. Parker, USN, "An Analysis of the Air Menace," *U.S. Naval Institute Proceedings* (hereafter *USNIP*) 58 (1932): 649.

57. Ibid.

58. Naval War College Staff Lecture, "The Employment of Aviation in Naval Warfare," serial 3429–1487/9–9-37 (September 1937), Record Group 14, Staff Presentations and Lectures, Naval Historical Collection, Naval War College, Newport, R.I. (hereafter cited as RG 14, NWC), 1.

59. "Employment of Aviation in Naval Warfare" (1937), 9. The listing from the staff lecture was : "Limitations: 1. Hampered by unfavorable weather. 2. Requires highly trained personnel. 3. Upkeep in field requires large transport tonnage. 4. Sphere of operation limited in radius and duration of effort. 5. Small effectiveness in single units. 6. Unable alone to accomplish unlimited military results. Capabilities: 1. Time: Can act quickly at great distances. 2. Space: Can not be denied

use of the air by direct attack. 3. Mass: Large numbers can be employed at a single point"; ibid.

60. Lieutenant Commander Logan C. Ramsey, USN, "Aerial Attacks on Fleets at Anchor," *USNIP* 63 (1937): 1126–32.

61. The records of the interwar fleet exercises are in *Records Relating to United States Navy Fleet Problems I to XXII, 1923–1941* (n. 16). In Fleet Exercise XIX, Admiral Kalbfus allowed King to operate the aircraft carriers independent of the Battle Force. King successfully launched air attacks against Pearl Harbor and Mare Island Naval Shipyard in California. Kalbfus held the carriers more closely in Fleet Exercise XX in the Caribbean the following year. See Buell, *Master of Sea Power,* 113–17.

62. On Richardson's relief, see Simpson, *Admiral Stark,* 54–61.

63. "Employment of Aviation" (1937), 10.

64. Ibid., 12.

65. Ibid., 17.

66. Ibid., 42.

67. Ibid., 27.

68. Ibid., 44.

69. Testimony of Rear Admiral A. B. Cook before the General Board on "Aircraft Building Program—1939," *General Board Hearings,* 1937, vol. 2, reel 11, 15.

70. Naval War College Staff Presentation. "Tactical Employment of the Fleet," 29 October 1937, 2–3, RG 14, NWC.

71. Naval War College Staff Presentation, "The Employment of Submarines," Synopsis, 3 October 1938, 2, RG 14, NWC.

72. Naval War College Staff Presentation, "The Employment of Submarines" (Confidential), 4 November 1937, 9, RG 14, NWC.

73. Testimony of Commander Charles A. Lockwood, USN, on "Characteristics of Submarines," 15 October 1937, *General Board Hearings,* roll 11, 324.

74. Testimony of Lieutenant Commander R. H. Smith, USN, on "Characteristics of Submarines," 15 October 1937, *General Board Hearings,* roll 11, 324–25.

75. Ernest J. King and Walter Muir Whitehill, *Fleet Admiral King: A Naval Record* (New York: Norton, 1952), 317.

76. Chairman, General Board to Secretary of the Navy, 31 May 1940, serial no. 1951, 31-May-40, E.J.K-LDB, on "Fleet Command and Administration," in Papers of Fleet Admiral Ernest J. King, USN, *General Board Hearings* (hereafter cited as "Fleet Command, 1940").

77. Quotations from "Fleet Command, 1940."

78. See King's letter to the chief of the Bureau of Navigation, quoted in Buell, *Master of Sea Power*, 109–10.

79. Quotations from "Fleet Command, 1940."

80. Reynolds's claims that the 27-knot battleships were unsuitable for operation with 33-knot carriers wrongfully assumed constant high-speed operations and ignored the reason for higher carrier speed: to rejoin the formation after maneuvering into the wind to launch and recover aircraft. Reynolds, *The Fast Carriers*, 39.

81. Friedman, *U.S. Carriers*, 207.

82. Admiral Richardson (commander-in-chief, U.S. Fleet) cited the low percentage of bomb hits during the Norway campaign in his opposition to armored flight decks during the design of the *Midway*-class large carriers. Friedman, *U.S. Carriers*, 212.

83. When the British navy lost contact, *Bismarck* was relocated by a British Catalina seaplane with a U.S. Navy co-pilot flying covertly in support of the British war effort; William H. Garzke and Robert O. Dulin, *Battleships: Axis and Neutral Battleships in World War II* (Annapolis: Naval Institute Press, 1986), 233. U.S. patrol planes also searched for *Bismarck* out of their base in Newfoundland; Buell, *Master of Sea Power*, 140.

84. Although wrecked by British battleship guns, *Bismarck*'s sinking was aided by her crew's efforts to scuttle her. Her armor design had warded off torpedo hits and resisted the 16-inch shells from HMS *Rodney* when fired from a point-blank range of 2,400 meters.

85. For a thorough, well-argued, revisionist treatment of the Italian naval war, see James Sadkovich, *The Italian Navy in World War II* (Westport, Conn.: Greenwood, 1994). Sadkovich's introduction addresses ethnic bias within military history. See the Roosevelt quote in Paul Fussell, *Wartime: Understanding and Behavior in the Second World War* (New York: Oxford University Press, 1989), 34.

86. Report of the British Ambassador in Japan to the Foreign Office, FO 371/27971, cited in Arthur J. Marder, *Old Friends, New Enemies: The Royal Navy and the Imperial Japanese Navy* (Oxford: Oxford University Press, 1981), 345.

87. Marder, *Old Friends, New Enemies*, 345.

88. Buell, *Quiet Warrior*, 72.

89. *The Reminiscences of Rear Admiral Arthur H. McCollum, U.S. Navy (retired)*, 2 vols., Oral History Office, U.S. Naval Institute (Annapolis: Naval Institute Press, 1973), vol. 1, 146–48, quoted in Marder, *Old Friends, New Enemies*, 354. Thomas G. Mahnken recently argued that U.S. Navy intelligence shortcomings regarding the Japanese were not the result of "consistent application of racial stereotypes," but rather due to the "failure of both conscious analytical constructs

and unconscious belief systems." Thomas G. Mahnken, "Gazing at the Sun: The Office of Naval Intelligence and Japanese Naval Innovation, 1918–1941," in *New Interpretations in Naval History: Selected Papers from the Twelfth Naval History Symposium*, ed. William B. Cogar (Annapolis: Naval Institute Press, 1997), 252.

90. Captain Charles D. Allen, USN, "Forecasting Future Forces," *USNIP* 108 (1982), quoted in Hughes, *Fleet Tactics*, 87.

91. Nimitz to King, quoted in Potter, *Nimitz*, 41.

92. On 28 October 1942, Vice Admiral John Towers, commander of Air Force, Pacific Fleet, recommended the transfer of a battleship division from Pearl Harbor to Admiral William F. Halsey to counter the "more extensive use by [the] Japanese of heavy surface forces." Towers memorandum to Admirals Chester Nimitz and Raymond Spruance, 28 October 1942, quoted in Reynolds, *Admiral Towers*, 408.

93. Hughes, *Fleet Tactics*, 91–92.

94. Treatment of aviator-surface officer relationships in the higher levels of the Pacific command are Reynolds, *Admiral Towers*, chaps. 14–15, and his *The Fast Carriers*, passim, especially chaps. 5 and 7; Potter, *Nimitz*, chap. 17, especially 267; Buell, *Master of Sea Power*, 365–75, and his *The Quiet Warrior*, 184.

95. The Act of 24 June 1926 limited command of aircraft carriers to naval aviators or naval aviation observers. See 44 Stat. 764–768, Ch. 668; Laws re Navy, 3491–93; and *United States Naval Aviation, 1910–1980*, NAVAIR 00–80P-1 (Washington, D.C.: U.S. Government Printing Office, 1981), 62.

96. Reynolds, *Admiral Towers*, 435–36 and 411. CinCPac is pronounced "sink-pack."

97. Reynolds, *The Fast Carriers*, 44; Sherman quoted on 46–47.

98. Admiral H. E. Yarnell, USN (Ret.), "Report on Naval Aviation," memorandum to Secretary of the Navy, Serial Op-50B, 6 November 1943, Admiral Harry E. Yarnell Papers, Naval Historical Center Library, Washington Navy Yard, Washington, D.C., Box 11.

99. Ibid.

100. Ibid.

101. For the decision to promote Towers, see Buell, *Master of Sea Power*, 374; and Reynolds, *Admiral Towers*, 453–54.

102. Yarnell, "Report on Naval Aviation."

103. For carrier accidents, see Captain Frank M. Seitz, USMM, and Captain Theodore F. Davis, USN (Ret.), "From Cockpit to Bridge," *USNIP* 114/1 (January 1988): 53–56.

104. Admiral Ernest J. King, USN to Secretary of the Navy, 29 January 1944, Letter Serial FF1/A9 516, "Commander in Chief, United States Fleet and Chief of

Naval Operations First Endorsement to Report of Admiral Yarnell on Naval Aviation dated 6 November 1943," Yarnell Papers, Box 11.

105. Ibid.

106. Ibid.

107. Buell, *Master of Sea Power*, King quote on 368, Mitscher quote on 370.

108. Figures from Buell, *Master of Sea Power*, 370.

109. Reynolds, *Admiral Towers*, 514–15. Malcolm Muir Jr., *Black Shoes and Blue Water: Surface Warfare in the United States Navy, 1945–1975* (Washington, D.C.: Naval Historical Center, 1995), 10.

110. Nimitz's testimony to Congress, quoted in Muir, *Black Shoes*, 9.

111. Naval War College Staff Presentation, "Battleships and Cruisers," 7 August 1947, RG 14, NWC, 3.

112. See O'Connell, *Sacred Vessels*, chap. 12.

113. See Jeffrey G. Barlow, *Revolt of the Admirals: The Fight for Naval Aviation, 1945–1950* (Washington, D.C.: Naval Historical Center, 1994).

CHAPTER 9.   CASTLES OF STEEL

1. William H. McNeill, *The Pursuit of Power: Technology, Armed Force, and Society since A.D. 1000* (Chicago: University of Chicago Press, 1982), 89; and Christopher Duffy, *Siege Warfare: The Fortress in the Early Modern World, 1494–1660* (London: Routledge, 1979).

2. Just as the castle was not embraced by warrior cultures worldwide in 1400, recall that the battleship and its strategy of guerre d'escadre was not ubiquitous circa 1890. A noteworthy exception in Europe was the French Jeune École, whose members advocated a strategy of guerre de course that avoided the acquisition cost, expensive upkeep, and necessity of a capital ship fleet to counter the British battleship fleet. Commerce raiding powers, characteristically "weaker" naval powers, relied on counterweapons such as torpedo boats to offset an attacking enemy battleship fleet.

3. As noted in chapter 6, "three-plane" warfare was elucidated by Admiral William Fullam during the 1924 Special Board hearings before the General Board of the Navy. See the testimony of Admiral William F. Fullam, USN, in *Hearings Before the Special Board* (1924), contained in *Hearings Before the General Board of the Navy, 1917–1950* , 716.

4. Battleships did have a definite advantage against the commerce-raider cruisers the Soviet Union produced after World War II. A November 1954 war game featured a single *Iowa*-class battleship "superior to all types in heavy weather." The battleship's guns outweighed the Soviet attacking force by 3:1; Malcolm Muir,

Black Shoes and Blue Water: Surface Warfare in the United States Navy, 1945–1975 (Washington, D.C.: Naval Historical Center [GPO], 1995), 49.

5. Or, as Naval War College students were told in 1950, "the key to victory over the submarine is to prevent it from emerging from its base." Naval War College Staff Presentation (Captain H. D. Felt, USN), "The Future Employment of the Carrier," Naval War College, Record Group 14, Staff Presentations and Lectures, Naval Historical Collection, Naval War College, Newport, R.I. (hereafter RG 14, NWC), 4

6. Naval War College Staff Presentation, "Submarine Warfare," 21 July 1947, RG 14, NWC, 12–16.

7. Muir, *Black Shoes*, 95.

8. Ibid., 31. For a strategic overview of the early postwar period, see George Baer, *One Hundred Years of Sea Power: The U.S. Navy, 1890–1990* (Stanford, Calif.: Stanford University Press, 1993), chap. 12.

9. For a recent treatment of the military-academic-industrial alliance, see Stuart W. Leslie, *The Cold War and American Science: The Military-Industrial-Academic Complex at MIT and Stanford* (New York: Columbia University Press, 1993).

10. Richard G. Alexander quoted in Muir, *Black Shoes*, 39.

11. See Michael S. Sherry's treatment of the "Age of Prophecy" in *The Rise of American Air Power: The Creation of Armageddon* (New Haven, Conn.: Yale University Press, 1987), chap. 2.

12. For Truman's relief of MacArthur and the role of atomic weapons in that relief, see Michael Schaller, *Douglas MacArthur: The Far Eastern General* (New York: Oxford University Press, 1989), 230–39.

13. Rear Admiral Bradley A. Fiske, USN (Ret.), to Rear Admiral William S. Sims, USN, 30 November 1920, Admiral William S. Sims Papers, Manuscript Division, Library of Congress, Washington, D.C., Box 57.

14. Richard Hegmann, "Reconsidering the Evolution of the US Maritime Strategy," *Journal of Strategic Studies* 14 (1991): 303.

15. Malcolm Muir, *Black Shoes*, 8. The July 1947 War College position was that carrier task forces were not "overly vulnerable" to atomic weapons and a doubling of the interval between ships would "limit the effectiveness of one atomic explosion to the complete loss of only one major ship"; Naval War College Staff Presentation, "The Carrier Task Force: Including the Aircraft Carrier — Its Capabilities and Limitations," 22 July 1947, RG 14, NWC, 20.

16. Rear Admiral C. R. Brown, USN, "The Role of the Navy in Future Warfare," presentation to the Naval War College, 1 February 1949, RG 14, NWC, 6–7.

17. According to Malcolm Muir, the combined surface ship and submarine cruise missile program (Regulus II) was in direct budget competition with *Forrestal*-class supercarriers being authorized one per year and with Polaris ballistic missile submarines — all within the very tight navy budget during the second Eisenhower administration; *Black Shoes*, 95. Chief of Naval Operations Admiral Arleigh Burke thought Regulus II would never have "amounted to anything much" but could have led to "a successor to Regulus II that would have been an effective weapon." Admiral Arleigh A. Burke, "A Sailor's Views on National Security," in *Into the Jet Age: Conflict and Change in Naval Aviation, 1945–1975, An Oral History*, ed. Capt. E. T. Wooldridge, U.S. Navy (Ret.) (Annapolis: Naval Institute Press, 1995), 134.

18. A list of the converted carriers is provided in Roy A. Grossnick, *United States Naval Aviation, 1910–1995* (Washington, D.C.: Naval Historical Center, 1996), 432.

19. Hegmann, "Maritime Strategy," 307.

20. Quoted in ibid., 311.

21. Quoted in ibid., 318.

22. John F. Lehman, *Command of the Seas* (New York: Charles Scribner's Sons, 1988), 344.

23. Ibid., 188. The S-3 was part of a larger problem during the 1970s. In a tight budget environment, that which is judged unimportant, like lighter-than-air ships during the late 1930s, has its funding shifted to what is considered important. Conventional aviation was more important in 1937, just as carrier-based antisubmarine efforts were considered marginal forty years later.

24. See Carl H. Builder, *The Masks of War: American Military Styles in Strategy and Analysis* (Baltimore: Johns Hopkins University Press, 1989), 29.

25. A mobility kill involves sufficient damage to a ship's propulsion or steering systems to prevent it from fulfilling its mission.

26. Muir, *Black Shoes*, 29.

27. Elmo Zumwalt Jr., *On Watch: A Memoir* (New York: Quadrangle, 1976), 81.

28. See Muir's discussion of nuclear-powered surface warships in *Black Shoes*, 126–29.

29. Kenneth J. Hagan, *This People's Navy: The Making of American Sea Power* (New York: The Free Press, 1991), 375–80.

30. The navy's two surface-effect test ships, *SES-200A* and *SES-200B*, proved inefficient and difficult to steer, and were fuel guzzlers. Another radical hull technology with more promise was SWATH (Small Waterplane Area Twin-Hull), basically two submerged propulsion hulls that looked like submarines, joined by an above-water hull — similar to a catamaran. SWATH remained too radical until

resurrected in the 1990s as the basis of the Lockheed/U.S. Navy stealth ship, *Sea Shadow.*

31. Lieutenant Commander Bruce R. Linder, USN, "FFG-7s: Square Pegs?" U.S. Naval Institute *Proceedings* (hereafter *USNIP*) 109 (June 1983): 38–43.

32. Baer, *Sea Power,* 410–11.

33. Ibid., 413.

34. Ibid., 419.

35. Ibid., 422.

36. A good insight into the debate on the Maritime Strategy is *Naval Strategy and National Security,* ed. Steven E. Miller and Stephen Van Evera (Princeton, N.J.: Princeton University Press, 1988). For a naval defense of the Maritime Strategy, see Ronald O'Rourke, "The Maritime Strategy and the Next Decade," 1988 Naval Institute Prize Essay, *USNIP* 114 (April 1988): 34–38, and the opposing essay "The Maritime Strategy — Bad Strategy? Or Global Deterrent" by William S. Lind and Colin Gray, *USNIP* 114 (February 1988): 52–59.

37. Hamlin A. Caldwell Jr., "Nuclear War at Sea," *USNIP* 114 (February 1988): 60–63, criticizes the navy "gloss over" of the probable Soviet use of nuclear weapons against the U.S. Navy.

38. Chief of Naval Operations Arleigh A. Burke recounted his successful efforts during the late 1950s to keep strike-at-source language within Joint Chiefs of Staff planning documents providing some linkage — albeit service-driven — between strategic plans; Hegmann, "Maritime Strategy," 309.

39. American naval strength hovered around nine hundred ships during the Vietnam War.

40. See Lieutenant Scott A. Hastings, USN, "A Maritime Strategy for 2038," *USNIP* 114 (July 1988): 30–35.

41. U.S. naval defense at Okinawa also had to deal with Japanese surface and submarine threats.

42. Admiral James D. Watkins, USN, "The Maritime Strategy," *USNIP* 112 (1986): 3–17.

43. Captain Gerald O'Rourke, USN (Ret.), "The End of the Submarine's Era?" *USNIP* 114 (February 1988): 64–68. In his advocacy of a "national 'naval system'" rather than emphasis on a single type of warship, O'Rourke was more in line with William Fullam.

44. For the development of kinematics at Merton College, Oxford, see David C. Lindberg, *The Beginnings of Western Science: The European Scientific Tradition in Philosophical, Religious, and Institutional Context, 600 B.C. to A.D. 1450* (Chicago: University of Chicago Press, 1992), 295.

45. Secretary Thomas quoted in Hegmann, "Maritime Strategy," 309.

46. H. Lawrence Garrett III, Admiral Frank B. Kelso, and General A. M. Gray, "The Way Ahead," *USNIP* 117 (1991): 36–47; Sean O'Keefe, Admiral Frank B. Kelso II, and General Carl E. Mundy, USMC, " . . . From the Sea: Preparing the Naval Service for the 21st Century," *USNIP* 118 (1992): 93–96; and John H. Dalton, Admiral J. M. Boorda, USN, and General Carl E. Mundy, USMC, "Forward . . . From the Sea," available at http://www.chinfo.navy.mil/navpalib/policy/fromsea/forward.txt.

47. Kenneth P. Werrell, *The Evolution of the Modern Cruise Missile* (Maxwell Air Force Base, Ala.: Air University Press, 1985).

48. Kenneth P. Werrell, "The Weapon the Military Did Not Want: The Modern Strategic Cruise Missile," *Journal of Military History* 53 (1989): 426. In January 1952, Lieutenant Colonel L. L. Frank, USMC, opined that "eventually the Guided Missile may and in all probability will take over many of the functions of our present day aircraft and artillery, but even the most rabid proponent of guided missiles must admit that the machine has never been made and probably never will be that can take the place of the grey matter between the ears of a human being"; Naval War College Staff Presentation, "Guided Missiles," 18 January 1952, RG 14, NWC, 21. Interestingly, Admiral Rickover was an early 1970s supporter of the SLCM since it might result in a new class of nuclear submarines.

49. Rear Admiral Walter M. Locke, USN (Ret.) and Kenneth P. Werrell, "Speak Softly and . . . ," *USNIP* 120/10 (October 1994): 30.

50. Ibid.

51. Ibid.

52. See "Navy Might Resurrect A-12; Defense Chief OK's New Version of $93 Billion Plane," *St. Louis Post-Dispatch*, 1 April 1991, and "Suit Blames Navy for A-12 Woes," *St. Louis Post-Dispatch*, 8 June 1991.

53. Locke and Werrell, "Speak Softly," 33.

54. Ibid.

55. Ibid.

56. Ibid., 35.

57. *The Sun* [Baltimore], 21 August 1998, A1.

58. See Builder regarding aviators' infatuation with their high-tech "toys"; *Masks of War*, 23. Information on the CNO's Strategic Studies Group is available at http://www.nwc.navy.mil/ssg/ssghist.htm.

59. *Navy Times*, "Navy Kills Futuristic Carrier Program," 8 June 1998.

60. Statement by the Chief of Naval Operations to the Senate Armed Services Committee, 21 May 1997, available at the Navy Office of Information web page at

http://www.chinfo.navy.mil/navpalib/people/flags/johnson_j/testimony/
cnoo521.txt and from the author's files.

61. Eighteen months before CVX's design delay, the house organ *Naval Aviation News* devoted a full issue to extolling the historical worth of the aircraft carrier
and the need for the CVX. *Naval Aviation News* 79/2 (January–February 1997).
The CVX Mission Statement is at http://www.navsea.navy.mil/cvx/teamcvx.html
and also available from the author's files.

62. See the comments regarding Soviet naval aviation in the Defense Department's *Soviet Military Power: An Assessment of the Threat, 1988* (Washington, D.C.,
1988), passim.

63. Peter Karsten, *The Naval Aristocracy: The Golden Age of Annapolis and
the Emergence of Modern American Navalism* (New York: The Free Press, 1972),
391–92.

64. Admiral Jay L. Johnson interview with the media at Naval Air Station
Oceana, Va., 9 August 1996, "Quotes from the CNO," available at the Navy Office
of Information web site at http://www.chinfo.navy.mil/navpalib/people/flags/
johnson_j/cnoquote.html and author's files.

65. See Paul M. Kennedy, *The Rise and Fall of the Great Powers: Economic
Change and Military Conflict from 1500 to 2000* (New York: Random House, 1987),
chaps. 7 and 8.

66. Thomas P. Hughes, "Technological Momentum," in *Does Technology Drive
History?: The Dilemma of Technological Determinism*, ed. Merritt Roe Smith and
Leo Marx (Cambridge, Mass.: The MIT Press, 1996), 111–12. The more energetic
dialectic within the naval profession is akin to Wiebe Bijker's micropolitics. See his
Of Bicycles, Bakelites, and Bulbs: Toward a Theory of Sociotechnical Change (Cam-
bridge, Mass.: The MIT Press, 1995), 288.

67. In refining the concept of technological trajectory, Henk van den Belt and
Arie Rip have built on Edward Constant's model of technological paradigms,
Kuhnian paradigms, and Neo-Schumpeterian economic theory. Van den Belt and
Rip, however, minimize the deterministic nature of technology selection and
its roots in biological evolution. Henk van den Belt and Arie Rip, "The Nelson-
Winter-Dosi Model and Synthetic Dye Chemistry," in *The Social Construction of
Technological Systems: New Directions in the Sociology and History of Technology*,
ed. Wiebe J. Bijker, Thomas P. Hughes, and Trevor Pinch (Cambridge, Mass.: The
MIT Press, 1994), 140–41.

68. Edward W. Constant II, *The Origins of the Turbojet Revolution* (Baltimore:
Johns Hopkins University Press, 1980), 17.

69. For the self-reinforcing nature of wargaming at the Naval War College, see

Robert L. O'Connell, *Sacred Vessels: The Cult of the Battleship and the Rise of the U.S. Navy* (Boulder, Colo.: Westview Press, 1991), 73–78. An assessment of more modern self-referential thinking is O'Connell's "A Useful Navy for 2017: What Can Naval History Tell Us?" in *New Interpretations in Naval History: Selected Papers from the Thirteenth Naval History Symposium*, ed. William M. McBride and Eric P. Reed (Annapolis: Naval Institute Press, 1998), 308–21.

70. See *Social Construction of Technological Systems*, especially the introduction and three essays that comprise Part I.

71. Bijker, *Of Bicycles, Bakelites, and Bulbs*, 276–77.

72. John Law, "Technology and Heterogenous Engineering: The Case of Portuguese Expansion," in *Social Construction of Technological Systems*, 112.

73. Ibid., 113.

74. Smith, *Military Enterprise*, 23.

75. In military cultures this involves the suppression of "counterweapons." See Robert L. O'Connell, *Of Arms and Men: A History of War Weapons and Aggression* (New York: Oxford University Press, 1989), passim and especially 7–11.

76. Michel Callon, "Society in the Making: The Study of Technology as a Tool for Sociological Analysis," in *The Social Construction of Technological Systems*, ed. Wiebe E. Bijker, Thomas P. Hughes, and Trevor Pinch (Cambridge, Mass.: The MIT Press, 1994), 83–87. In addition to paradigmatic filtration, equipment tests provided "one of the main avenues by which the military influences industrial design and production"; Smith, *Military Enterprise*, 18.

77. Howard Margolis, *Paradigms and Barriers: How Habits of Mind Govern Scientific Beliefs* (Chicago: University of Chicago Press, 1993); and Bijker, *Of Bicycles, Bakelites, and Bulbs*, 276.

78. Chief of Naval Operations (Planning Division) Secret Memorandum to the General Board, Serial Op-12-D, P.D.138–15, 30 October 1920, included in *Hearings of the General Board of the Navy, 1917–1950*, microfilm, Nimitz Library, U.S. Naval Academy, reel 3, 926ff. Quote from Captain Yates Stirling, USN, "Some Fundamentals of Sea Power," *USNIP* 51 (1925): 889–918, see 913–14.

79. Bijker, *Of Bicycles, Bakelites, and Bulbs*. The different technical frames appeared in a satirical article in 1932 in which battleship supporters are "cave men" defeated by superior "Neanderthal" aviators; Commander Ralph C. Parker, USN, "An Analysis of the Air Menace," *USNIP* 58 (1932): 649.

80. Smith, *Military Enterprise*, 27.

81. For an account of the pejorative treatment of the Italian navy during World War II, start with James J. Sadkovich, *The Italian Navy in World War II* (Westport, Conn.: Greenwood, 1994).

82. See Bradley A. Fiske, "Air Power," *USNIP* 43 (1917): 1702.

83. Admiral William A. Owens, USN, *High Seas: The Naval Passage to an Uncharted World* (Annapolis: Naval Institute Press, 1995), 126.

84. For the jargon-laden "centric" warfare, see Admiral Jay L. Johnson, USN, chief of naval operations, "Remarks to Surface Navy Association, Washington, D.C., 30 May 1997," available at the Navy Office of Information web site http://www.chinfo.navy.mil/navpalib/people/flags/johnson_j/speecs/sna0530. txt and from the author's files.

85. "Statement of Admiral William D. Leahy, United States Navy, Chief of Naval Operations," during "Hearings on H.R. 9218 to Establish the Composition of the United States Navy, to Authorize the Construction of Certain Naval Vessels, and for Other Purposes," *Hearings before Committee on Naval Affairs of the House of Representatives on Sundry Legislation Affecting the Naval Establishment, 1937-1938, Seventy-fifth Congress, Second and Third Sessions* (Washington, D.C.: U.S. Government Printing Office, 1938), 1937–2100; quote on 1943.

86. See the partial transcript of Admiral Johnson's question-and-answer session at the Naval Institute's Seventh Annapolis Seminar, reprinted in the *USNIP* 123/6 (June 1997): 9. For the viability of the CVX into the latter half of the twenty-first century, see "Statement by the Chief of Naval Operations to the Senate Armed Services Committee, 21 May 1997" available at the Navy Office of Information web page at http://www.chinfo.navy.mil/navpalib/people/flags/johnson_j/testimony/ cno0521. txt and from the author's files.

87. Hagan, *This People's Navy*, 60–61.

88. For another application of a paradigmatic framework to war, see Andrew F. Krepinevich Jr., *The Army and Vietnam* (Baltimore: Johns Hopkins University Press, 1986).

89. I. B. Holley Jr., *Ideas and Weapons: Exploitation of the Aerial Weapon by the United States during World War I: A Study in Technological Advance, Military Doctrine, and the Development of Weapons* (New Haven, Conn.: Yale University Press, 1953 reprinted by the Office of Air Force History, 1983), 178.

90. The intellectual differences among the services runs throughout Builder, *Masks of War.* For a defense of the humanities in the navy, see John H. Mitchell, "Bull? Or the Real Thing?" *USNIP* 118/4 (April 1992): 40–46. The simplest measure of support for intellectual development is the support for officer doctorates in nontechnical fields, such as history. The army and air force consider it important and have more. The navy seems to care little.

91. For example, "The way we are going to take ourselves and keep ourselves relevant as we punch our way into the next century is with a vision that has us look-

ing out ahead and steering by the stars that are up there instead of looking over our shoulders and steering by the wake that is behind us." Quote from Admiral Johnson's speech at the Sea-Air-Space Luncheon, 26 March 1997, available at "Quotes from the CNO" at the Navy Office of Information web page at http://www.chinfo.navy.mil/navpalib/people/flags/johnson_j/cnoque.html and from the author's files. Kettering's comment from Stuart W. Leslie, *Boss Kettering: Wizard of General Motors* (New York: Columbia University Press, 1983), ix.

Note on Sources

The documents dealing with the U.S. Navy during the period of this study are extensive. The secondary literature on general naval history is sizable, but the quality of its scholarship varies widely. Although there are significantly fewer works dealing with the navy within the framework of the history of technology, their scholarship makes up for their scarcity.

The primary documents on which this study was based include manuscript sources, official papers, and official documents. The most useful manuscripts were in the papers of naval officers, and those involved with the navy, held by the Manuscript Division of the Library of Congress in Washington, D.C. These include the interesting, informative papers of Admiral William S. Sims, including his wide-ranging correspondence with Rear Admiral Bradley Fiske; the papers of relevant chiefs of naval operations, including Fleet Admiral Ernest J. King, Fleet Admiral William D. Leahy, Admiral William S. Benson, Admiral William H. Standley, and Admiral William V. Pratt; the papers of influential personages within the Department of the Navy, such as Secretary of the Navy Josephus Daniels, Admiral of the Navy George Dewey, and Rear Admiral George W. Melville; other flag officers who affected naval policy, including Admiral Harry E. Yarnell, Admiral Marc A. Mitscher, Admiral Hillary P. Jones, Admiral Montgomery M. Taylor, Vice Admiral Emory S. Land, Rear Admiral Henry C. Taylor, and Admiral Claude C. Bloch; the papers of the first historian of the navy's Engineering Corps, Captain Frank M. Bennett; and the papers of William Conant Church, editor of the *Army and Navy Journal,* a periodical devoted to popular treatment of issues relevant to the army and navy.

For the interwar period I relied on the papers of Presidents Herbert C. Hoover and Franklin D. Roosevelt. Worthwhile papers at the Franklin D. Roosevelt Presidential Library in Hyde Park, New York, include Roosevelt's papers as assistant secretary of the navy (1913–20); Roosevelt's papers as president, including the Official File, Personal File, President's Secretary's File, and Alphabetical File; the papers of the naval aides to the president, including Wilson Brown, John L. McCrea, and William Rigdon; and the papers of Roosevelt's mentor and chief adviser, Louis M. Howe, and of Harry L. Hopkins, an important figure within the Roosevelt administration.

Relevant collections at the Herbert Hoover Presidential Library in West Branch, Iowa, include Hoover's official and personal papers as secretary of commerce, 1921–28. Important material on the navy, defense matters, and arms control can be found in the Presidential Period Collection and includes the Cabinet Offices File, Subject File, Foreign Affairs (Disarmament) File, Individual Name File, President's Personal File, and the Taylor/Gates Collection. Helpful information can also be found in the Post-Presidential Period Collection (1933–63), including the General Correspondence File, Individual Correspondence File, and Subject File. Other relevant collections at the Hoover Library include the papers of William R. Castle Jr., who was ambassador to Japan and undersecretary of state, the papers of William P. MacCracken, assistant secretary of commerce for aviation, 1926–29, and the papers of Hoover friend and Republican Party functionary, John Callan O'Laughlin, owner and publisher of the *Army and Navy Journal* from 1925 to 1949.

Three other archives provided manuscript sources for this study. The account of the establishment of the navy's graduate program in naval architecture was based on material from the Massachusetts Institute of Technology's Archives and Special Collections, specifically Collection AC13. The Vice Admiral Harold G. Bowen Papers at the Seeley G. Mudd Manuscript Library of Princeton University provided material on battleship design during the 1930s. I found information on naval-industrial relations in the Elmer A. Sperry Papers and Sperry Gyroscope Company Records at the Hagley Museum and Library in Wilmington, Delaware.

The official records of the U.S. Navy are sizable and located primarily in Washington, D.C. The most comprehensive records regarding the administration and operation of the navy are found in Record Group 80, General Records of the Secretary of the Navy, maintained by the National Archives and Records Administration (NARA). In addition to documents originating from the office of the secretary of the navy, Record Group 80 also contains a large number of reports, correspondence, and documents from subordinate commands and bureaus within the navy. The material contained in NARA Record Group 19, Records of the Bureau of Construction & Repair, illuminated many of the technological aspects of U.S. warship designs and the process of technological change within the scope of this study.

Those interested in the history of the Naval Academy, including the programs focusing on undergraduate and graduate technical education of cadets, midship-men, and commissioned officers, will be well served by the material in Record Group 405, Records of the U.S. Naval Academy and Correspondence of the Su-perintendent, maintained in the Special Collections, Nimitz Library, U.S. Naval Academy, Annapolis, Maryland.

Another archive of particular worth is the Naval Historical Collection main-tained by the U.S. Naval War College in Newport, Rhode Island. I found the ma-terial contained in Record Group 6, Technology and Intelligence; Record Group 14, Staff Presentations and Lectures; Records of the 1908 Battleship Conference; and the files of Office of Naval Intelligence Publications helpful in gauging the strategic dialectic within the navy.

Official documents regarding the navy were published by the executive and Congress. Probably the most useful detailed overview of the operation of the De-partment of the Navy and its subordinate commands and bureaus is the *Annual Report of the Secretary of the Navy*. Typically, each report also included the reports made to the secretary of the navy by the chiefs of the navy's bureaus, operational commands, and the chiefs of naval operations after the creation of that post. A de-tailed picture of the official discussions on strategy, tactics, organization, and tech-nology can be found in the voluminous *Hearings Before the General Board of the Navy, 1917–1950*, available on microfilm. In addition to often candid, confidential testimony and discussions, the hearings also include studies, reports, and various correspondence relevant to the development of naval policy. Another useful pub-lication, also in microfilm, *Records Relating to United States Navy Fleet Problems I to XXII, 1923–1941*, pertains to the interwar fleet exercises.

In addition to publications by the executive, congressional hearings, testimony, and legislation provide useful insights into the development of naval policies and its role within national policies. I found the annual *Hearings Before the Commit-tee on Naval Affairs of the House of Representatives* and *Hearings Before the Com-mittee on Naval Affairs, United States Senate* of value. In addition, *Hearings Before the Committee on Naval Affairs, United States Senate on the Senate Bill (S.3335) to Increase the Efficiency of the Personnel of the Navy and Marine Corps of the United States* and *Report Concerning Certain Alleged Defects in the Vessels of the United States Navy, by Washington Lee Capps, Chief Constructor, U.S. Navy, and Chief of the Bureau of Construction & Repair, Senate Document No. 297, 60th Con-gress, 1st Session*, both published in 1908, aid in understanding the events sur-rounding William Sims's attacks on the bureaus and American battleship designs.

Probably the single most important contemporary publication, especially in un-derstanding the process of technological change during the period of this study, is the Naval Institute *Proceedings.* Prior to World War II, the *Proceedings* reprinted newspaper and journal articles on naval and maritime affairs. The value of the *Proceedings* also lies in its articles reflecting the dynamic dialectic relating strategy, tactics, and technology within the American naval profession.

Additional insights into technological developments can be found in the *Journal of the American Society of Naval Engineers,* the *Transactions* of the Society of Naval Architects and Marine Engineers, and the *Transactions, (Royal) Institution of Naval Architects.* Periodically, *Scientific American* took an interest in naval affairs and the relevant discussions on its editorial pages often proved illuminating. Additional information can be found in the periodicals *Army and Navy Journal* and *Army and Navy Register.* Other useful accounts of the navy can be found in the *New York Times* and *The Sun* (Baltimore).

A large number of general histories of the U.S. Navy have been written, and their scholarship and usefulness vary. Most naval histories tend to be "data streams," sometimes punctuated by tactical diagrams, with no analysis and no discernible thematic focus. They fall into a category a former colleague termed "'Dragnet' history"—presenting "just the facts" in an Orwellian tale with no sense that some facts deserve to be more "equal" than others. Recent exceptions include Kenneth Hagan's *This People's Navy: The Making of American Sea Power* (1991) and George Baer, *One Hundred Years of Sea Power: The U.S. Navy, 1890–1990* (1993).

While the field of general naval history has its shortfalls, there are many excellent studies dealing with discrete aspects of naval history. Older works which have retained their relevance include Peter Karsten's *The Naval Aristocracy: The Golden Age of Annapolis and the Emergence of Modern American Navalism* (1972); William Reynolds Braisted's *The United States Navy in the Pacific, 1897–1909* (1958); Walter R. Herrick's *The American Naval Revolution* (1966); Robert G. Albion's *Makers of Naval Policy, 1798–1947* (1980); Gerald E. Wheeler's *Prelude to Pearl Harbor: The United States Navy and the Far East, 1921–1931* (1963); Roger Dingman's *Power in the Pacific* (1979); Elting Morison's *Admiral Sims and the Modern American Navy* (1942); Paolo Coletta's *Admiral Bradley A. Fiske and the American Navy* (1979); Stephen E. Pelz's *Race to Pearl Harbor: The Failure of the Second London Naval Conference and the Onset of World War II* (1974); Raymond G. O'Connor's *Perilous Equilibrium: The United States and the London Naval Conference of 1930* (1969); Passed Assistant Engineer Frank M. Bennett's *The Steam Navy of the United States: A History of the Growth of the Steam Vessel of War in the U.S. Navy, and of the Naval Engineer Corps* (1896); and Thomas B. Buell's *The Quiet Warrior: A*

Biography of Admiral Raymond A. Spruance (1974) and his *Master of Sea of Sea Power: A Biography of Fleet Admiral Ernest J. King* (1980).

More recent works include Christopher Hall's *Britain, America, and Arms Control, 1921–1937* (1987); B. Mitchell Simpson III's *Admiral Harold R. Stark: Architect of Victory, 1939–1945* (1989); Malcolm Muir's *Black Shoes and Blue Water: Surface Warfare in the United States Navy, 1945–1975* (1995); Clark G. Reynolds's *Admiral John H. Towers: The Struggle for Naval Air Supremacy* (1991); Captain Wayne P. Hughes's *Fleet Tactics: Theory and Practice* (1986); and William F. Trimble's *Admiral William A. Moffett: Architect of Naval Aviation* (1994).

Informative histories of Japanese naval policy include *Pearl Harbor as History: Japanese-American Relations 1931–1941*, ed. Dorothy Borg and Shumpei Okamoto, (1973); *Japan Erupts: The London Naval Conference and the Manchurian Incident, 1928–1932*, ed. James William Morley (1984); and Sadao Asada's "Japanese Admirals and the Politics of Naval Limitation: Katō Tomosaburo and Katō Kanji," in *Naval Warfare in the Twentieth Century, 1900–1945: Essays in Honour of Arthur Marder*, ed. Gerald Jordan (1977).

Those interested in the debate surrounding the navy's strategic developments during the 1980s should start with *Naval Strategy and National Security*, ed. Steven E. Miller and Stephen Van Evera (1988) and Richard Hegmann's "Reconsidering the Evolution of the US Maritime Strategy" in the 1991 *Journal of Strategic Studies*. Also of interest to students of the cold war military is Carl H. Builder's *The Masks of War: American Military Styles in Strategy and Analysis* (1989).

The theoretical framework for this study is based in Edward W. Constant II, *The Origins of the Turbojet Revolution* (1980); Thomas Kuhn, *The Structure of Scientific Revolutions*, 2nd ed. (1970); Ludwik Fleck, *Genesis and Development of a Scientific Fact*, ed. Thaddeus K. Trenn and Robert K. Merton (1979); *The Social Construction of Technological Systems: New Directions in the Sociology and History of Technology*, ed. Wiebe J. Bijker, Thomas P. Hughes, and Trevor Pinch (1994); and *Does Technology Drive History?: The Dilemma of Technological Determinism*, ed. Merritt Roe Smith and Leo Marx (1996). Also relevant are Wiebe J. Bijker's *Of Bicycles, Bakelites, and Bulbs: Toward a Theory of Sociotechnical Change* (1995); Merritt Roe Smith's *Harpers Ferry Armory and the New Technology: The Challenge of Change* (1977); and John M. Staudenmaier's *Technology's Storytellers: Reweaving the Human Fabric* (1989).

Studies within the history of technology relevant to this work include Robert L. O'Connell's *Sacred Vessels: The Cult of the Battleship and the Rise of the U.S. Navy* (1991) and his *Of Arms and Men: A History of War Weapons and Aggression* (1989);

Alex Roland's *Underwater Warfare in the Age of Sail* (1978); Jon Tetsuro Sumida's *In Defence of Naval Supremacy: Finance, Technology, and British Naval Policy, 1889–1914* (1989); Elting Morison's *Men, Machines, and Modern Times* (1966); Monte A. Calvert's *The Mechanical Engineer in America, 1830–1910: Professional Cultures in Conflict* (1967); Edward William Sloan III's *Benjamin Franklin Isherwood, Naval Engineer: The Years as Engineer in Chief, 1861–1869* (1965); Lance C. Buhl's "Mariners and Machines: Resistance to Technological Change in the American Navy, 1865–1869," *Journal of American History* (1974); and Kenneth P. Werrell's *The Evolution of the Modern Cruise Missile* (1985) and his "The Weapon the Military Did Not Want: The Modern Strategic Cruise Missile," *Journal of Military History* (1989).

Other applications of paradigmatic analysis in military history include Andrew F. Krepinevich Jr.'s *The Army and Vietnam* (1986) and Tim Travers's *The Killing Ground: The British Army, The Western Front and the Emergence of Modern Warfare, 1900–1918* (1987).

Index

Related Books in the Series

The American Railroad Passenger Car   John H. White, Jr.

Neptune's Gift: A History of Common Salt   Robert P. Multhauf

Electricity before Nationalisation: A Study of the Development of the Electricity Supply Industry in Britain to 1948   Leslie Hannah

Alexander Holley and the Makers of Steel   Jeanne McHugh

The Origins of the Turbojet Revolution   Edward W. Constant

Engineers, Managers, and Politicians: The First Fifteen Years of Nationalised Electricity Supply in Britain   Leslie Hannah

Stronger Than a Hundred Men: A History of the Vertical Water Wheel   Terry S. Reynolds

Authority, Liberty, and Automatic Machinery in Early Modern Europe   Otto Mayr

Inventing American Broadcasting, 1899–1922   Susan J. Douglas

Edison and the Business of Innovation   André Millard

What Engineers Know and How They Know It: Analytical Studies from Aeronautical History   Walter G. Vincenti

Alexanderson: Pioneer in American Electrical Engineering   James E. Brittain

Steinmetz: Engineer and Socialist   Ronald R. Kline

A Nation of Steel: The Making of Modern America, 1865–1925   Thomas J. Misa

The Machine in the Nursery: Incubator Technology and the Origins of Newborn Intensive Care   Jeffrey P. Baker